Three-Dimensional Ultrasound

Three-Dimensional Ultrasound

Thomas R. Nelson, Ph.D
Professor
Division of Physics
Department of Radiology
University of California, San Diego
La Jolla, CA

Dónal B. Downey, M.B., B.Ch, F.R.C.P.C., F.R.C.R., F.F.R.(R.C.S.I.)
Associate Professor
Department of Radiology and Nuclear Medicine
University of Western Ontario
London, Ontario, Canada

Dolores H. Pretorius, M.D.
Professor
Division of Ultrasound
Department of Radiology
University of California, San Diego
La Jolla, CA

Aaron Fenster, Ph.D, F.C.C.M.P.
Director
Imaging Research Laboratories
The John P. Robarts Research Institute
Professor
Department of Radiology and Nuclear Medicine
University of Western Ontario
London, Ontario, Canada

LIPPINCOTT WILLIAMS & WILKINS
PHILADELPHIA · NEW YORK · BALTIMORE

Acquisitions Editor: James Ryan
Developmental Editor: Brian Brown
Manufacturing Manager: Tim Reynolds
Production Manager: Liane Carita
Production Editor: Patrick Carr
Cover Designer: Mark Lerner
Indexer: Sandi Frank
Compositor: Maryland Composition
Printer: Maple Vail

© 1999, by Lippincott Williams & Wilkins. All rights reserved. This book is protected by copyright. No part of it may be reproduced, stored in a retrieval system, or transmitted, in any form or by any means—electronic, mechanical, photocopy, recording, or otherwise—without the prior written consent of the publisher, except for brief quotations embodied in critical articles and reviews. For information write **Lippincott Williams & Wilkins, 227 East Washington Square, Philadelphia, PA 19106-3780.**

Materials appearing in this book prepared by individuals as part of their official duties as U.S. Government employees are not covered by the above-mentioned copyright.

Printed in the United States of America

9 8 7 6 5 4 3 2 1

Library of Congress Cataloging-in-Publication Data

Three-dimensional ultrasound / Thomas R. Nelson . . . [et al.].
p. cm.
Includes bibliographical references and index.
ISBN 0-7817-1997-6
1. Ultrasonic imaging. 2. Three-dimensional imaging in medicine.
I. Nelson, Thomas, Ph.D. II. Title: Three dimensional ultrasound.
[DNLM: 1. Ultrasonography—methods. 2. Ultrasonography—trends.
WN 208 T531 1999]
RC78.7.U4T48 1999
616.07′543—dc21
DNLM/DLC
for Library of Congress 98-32024
CIP

Care has been taken to confirm the accuracy of the information presented and to describe generally accepted practices. However, the authors, editors, and publisher are not responsible for errors or omissions or for any consequences from application of the information in this book and make no warranty, expressed or implied, with respect to the contents of the publication.

The authors, editors, and publisher have exerted every effort to ensure that drug selection and dosage set forth in this text are in accordance with current recommendations and practice at the time of publication. However, in view of ongoing research, changes in government regulations, and the constant flow of information relating to drug therapy and drug reactions, the reader is urged to check the package insert for each drug for any change in indications and dosage and for added warnings and precautions. This is particularly important when the recommended agent is a new or infrequently employed drug.

Some drugs and medical devices presented in this publication have Food and Drug Administration (FDA) clearance for limited use in restricted research settings. It is the responsibility of the health care provider to ascertain the FDA status of each drug or device planned for use in their clinical practice.

to my parents for their support, encouragement, and warmth from the beginning; and to my co-author and wife, Dolores Pretorius, for her friendship, collegiality, and enthusiasm for life.

TRN

to my parents, who have supported my career since I was an inquisitive child, and to my co-author and husband, Thomas Nelson, for his boundless support of our family and our careers together.

DHP

to our children, Novella and Tanya, who have survived having parents working on this book and who have been able to share the fun and excitement of our lives together.

DHP and TRN

to my family–David, Jennifer, and Pam–for the selfless love, patience, support, and kindness you showed to me during this project and that you show me always.

DBD

to my family–Katherin and Jade–for their inspiration, and for making my achievements worth while.

AF

Contents

Preface

Ultrasound is one of those marvelous imaging techniques that continues to amaze. While it is true that one cannot hear ultrasound, ultrasound can reveal the hidden world within objects often with astounding clarity. The scope of ultrasound imaging applications extend far beyond medicine to the realm of non-destructive testing, often on scales from microscopic to macroscopic. Yet within medicine, ultrasound continues to make important and significant contributions to reassuring patients and enhancing their quality of life. The intrinsic ability of ultrasound to image deep within the patient can help physicians understand the anatomy of the patient in ways not possible with other techniques. Ultrasound is unique in its ability to image in real-time not only the patient's anatomy but also the patient's physiology, providing an important, rapid, and non-invasive means of evaluation.

Early applications of ultrasound in medicine began shortly after World War II and derived from knowledge gained in developing sonar for submarines. Progress has continued during the past half-century with advances in clinical applications and equipment performance. As a result, ultrasound has gained an ever larger role in the evaluation of patients in radiology, obstetrics, cardiology, pediatrics, and emergency departments, to name a few. With the continuing development of improved scanners and new techniques, and the increased importance of cost containment, ultrasound continues to expand its role in health care.

Of the many areas of ultrasound development currently under investigation, none is of greater interest than three-dimensional ultrasound (3DUS). 3DUS exploits the real-time capability of ultrasound to build a volume, often before the operator's eyes, that can then be explored using the power of modern computer viewing workstations. 3DUS represents a logical and natural extension of previous ultrasound developments and also complements other parallel developments underway at this time, including harmonic imaging and the development of ultrasound contrast materials.

When the authors first became interested in 3DUS in the late 1980's a body of literature extending back to the 1960's already existed. But early efforts often were hampered by limitations in the technology available to implement clinically viable scanning equipment. It is no understatement to say that, as with so many other aspects of modern life, the advent of modern ultrasound imaging has resulted from the availability of increasingly powerful and affordable computers in packages small enough to make them portable. By the late 1980's, computer performance had progressed to the point where affordable systems could be used in the clinical setting. Some of Drs. Nelson and Pretorius' first work began using the computer resources of the San Diego Supercomputer Center to perform initial feasibility studies before transferring the methods to the clinical setting on smaller graphics workstations.

Since acquisition of a volume of data required not only a powerful computer to process the data rapidly but also required a means of registering images acquired from a volume, position sensing systems also were needed. Many approaches have been tried, including spark gap, electromagnetic, and mechanical devices, and methods continue to evolve to address this central issue. More recent research using image processing and multi-dimensional arrays promises to bring real-time volume imaging to the patient's bedside in the near future. The combination of computers and a way to register images made it possible to produce volume data; the next challenge has been extracting meaningful clinical information from the volume. Ultrasound image data presents a major challenge for image processing techniques, in part due to its speckle, signal-to-noise properties, and the fundamental physics of how acoustic images are formed. Thus, image processing and visualization remain areas requiring significant research dependent on powerful computing resources. Visualization of underlying patient anatomy including blood flow offers new ways to visualize complex anatomy and enhance the visibility of structures that previously were difficult to appreciate.

With improved visualization tools, from reslicing the volume to rendering in a single image the complex anatomy of a fetal face or vascular bed, clinical applications of 3DUS are expanding rapidly. One of the most important requirements for clinical scanning systems is the ability to provide superior image quality and interactivity to manipulate the data in real-time as part of reviewing the patient study. Increasingly, clinical scanners are becoming available that can accomplish this amazing feat at the patient's bedside. 3DUS also offers the physician the capability to image patients after they have left the scanning suite and reevaluate diagnoses with experts across networks at remote locations.

This book reviews areas of clinical 3DUS research to date, and speculates on other areas that may become important in the future. However, as we have explored and developed this imaging technique, we have gained experience in new applications and areas we had not anticipated as often as we have confirmed our hypothesis regarding an important clinical question. Indeed, the very presence of real-time 3DUS imaging equipment in the clinical setting will expand and stimulate new areas of investigation and identify areas where 3DUS can enhance clinical care.

Ultimately, while 3DUS technology will provide a central integrating focus for ultrasound imaging, the success of 3DUS methods will depend on providing performance that equals or exceeds that of 2D sonography, including real-time capability and interactivity. Although the eventual role for 3DUS has yet to be determined, there is little doubt that its impact will be broad and substantial.

Future 3DUS equipment will integrate conventional 2DUS imaging while seamlessly expanding capability, making volume acquisition and display central to obtaining the patient diagnosis. Focused clinical application of 3DUS methods will extend beyond the applications in which ultrasound is currently the predominate imaging modality. Through rapid transmission of volume data to specialists at distant locations, patient care will benefit from improved diagnosis and treatment. Although it is difficult to predict future applications, we all anticipate the rapid progress of this exciting new area of the medical imaging field and expect to see continued progress and development beyond our expectations. We hope that this book will serve as a starting point to interest and inform our readers and open this field to clinical application.

Three-Dimensional Ultrasound

Acknowledgments

No book would be possible without the support and encouragement of friends, colleagues, and collaborators across the world. To all those who were willing to listen to our hopes and aspirations as we have explored this exciting and rapidly developing new field, we greatly appreciate your support and encouragement.

Drs. Nelson and Pretorius appreciate the support of our research efforts from Drs. George Leopold, Barbara Gosink, and our colleagues in the Radiology Department, and Dr. O.W. Jones in the Medicine Department at the University of California, San Diego. Dr. Pretorius also would like to thank Dr. Michael Manco-Johnson for his mentorship and friendship throughout her journey discovering the secrets of ultrasound. Drs. Nelson and Pretorius also wish to acknowledge Dr. Sam Maslak and his colleagues at Acuson Corporation (Mountainview, CA) whose support early on was essential to begin our exploration into 3DUS. We have had the good fortune to have had the support and encouragement of the groups at Kretztechnique (Zipf, Austria) and Medison (Seoul, Korea and Pleasanton, CA) in helping advance the clinical applications of 3DUS. Certainly, those individuals and corporations who have supported our work financially (the Society of Radiologists in Ultrasound, Acuson Corporation, Medison America, Inc., The California Tobacco Research Program, and the University of California, Academic Senate Grants) have greatly assisted the progress we have made.

Drs. Downey and Fenster wish to acknowledge the tremendous support they received from their Medical and Research colleagues and students at the Robarts Research Institute, London, Ontario, and the London Health Sciences Center, London Ontario. Drs. Downey and Fenster also wish to acknowledge funding received from The Medical Research Council of Canada, The Canadian Association of Radiologists, The Heart and Stroke Foundation of Ontario, and The University Hospital Research Fund, London, Ontario.

The authors wish to acknowledge the assistance of their colleagues in providing images and perspective during the preparation of this book: Dr. David Nicolle, London Health Sciences Center, London, Ontario; Drs. Derek Boughner and Peter Pflugfelder, London Health Sciences Center, London, Ontario; Dr. Ji-Bin Liu, Thomas Jefferson University, Philadelphia, PA; Dr. Mark Sklansky, University of California, San Diego, CA; Dr. Nancy Budorick, Columbia University, New York, NY; Dr. Michael Riccabona, University Hospital, Graz, Austria; Karen Lou, RDMS, San Francisco, CA; Michael Tartar RDMS, San Diego, CA; Dr. Kathy Spaulding, Raleigh, NC; and Sandy Hagen-Ansert, Memphis, TN.

The authors wish to acknowledge the use of equipment and images received from: Life Imaging Systems Inc., London, Ontario; Advanced Technologies Limited, Bothell, WA; Siemens Ultrasound Group, Issaquah, WA; Aloka, USA; General Electric Medical Systems, Milwaukee, WI; Diasonics-Vingmed, TomTec, Inc., Boulder, CO; and Medison-America, Pleasanton, CA.

PART I

Overview

1

Overview of Medical Visualization

OVERVIEW

This chapter will provide a brief overview of medical visualization and the growing role of ultrasound in medical diagnosis and patient management.

KEY CONCEPTS

- The primary role of visualization in medicine is to provide the physician with information.
- Physicians use visualization to gain insight and share observations with colleagues and patients.
- Medical visualization is used with x-ray, computed tomography, magnetic resonance imaging, nuclear, and ultrasound modalities.
- Ultrasound is flexible, has moderate cost, is real-time, can measure physiology, has no bio-effects, and is available in most hospitals and clinics.
- Advanced computer technology facilitates interactive review of 3DUS data.
- The eventual role for 3DUS will be broad and substantial and become a routine part of patient diagnosis and management.

MEDICAL VISUALIZATION

The primary role of visualization in medicine is to provide the physician with information needed to make an accurate diagnosis of the patient's condition and assess the patient's response to therapy. Medical visualization methods increasingly rely on computer graphics techniques to assist physicians in understanding patient anatomy. Visualization helps physicians extract meaningful information from numerical descriptions of complex phenomena using interactive systems. Physicians need these imaging systems for their own insight and to share their observations with clinical colleagues and patients. Few medical articles are published without some type of data visualization.

Medical visualization resources encompass a broad array of different modalities, including x-ray imaging, computed tomography (CT), magnetic resonance imaging (MRI), nuclear medicine, and ultrasound (US). Sensor technologies utilize a variety of physical mechanisms (e.g., electron density, isotope distribution, magnetic coupling, and acoustic impedance) to produce image data. Increasingly, images are developed from more sophisticated sensors that require physicians to comprehend information derived from technology neither developed nor discussed during their medical training.

The inherent flexibility of ultrasound imaging, its moderate cost, and advantages that include real-time imaging, physiologic measurement, use of nonionizing radiation, and no known bio-effects give ultrasound a vital role in the diagnostic process and important advantages compared with MRI and CT. Ultrasound visualization is in routine use in nearly all hospitals and in many physician offices and clinics, and currently is used to diagnose many different pathologies.

ULTRASOUND IMAGING

Over the past few years ultrasound imaging has made tremendous progress in obtaining important diagnostic information from patients in a rapid, noninvasive manner and has benefitted from significant improvements in image quality and visualization clarity (Figure 1-1). Much of this progress has been derived from utilizing information present in the ultrasound signal, such as acoustic scatter and Doppler shift, to extract useful physiologic and tissue structure information. Ultrasound has the advantage in that it offers tomographic imaging capability at fast update rates (10–100 images/sec). As ultrasound equipment has benefitted from increasingly sophisticated computer technology, systems integration has ensured better image quality, data acquisition, analysis, and display. Advances in technology, particularly high-speed computing and storage hardware, have further expanded the possibilities for maximizing patient diagnostic information.

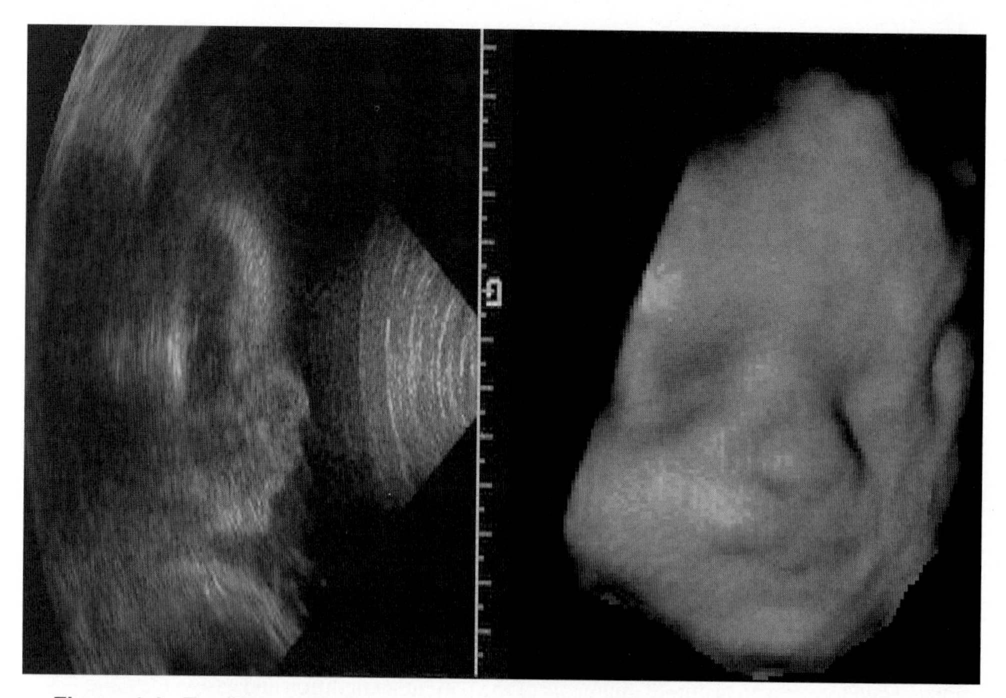

Figure 1-1. Traditional 2DUS produces planar tomographic slices through the object of interest, in this case a 2DUS image of a fetal head and face seen in profile on the left. 3DUS visualization methods provide a means of displaying the entire structure in a single image significantly improving comprehension of spatial relationships as is seen in the fetal face on the right where the eyes, nose, and lips are readily visible.

Applications of ultrasound imaging in medicine are ubiquitous (1). With the advent of specialized intravascular, endotracheal, and endocavitary imaging probes nearly every organ system is accessible to ultrasound scanning. Ultrasound contrast agents will further extend the range of diagnostic application. Although diagnostic ultrasound power levels have not been shown to produce bio-effects, recent work using high-intensity focused ultrasound has been shown to have potentially useful therapeutic value in producing localized tumor hyperthermia (2). Three-dimensional ultrasound (3DUS) can play an important role in assessing the extent of hyperthermia response and optimization of treatment.

Two-dimensional ultrasound (2DUS) traditionally has relied on acquisition of images from a variety of orientations in which the operator has a good eye–hand linkage to assist in feature recognition. As a result, ultrasound imaging has been one of the few areas of medical imaging that has not routinely used standardized viewing orientations, relying instead on the interactivity of the imaging process to optimize visualization of patient anatomy. Because viewing of important landmarks is essential for interpretation and identification of anatomy, particularly in less skilled practitioners, one benefit that 3DUS brings to patient diagnosis is that data review may be carried out at the console after the patient has left the clinic. Because patient data can be reoriented to standard anatomic position, viewer comprehension and recognition of anatomy are enhanced.

An important part of 3DUS is the ability to review patient data interactively. The flexibility to rotate, scale, and view objects from perspectives that optimize visualization of the anatomy of interest is critical. Physician involvement in optimizing and enhancing the tools to accomplish this is essential in the ongoing evaluation of all these techniques.

At present, real-time 2DUS makes it possible for physicians to make important contributions to patient management. However, there are occasions when it is difficult to develop a three-dimensional impression of the patient anatomy, particularly with curved structures, when there is a subtle lesion in an organ, when a mass distorts the normal anatomy, or when there are tortuous vessels or structures not commonly seen that can be difficult to visualize with 2DUS methods. The typical approach to overcome this problem is to scan repeatedly through the region of interest to clarify the exact spatial relationships. Complex cases often make it difficult even for specialists to understand three-dimensional anatomy based on two-dimensional images and can be time-consuming. Abnormalities may be difficult to demonstrate with 2DUS because of the particular planes that must be imaged to develop the entire three-dimensional impression.

Advanced technology permits volume imaging methods to be applied to diagnostic ultrasound incorporating interactive manipulation of volume data using rendering, rotation, and zooming in on localized features. Integration of views obtained over a region of a patient with 3DUS may permit better visualization in these situations and allow a more accurate diagnosis (3–5). Although the eventual role for 3DUS has yet to be determined, there is little doubt that its impact will be broad and substantial. 3DUS also has an important role in demonstrating normalcy and reassuring patients. Ultimately, the success of 3DUS methods will depend on providing performance and capabilities that exceed those of 2D sonography. In the near future 3DUS will be a routine part of patient diagnosis and management.

Ultrasound imaging and volume visualization methods can assist this process by presenting the entire volume in a single image. Currently, most volume visualization is performed using graphics workstations to provide interactive performance. Clinical scanners such as the Voluson 530D (Medison, Korea) are increasingly integrating workstation performance in the scanner to provide volume visualization at the patient's bedside. Volume imaging provides the physician with the capability to image the patient after he or she has left the scanning suite and to reevaluate the diagnosis with experts across networks at remote locations. 3DUS offers new opportunities in patient visualization and has sparked interest both in the academic community and in commercial industry (5,6).

Early clinical experience with 3DUS suffered from the limitations of available technology. Nonetheless, early work clearly demonstrated the potential of using ultrasound for volume imaging (7–11). The explosive growth of 3DUS over the past few years is readily

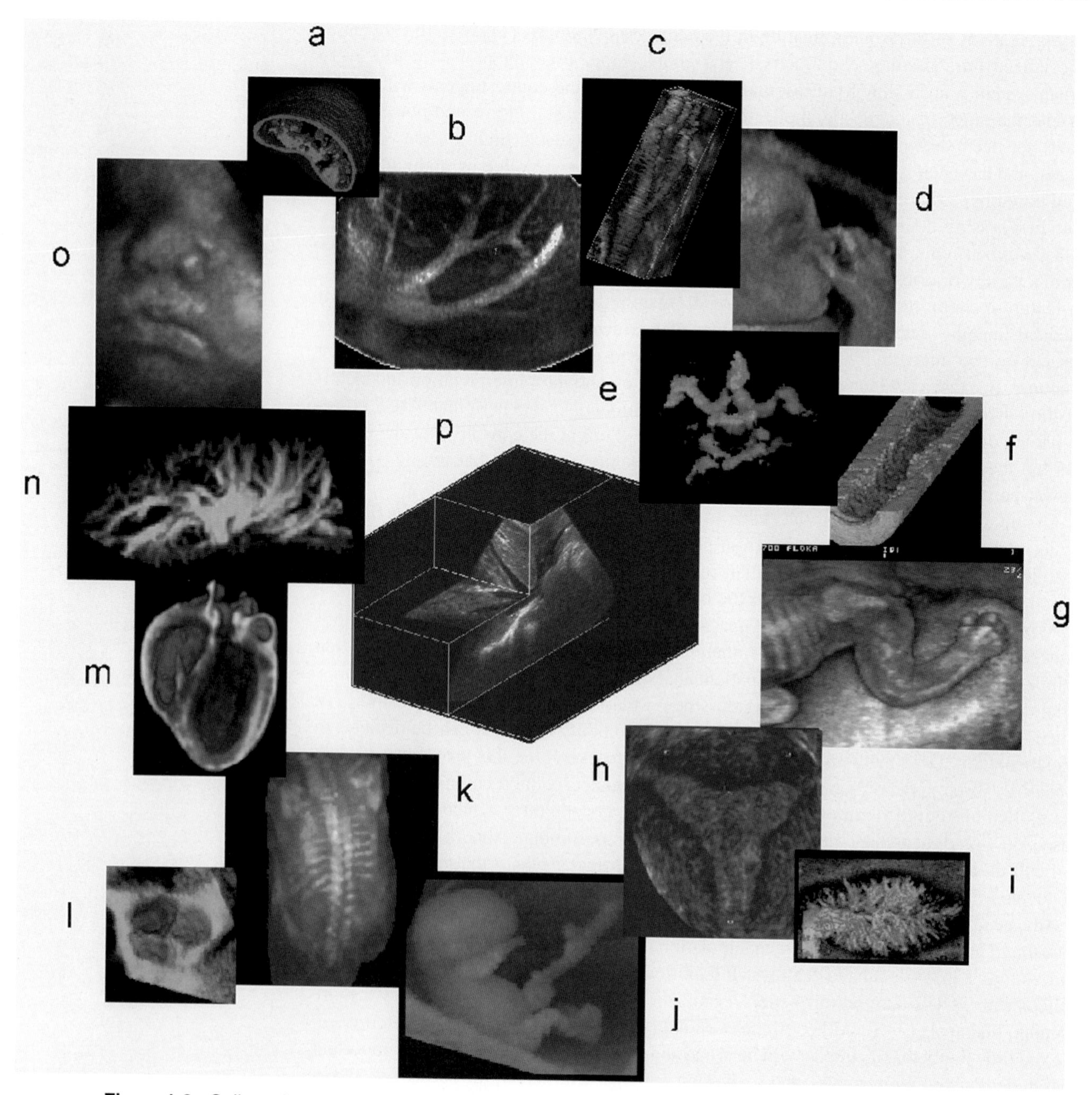

Figure 1-2. Collage demonstrating several applications of 3DUS to image vascular, cardiac, and obstetric patients. **a**, prostate cancer; **b**, liver vasculature; **c**, carotid arteries; **d**, fetal face; **e**, adult circle of Willis; **f**, coronary artery; **g**, fetal skeleton; **h**, uterus; **i**, lymph node; **j**, 10-week embryo; **k**, fetal spine; **l**, aortic valve; **m**, dog heart; **n**, kidney vasculature; **o**, fetal face; **p**, liver anatomy. See color version of figure.

appreciated by reviewing the number of published articles, which reflects the recent availability of improved clinical 3DUS imaging equipment. Although the areas of clinical applicability for 3DUS methods continue to expand, the earliest work has been in the field of cardiology, with more recent efforts focused on obstetric, vascular, and gynecologic areas (Figure 1-2). Identification of critical clinical areas in which 3DUS contributes to the patient management is a central focus of much research in 3DUS worldwide.

Among the many potential advantages of 3DUS compared with current 2DUS methods are improved visualization of normal and abnormal anatomic structures, and evaluation of complex anatomic structures for which it is difficult to develop a 3D understanding. Reduced patient scanning times compared with current 2DUS techniques could increase the number of patients scanned, thereby increasing operational efficiency and providing more cost-effective use of sonographers and equipment. Standardization of the ultrasound examination protocols can lead to uniformly high-quality examinations and decreased health care costs.

Ultimately, the improved understanding of patient anatomy offered by 3DUS may make it easier for primary-care physicians to understand complex patient anatomy. Tertiary-care physicians specializing in ultrasound can further enhance the quality of patient care by using high-speed computer networks to review volume ultrasound data at specialization centers. Access to volume data and expertise at specialization centers afford more sophisticated analysis and review, further augmenting patient diagnosis and treatment.

3DUS ultimately will increase patient throughput and reduce health care costs through decreased patient imaging time, improved understanding of complex anatomy and physiologic spatial relationships, and, when necessary, rapid computer network transfer of data to more experienced physicians. Shared high-speed networks will connect examination rooms, with specialists around the world sharing volume data easily. Physicians in different cities will be able to investigate three-dimensional patient data collaboratively, which will lead to instantaneous data analysis and diagnosis that will improve health care delivery.

THIS BOOK

The field of 3DUS is advancing rapidly. This book is intended to provide clinicians, scientists, and engineers with an up-to-date overview of the rapidly emerging field of 3DUS or volume sonography and to review the current state of development of 3DUS. It is intended to serve as a reference for those who wish to learn more about 3DUS imaging, and it will unify much of the work published to date.

The following chapters review the state of the art with respect to 3DUS imaging, methods of data acquisition, analysis, and display approaches. Clinical chapters summarize patient research study results to date with discussion of applications by organ system and areas of greatest current and future diagnostic importance. The basic algorithms and approaches to visualization of 3DUS and 4DUS are reviewed, including issues related to the interaction of the user with 3DUS data and different user interfaces. The implications of recent developments for future ultrasound imaging/visualization systems are considered. Emerging technologies for volume imaging will be discussed in the context of real-time volume imaging, commercialization, and clinical use. The design of clinical efficacy studies will be important to the acceptance of volume sonography into the clinical arena.

REFERENCES

1. Wells PNT. The present status of ultrasonic imaging in medicine. *Ultrasonics* 1993;31:345–353.
2. ter Haar G. Ultrasound focal beam surgery. *Ultrasound Med Biol* 1995;21:1089–1100.
3. Baba K, Jurkovic D. *Three-dimensional ultrasound in obstetrics and gynecology.* New York: Parthenon Press, 1997.
4. Baba K, Okai T, Kozuma S, Taketani Y, Mochizuki T, Akahane M. Real-time processable three-dimensional US in obstetrics. *Radiology* 1997;203(2):571–574.
5. Pretorius DH, Nelson TR. 3-Dimensional ultrasound imaging in patient diagnosis and management: the future. *Ultrasound Obstet Gynecol* 1991;1(6):381–382.
6. Kossoff G. Three-dimensional ultrasound—technology push or market pull? *Ultrasound Obstet Gynecol* 1995;5:217–218.
7. Blankenhorn DH, Chin HP, Strikwerda S, Bamberger J, Hestenes JD. Common carotid artery contours reconstructed in three dimensions from parallel ultrasonic images. Work in progress. *Radiology* 1983;148: 533–537.
8. Brinkley JF, McCallum WD, Muramatsu SK, Liu DY. Fetal weight estimation from ultrasonic three-dimensional head and trunk reconstructions: evaluation *in vitro*. *Am J Obstet Gynecol* 1982;144:715–721.
9. Greenleaf JF. Three-dimensional imaging in ultrasound. *J Med Systems* 1982;6:579–589.
10. Matsumoto M, Inoue M, Tamura S, et al. Three-dimensional echocardiography for spatial visualization and volume calculation of cardiac structures. *J Clin Ultrasound* 1981;9:157–165.
11. Robinson DE, Display of three dimensional ultrasonic data for diagnosis. *J Acoust Soc Am* 1972;52(2): 673–687.

PART II

Three-Dimensional Ultrasound Technology

2

Acquisition Methods

OVERVIEW

This chapter will provide an overview of volume acquisition methods used in three-dimensional ultrasound (3DUS) imaging as well as a discussion of the advantages and disadvantages of each technique.

KEY CONCEPTS

- Volume acquisition factors to be considered and optimized in 3DUS imaging:
  - Fast (or gated) scanning to avoid artifacts from involuntary, respiratory, or cardiac motion.
  - Accurate and calibrated scanning geometry to avoid geometric distortions and make accurate measurements.
  - Unified scanning techniques integrated with exam procedure.
- Volume acquisition scanning approaches:
  - Mechanical scanning devices (scan geometry [linear, tilt, and rotational] defined and controlled by motorized assembly):
    - Integrated device with transducer in scanning probe.
    - External device with attachment to transducer.
  - Free-hand scanning with position sensors (routine scanning procedure with position and orientation monitored by sensing method [acoustic sensor, articulated arm, magnetic field sensor, image correlation]).
  - Free-hand scanning without position sensing (routine scanning procedure with no position data—assumed uniform and steady scanning geometry).
  - Two-dimensional array transducers.
- Volume reconstruction (feature based and voxel based):
  - Feature-based reconstruction:
    - Analyze two-dimensional ultrasound images, identify and classify desired features in image, outline and assign a color to the feature, reconstruct boundaries into three-dimensional volume.

KEY CONCEPTS, *continued*

- Voxel-based reconstruction:
  - Each pixel in the acquired 2DUS images is placed into its appropriate position in the volume based on geometric information to generate a three-dimensional grid of volume elements (voxels) and to fill gaps between acquired two-dimensional planes by interpolation.

Acquisition of the information needed to produce a three-dimensional ultrasound (3DUS) image depends on obtaining echo data throughout the volume of interest. The general features of a 3DUS acquisition system are shown in Figure 2-1. Numerous efforts have focused on the development of various types of three-dimensional (3D) imaging techniques using ultrasound's positioning flexibility and data acquisition speed (1–12). Much of the effort has focused on integrating transducer position information with the gray-scale sonogram. Over the past few decades, numerous investigators have attempted to produce clinically useful systems; however, because of the enormous computing demands needed to produce clinically useful systems, most attempts have not succeeded. Only in the last few years has technology progressed sufficiently to make 3DUS imaging viable. Reviews of various aspects of this emerging technology have been published (3,7,12–15).

The challenges and major differences in the various 3DUS imaging approaches come from the specific method used to locate the position of the 2DUS image within the tissue volume under investigation. The performance requirements for a positioning system are quite exacting and are becoming more so with improvements in ultrasound scanners. Clearly, the choice of the scanning technique used to produce the 3DUS data is crucial to produce accurate representation of the anatomy.

In general, current 3DUS imaging systems are based on commercially available, one-dimensional, or annular transducer arrays whose position is known accurately or monitored by a position-sensing device. Position data may be obtained from stepping motors in the scan head, a translation or rotation device, or a position sensor that may be electromagnetic, acoustic, or mechanical. During acquisition, 2DUS images and position data are stored in a computer for subsequent reconstruction into 3DUS data. Depending on the type of acquisition used, the acquisition slices may be in the pattern of a wedge, a series of parallel slices, a rotation around a central axis (e.g., from an endocavitary probe), or arbitrary orientations, as discussed in detail in the subsequent sections.

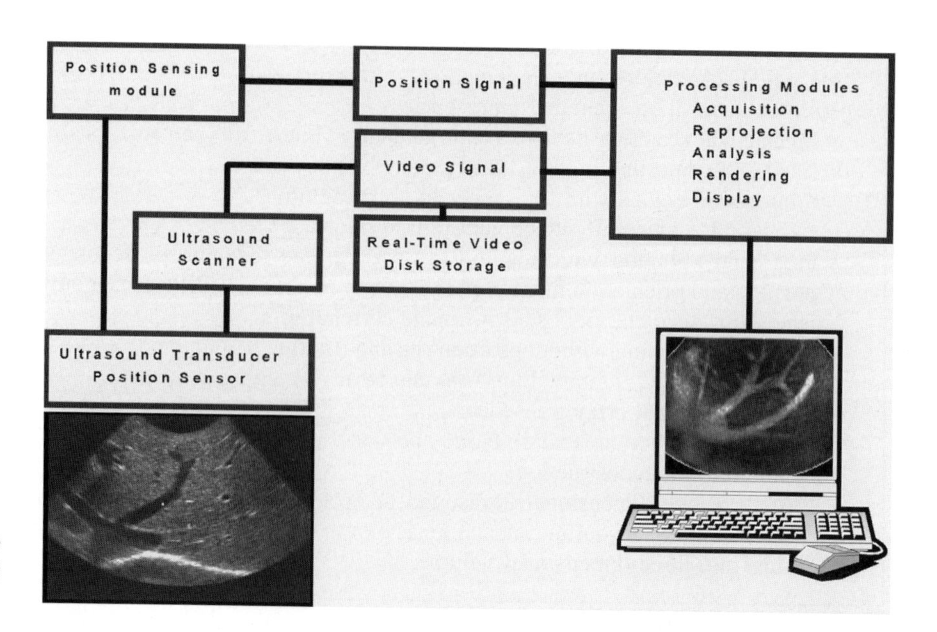

Figure 2-1. Block diagram showing the general features of a 3DUS imaging system.

Table 2-1. *Factors necessary to optimize 3DUS images*

Factor	Reason
Fast or gated	To avoid artifacts due to motion
Known geometry	To avoid distortions and allow accurate measurements
Convenient	To avoid interference with the exam and stresses on the operator

To produce proper images, three factors must be considered and optimized (Table 2-1). First, the scanning technique must be fast or gated to avoid artifacts and distortions caused by involuntary, respiratory, or cardiac motion. Second, the geometry of the scanning technique must be known accurately or calibrated correctly to avoid geometric distortions and measurement errors. Third, the scanning technique must be easy to use, so that it does not interfere with the patient exam or is not inconvenient to use. Four different approaches are currently being pursued: mechanical scanners, sensed free-hand techniques, free-hand without location sensing, and two-dimensional (2D) arrays.

MECHANICAL POSITIONING

Real-time 3DUS systems using 2D arrays are still unavailable for routine imaging, which means that most of the approaches have focused on using conventional one-dimensional (1D) transducer arrays. In this scanning approach, a series of 2DUS images is recorded rapidly while the conventional transducer is manipulated over the anatomy using a variety of techniques. These digitally recorded 2DUS images are then reconstructed into 3DUS data and made available for viewing. To avoid geometric distortions and inaccuracies, the relative position and angulation of each 2D image must be known accurately. One way to accomplish this is to use mechanical means to move the conventional transducer over the anatomy in a precise, predefined manner. As the transducer is moved, the 2DUS images generated by the ultrasound machine are acquired at predefined positional intervals (either angle or distance).

In the early days of development, images were stored on videotape for later processing, but with advances in computer technology and ultrasound instrumentation, they are now more commonly digitized into an external computer immediately or stored in original digital format in the ultrasound system's computer. Either the ultrasound machine's computer or the external computer reconstructs the 3DUS data using the predefined geometric information, which relates the digitized 2DUS images to each other. To ensure that the scanning procedure avoids missing any regions in the body, the spacing interval between the digitized 2DUS images is precomputed and usually is made adjustable to minimize the scanning time while optimally sampling the volume.

Various kinds of mechanical scanning systems have been developed, requiring the conventional transducer to be mounted in a special assembly and rotated or translated by a motor (Table 2-2). When the motor is activated under computer control, the mechanical

Table 2-2. *Mechanical scanning approaches used in 3DUS imaging*

Scanning approach	Advantages	Disadvantages
Integrated 3D probe	• Small size • Interface to ultrasound machine optimized for 3D	• Requires the purchase of a special ultrasound machine • Advances in ultrasound technology not immediately available in 3D
External fixture	• Does not require special ultrasound machine • Can interface to many machines in department • Advances in ultrasound technology available immediately in 3D	• Bulkier and heavier fixture • Systems not optimized for 3D without special calibration

Table 2-3. *Summary and major uses of mechanical scanning types*

Scanning types	Arrangements of images	Major uses
Linear	Parallel to each other with equal spacing	Vascular (carotids and peripheral), breast
Tilt	Fanlike geometry with equal angular spacing	Obstetrics, abdominal, prostate
Rotation	Propeller-like geometry with equal angular spacing	Cardiac, prostate, gynecology

assembly rotates or translates the transducer rapidly over the region being examined. Assemblies have been developed that vary in size from small integrated 3D probes housing the mechanical mechanism and motor to ones employing an external fixture connected to the conventional 2DUS transducer.

In general, the integrated mechanical scanning 3D probes are small, allowing easier use by the operator, and can be optimized for 3D imaging; however, they require the purchase of a special ultrasound machine capable of interfacing to these probes. The external fixture approach results in bulkier assemblies, but they can accommodate any conventional ultrasound machine's transducers, obviating the need to purchase a special-purpose 3DUS machine. In this way, improvements in image quality and imaging techniques (e.g., power Doppler) afforded by advances in the ultrasound machine also can be achieved in 3D. In addition, because the external fixture can house a transducer from any ultrasound machine, the 3D capability can be extended to a number of ultrasound machines in a department.

A number of investigators and commercial companies have developed different types of mechanical assemblies used to produce 3DUS data. These can be divided into three basic types of motion: linear, tilt, and rotation. The clinical applications of each type are shown in Table 2-3. The advantages and disadvantages of each type are shown in Table 2-4.

Linear Scanning

In linear scanning, the transducer translates linearly over the patient's skin, so that the acquired 2DUS images are all parallel to each other. This is accomplished by mounting the conventional linear or curved transducer in an assembly housing a motor and drive mechanism, such as a lead screw as shown in Figure 2-2. When the motor is activated, the lead screw rotates, moving the transducer parallel to the skin surface. The 2DUS images can be acquired at regular spatial or temporal intervals and made available for reconstruction as shown in Figure 2-2B. If 2DUS images are acquired with uniform linear translation motion at regular temporal intervals (e.g., 30 images/sec), then a set of parallel 2DUS images separated by a regular spatial interval will be obtained. If the temporal

Table 2-4. *Advantages and disadvantages of mechanical scanners*

Scanning type	Advantages	Disadvantages
Linear	• Simple geometry easy to reconstruct in 3D • Angulation easy to interpret for Doppler imaging	Bulky mechanism
Tilt	• Relatively compact mechanism • Simple geometry easy to reconstruct in 3D	Resolution degrades with depth
Rotation	• Ideal for transrectal and transesophageal imaging • Compact mechanism	Any motion of axis of rotation will generate artifacts in the center of the image

A

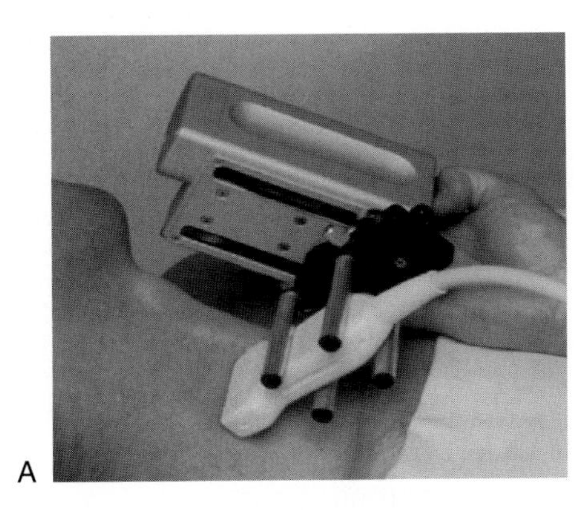

B

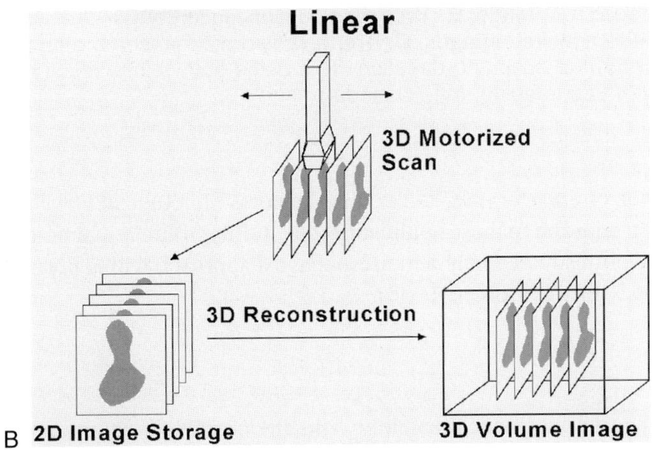

Figure 2-2. A: Photograph of a linear scanning mechanism using an external fixture approach. A transducer is inserted into the fixture and positioned over the carotid arteries. When the motor is activated, the transducer is scanned over the vessels as 2DUS images are digitized and stored in a computer. (Image courtesy of Life Imaging, Inc., Canada.) **B:** Schematic diagram showing a transducer scanning in a linear fashion over the skin. The acquired images are parallel to each other, with a fixed distance between them.

sampling interval is fixed, then the speed of translation can be varied to change the image sampling distance of the acquisition. The ability to vary the sampling interval is important because it always allows imaging the organ in 3D with an appropriate sampling interval for the particular elevational resolution of the transducer and examination depth.

Because the acquired 2DUS images are parallel to each other (see Figure 2-2B) and the spatial sampling interval is known prior to the scan, many of the parameters required for the reconstruction can be precomputed so that the reconstruction time can be made very short. Three-dimensional reconstruction obtained from linear scanning has been shown to require less than 0.5 sec after acquisition of 200 images of 336 × 352 pixels each (16).

Because of the simplicity of the scanning motion, the resolution in the final 3DUS data can be easily understood. In the planes of the 3DUS data corresponding to the acquired 2DUS images (Figure 2-3A) the resolution will be the same as or similar to the original images. In the plane that incorporates the scanning direction and the axial direction of the original images (Figure 2-3B), the resolution in the scanning direction will be worse because of the poor elevational resolution of the transducer. Similarly, in the plane that incorporates the lateral direction of the original images and the scanning direction (Figure 2-3C), the resolution will be worse in the scanning direction. Clearly, for optimal results, a transducer with good elevational resolution should be used.

The linear scanning approach has been used successfully in a number of vascular imaging application using B-mode (17–21), color Doppler imaging of the carotid arteries

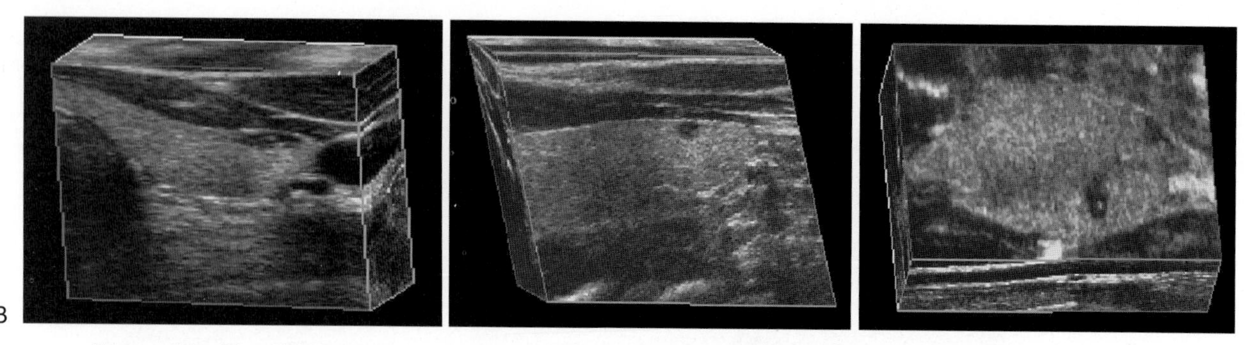

Figure 2-3. The 3DUS data of a thyroid acquired by the linear scanning approach have been sliced in three orthogonal directions to demonstrate the different resolution in each of the directions. **A:** The revealed plane corresponds to an acquired 2DUS image. **B:** The revealed plane incorporates the scanning direction and the axial direction of the acquired images. **C:** The revealed plane incorporates the lateral direction of the acquired images and the scanning direction. This plane shows less resolution than those in **A** and **B**. (Image courtesy of Life Imaging, Inc., Canada.)

(18,22–26), tumor vascularity (19,27–29), test phantoms (23,30), and power Doppler imaging (18,19,23). In addition to these applications, the utility of linear scanning in echocardiology has been demonstrated using a transesophageal approach with a horizontal scanning plane and pullback of the probe (31–33).

Tilt Scanning

In tilt scanning, the mechanical assembly tilts the transducer about an axis parallel to the axis of the transducer as shown in Figure 2-4A. This type of motion using the external fixture approach allows the transducer face to be placed at a single location on the patient's skin. Activating the motor causes the transducer to pivot on the point of contact of the skin (Figure 2-4B). Use of the integrated 3D probe approach, which incorporates the transducer inside a housing, allows the axis of the tilting motion to be located away from the patient's surface. In this way the housing is placed against the skin, while the transducer is tilted and slides against the housing to produce an angular sweep as shown in Figure 2-4C.

The process of acquiring the 2DUS images at fixed temporal intervals during the tilting motion or at regular angular intervals results in the computer storage of a set of 2D image planes arranged in a fanlike geometry. This approach sweeps out a large region of interest with a fixed predefined angular separation as shown schematically in Figures 2-4B and 2-4D (22,34–37).

This simple approach lends itself to compact designs for both integrated 3D probes and external fixtures. Medison (Korea—also known as Kretztechnik [Zipf, Austria]) and Aloka (Japan) have demonstrated an integrated 3D transducer for use in abdominal and obstetric imaging that is coupled to a special 3DUS system, whereas Life Imaging Systems, Inc. (London, Canada) has demonstrated an external fixture system for use in abdominal imaging that can couple to any manufacturer's ultrasound machine.

This type of scanning can also be used for transesophageal and transrectal transducers (Figures 2-4E and 2-4F). In this approach, a side-firing linear transducer array is used for both the external fixture approach and the integrated 3D probe. When the motor is activated, the transducer rotates about its axis while 2DUS images are digitized. After a rotation of about 90 degrees, the acquired 2DUS images are arranged in a fanlike geometry in the same way as that obtained with the tilting scan used for abdominal imaging. This approach has been used successfully in prostate imaging (18,19,22,38,39) and in 3DUS guided cryosurgery (16,40,41), and has been commercialized by Life Imaging Systems, Inc. (London, Canada). Successful applications in echocardiology have been demonstrated by TomTec, Inc. (Munich, Germany) using a transesophageal approach in which the imaging plane is vertical (i.e., parallel to the axis of the probe) and the probe is made to rotate by

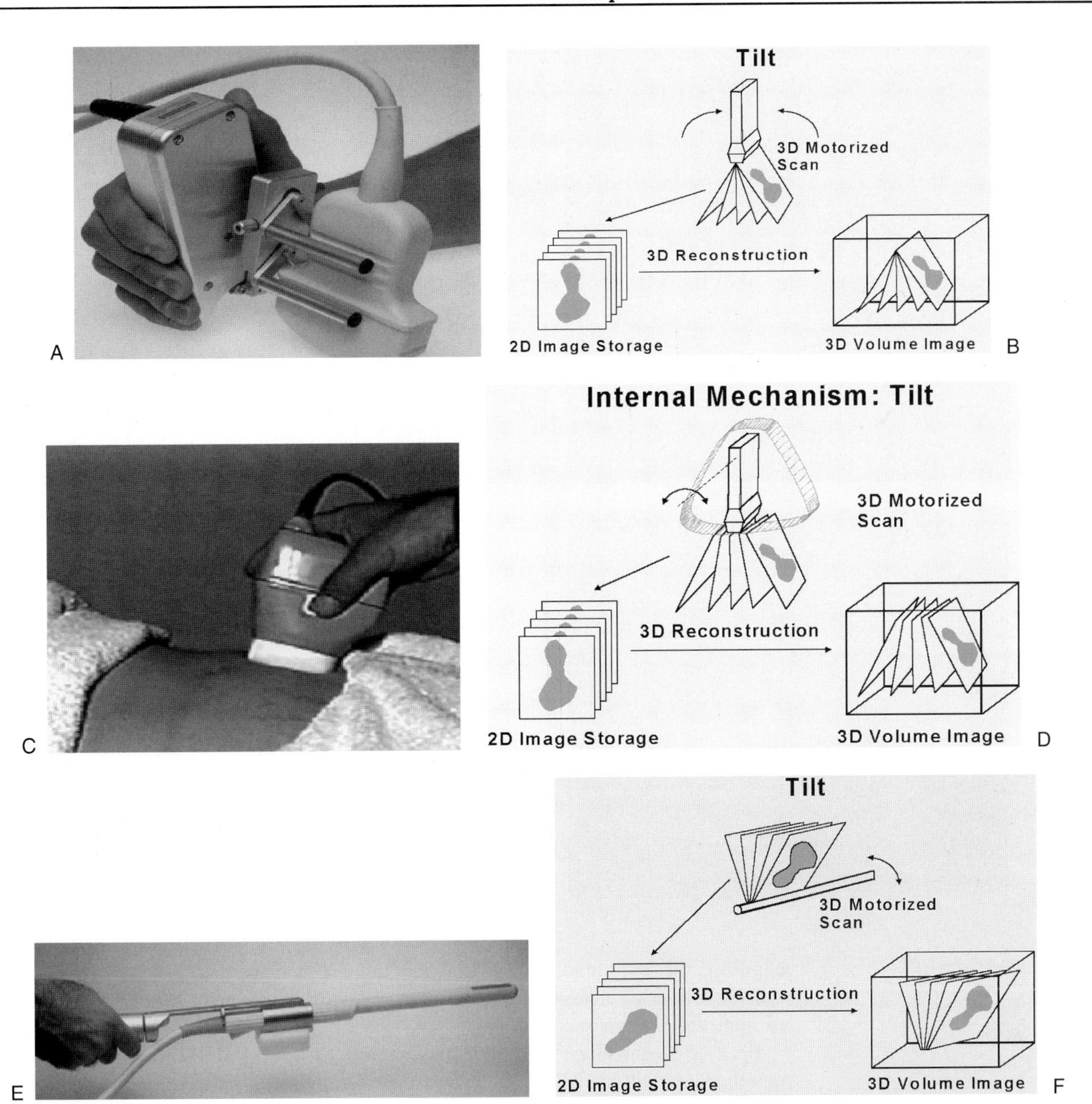

Figure 2-4. A: Photograph of a tilt scanning mechanism using an external fixture approach. When the motor is activated, the transducer is held against the skin and tilts about its front face with the probe sensor held against the skin. (Image courtesy of Life Imaging, Inc., Canada.) **B:** Schematic diagram showing a transducer scanning in a tilting fashion with its front face fixed. The acquired images are arranged in a fanlike geometry, with a fixed angular separation between images. **C:** Photograph of a tilt scanning mechanism integrated inside an ultrasound probe housing. **D:** Schematic diagram showing a tilt scanning mechanism inside an ultrasound probe housing. **E:** Photograph of a transrectal tilting scanning mechanism housed in an external fixture. When the motor is activated, the transducer rotates about the long axis of the probe, which must be held stationary during the scan. (Image courtesy of Life Imaging, Inc., Canada.) **F:** Schematic diagram showing the transrectal tilt scanning mechanism. The acquired images are arranged in a fanlike geometry, with a fixed angular separation between images.

an external motor assembly with the axis of rotation along the central axis of the probe (42,43).

The main advantage of the tilt scanning technique, which generates a set of planes arranged in a fan, is that the mechanism can be made compact to allow easy hand-held manipulation with typical scanning times of about 5–10 sec. As with the linear scanning

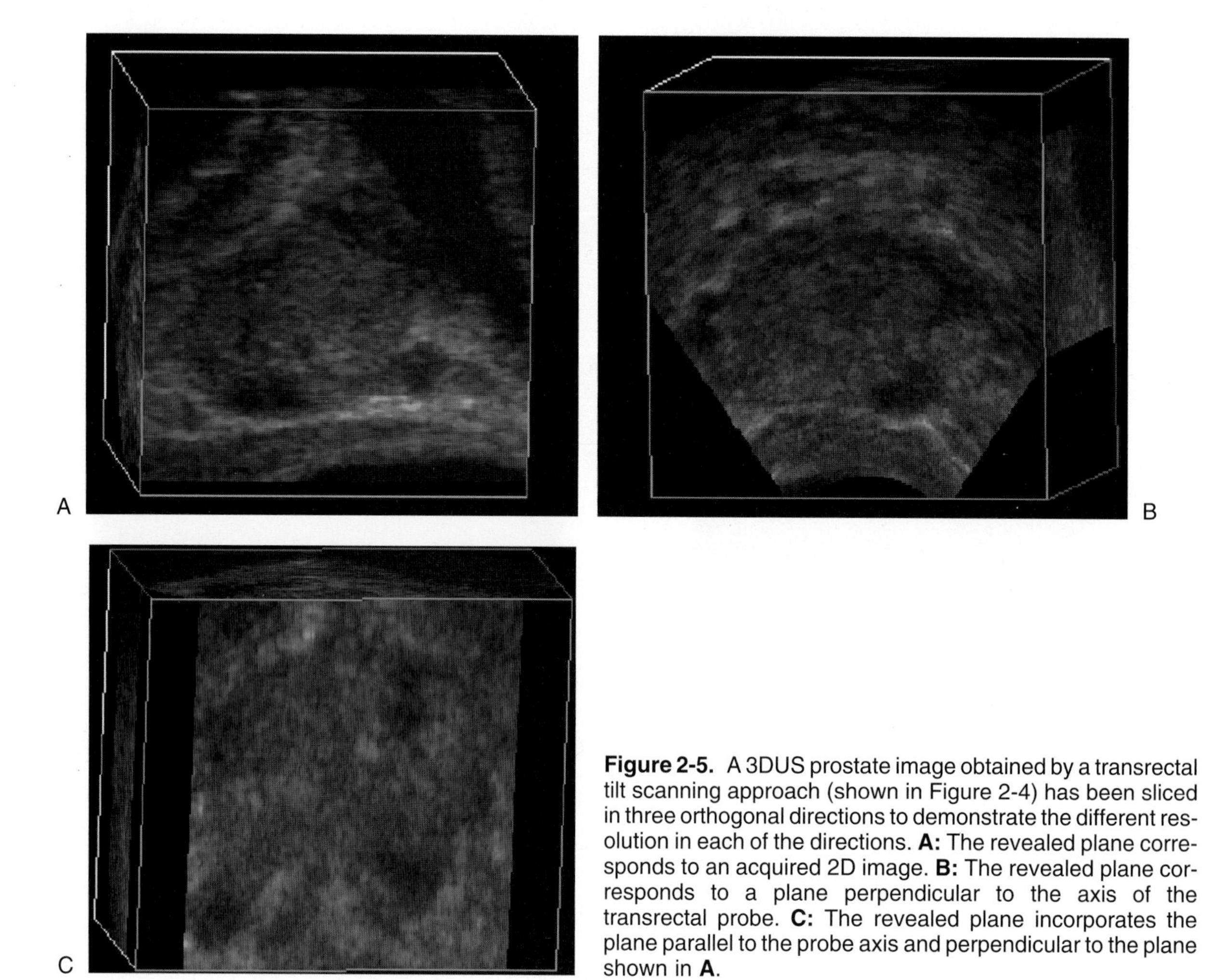

Figure 2-5. A 3DUS prostate image obtained by a transrectal tilt scanning approach (shown in Figure 2-4) has been sliced in three orthogonal directions to demonstrate the different resolution in each of the directions. **A:** The revealed plane corresponds to an acquired 2D image. **B:** The revealed plane corresponds to a plane perpendicular to the axis of the transrectal probe. **C:** The revealed plane incorporates the plane parallel to the probe axis and perpendicular to the plane shown in **A**.

approach, the set of planes is acquired with a predefined geometry but with equal angular spacing. Therefore many of the geometric parameters can be precalculated, allowing short or immediate reconstruction times.

Because the angular step between acquired images is fixed, the distance between acquired planes will increase with depth. Near the transducer, the sampling distance will be small, whereas far from the transducer the distance will be large (see Figures 2-4B and 2-4C and Figure 2-4F). This change in sampling distance with depth coincidentally matches the elevational resolution degradation of 2DUS transducers with depth. Using an appropriate choice of the angular steps (0.5–2.0 degrees), the sampling can be made to match the elevational resolution so that no additional degradation of resolution is apparent. However, the resulting resolution in the 3DUS data will not be isotropic. The resolution will be worse in the scan (tilting) direction and far from the transducer, because of the combined effects of poor elevational resolution and sampling (see Figure 2-5).

Rotational Scanning

In rotational scanning, the ultrasound transducer is placed into an external assembly or incorporated inside a 3D mechanical probe. In both cases, a motor rotates the transducer array by more than 180 degrees around an axis that is perpendicular to the array and bisects it as shown in Figure 2-6. This rotation geometry allows the axis of rotation to be fixed, while the acquired images sweep out a conical volume in a propeller-like fashion. As with the other mechanical scanning approaches, a fixed angular or temporal sampling interval will result in a set of acquired 2DUS images arranged as shown in Figure 2-6B.

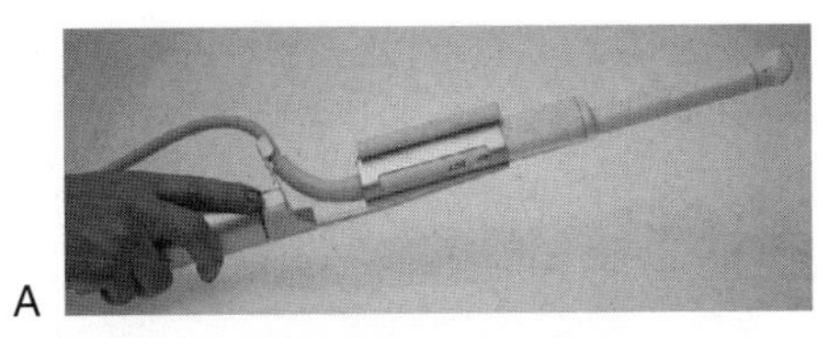

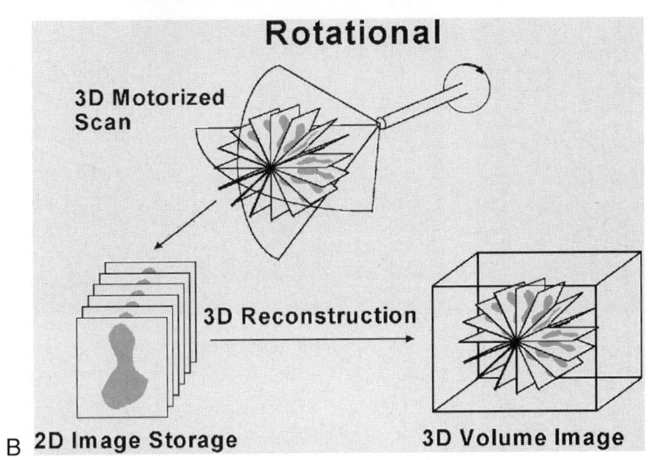

Figure 2-6. A: Photograph of a rotational scanning mechanism using an external fixture approach. When the motor is activated, the transducer rotates about the long axis of the probe, which must be held stationary during the scan. (Image courtesy of Life Imaging Systems, Inc., Canada.) **B:** Schematic diagram showing a transducer rotational scanning approach. The acquired images are arranged in a propeller-like geometry, in which the images intersect along the scanning axis, and the angular separation between images is fixed.

In rotational scanning, the images intersect along the central rotation axis, with the highest spatial sampling being near the axis and poorest away from it (Figure 2-6B). Thus the resolution in the 3DUS data will vary in a complicated fashion. The resolution will be highest closer to the transducer, degrading as one moves further from the transducer because of resolution loss in the axial direction in the original 2DUS images and the reduced elevational resolution with depth. Moreover, the angular sampling may impose additional resolution degradation away from the central scanning axis if the sampling is not optimized (see Figure 2-7).

Rotational scanning has been used successfully by Life Imaging Systems, Inc. (London, Canada) for transrectal imaging of the prostate and endovaginal imaging. In this system, 200 images (320 × 320 pixels each) are acquired at 15 images/sec over a 200-degree angular sweep. Investigators have used rotational scanning for imaging the prostate (19) and imaging the heart using a multiplane transesophageal transducer (6,44–46).

In the rotational scanning geometry, the acquired images intersect along the rotational axis in the center of the volume. Thus any motion of the axis of rotation or of the patient during the acquisition will cause the acquired images to be inconsistent. For example, if motion were to occur during scanning, the images acquired at 0 degrees and 180 degrees would not mirror each other. Thus if the reconstruction does not compensate for this, artifacts will be present in the center of the image. In addition to the sensitivity to motion, the relative geometry of the acquired images must be accurately known. Thus any image tilt or offset from the rotation axis must be known and the reconstruction must compensate for these to avoid artifacts in the center of the 3DUS data.

FREE-HAND SCANNING WITH POSITION SENSING

Although the mechanical scanning approach to ultrasonography offers speed and accuracy, at times the bulkiness and weight of the devices hinder the scan, particularly when

A

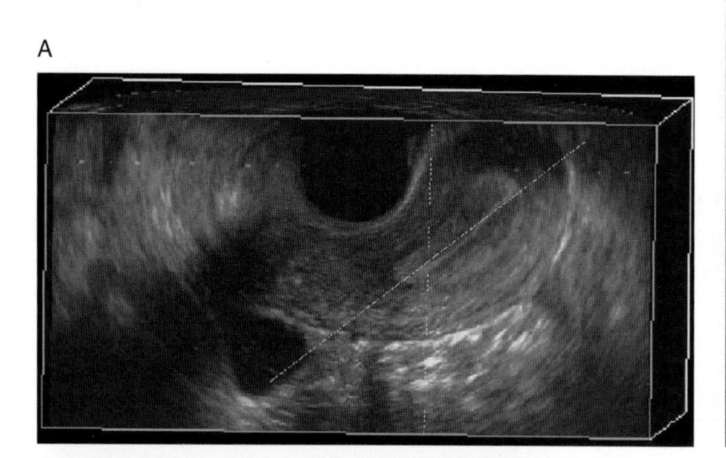

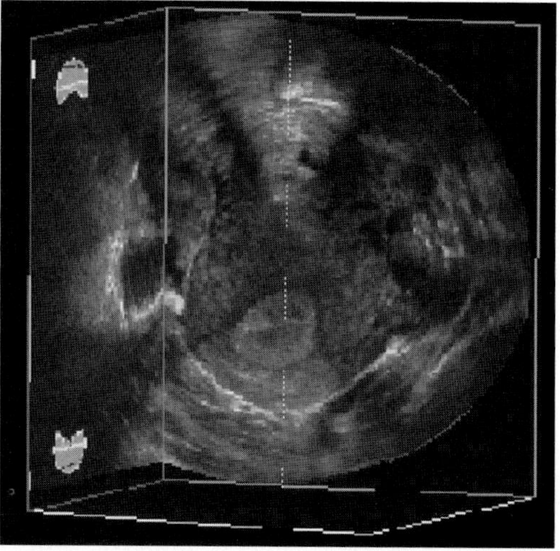

B

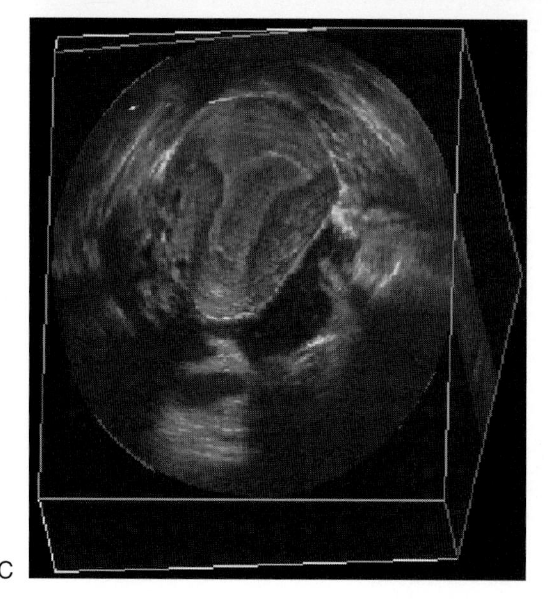

C

Figure 2-7. 3DUS study of a normal anteflexed uterus in a 19-year-old woman. Images obtained by mechanically rotating an end-firing endovaginal transducer. **A:** Longitudinal perspective of the uterus with prominent endometrium in the fundus. **B:** The 3DUS image has been sliced orthogonal to the axis of the transducer. The echogenic fundus is well seen again toward the right side of the fundus. The endometrium in the mid-uterus is not seen at all. However, on the left side of the image, the endometrium is seen containing some nabothian cysts. **C:** The 3DUS image has been sliced at an angulated plane to put the fundus at the top of the image. By slicing obliquely into the uterus, an improved perception of the anatomy is possible. The *dotted lines* show the locations of the intersecting planes. The *solid line* shows the orientation of the angulated plane.

imaging large structures. To overcome this problem many investigators have attempted to develop various free-hand scanning techniques in which the operator can hold the transducer with a position sensor attachment and manipulate it over the anatomy in the usual manner. While the transducer is being manipulated, the conventional 2DUS images generated by the ultrasound machine are recorded by a computer, as are their position and orientation. The operator is free to select the optimal views and orientation to best view the anatomy and accommodate complex surfaces of the patient. Since geometric information about the transducer's location is not predefined as with the mechanical scanning approaches, the exact relative position and angulation of the digitized 2DUS images must be known accurately. This information is then used in the reconstruction algorithm to produce a 3DUS volume in a manner that avoids distortions and inaccuracies. In addition, because the relative locations of the acquired images are not predefined, the operator must ensure that no significant gaps are left when scanning the anatomy. This generally can be achieved with training and by using careful scanning techniques. Over the past two decades, a number of free-hand scanning approaches have been developed that make use of four basic position sensing techniques: acoustic, articulated arm, magnetic field, and image correlation methods (Table 2-5).

Table 2-5. *Summary of free-hand scanning with position sensing methods*

Scanning method	Required equipment	Measurement method
Acoustic sensor	Spark gap sound emitters on the transducer and microphones above the patient	Measure speed of sound between emitters and microphones
Articulated arm	Multiple jointed arms assembly attached to transducer	Measure angles between movable arms
Magnetic sensor	Magnetic sensor on transducer and electromagnetic emitter beside the patient	Measure AC or DC magnetic field strength using three orthogonal coils
Image correlation	Fast processor for on-the-fly image correlation calculations and calibration for transducer	Measure image-to-image correlation loss to estimate position

Acoustic Sensors

Acoustic sensing was one of the first free-hand scanning methods for producing 3DUS images. This method uses three sound-emitting devices, such as spark gaps, mounted on the transducer, and an array of fixed microphones, which are typically mounted above the patient. To produce 3DUS data, the operator moves the transducer freely over the patient's anatomy, scanning the volume of interest. During the scan, the sound-emitting devices are active and the microphones continuously receive sound pulses while the 2DUS images are recorded. The position and orientation of the recorded images are determined from knowledge of the speed of sound in air and the time of flight of the sound pulses from the emitters on the transducers to the fixed microphones. To be able to determine the position and orientation of the recorded images, the fixed microphones must be placed over the patient in a way that provides unobstructed lines of sight to the emitters and to be sufficiently close to allow detection of the sound pulses. Because the speed of sound varies with temperature and humidity, corrections must be made to avoid distortions in the 3DUS data (2,47–53).

Articulated Arms

A transducer mounted on a multiple-jointed mechanical arm system is the simplest approach for free-hand scanning. Potentiometers located at the joints of the movable arms provide the information about their relative angulation during the scan. This arrangement allows the operator to manipulate the transducer and scan the desired patient anatomy, whereas the computer records the 2DUS images and the relative angulation of all the arms. From this information the position and angulation of the transducer and each recorded 2DUS image can be computed and continuously monitored. The 3DUS volume is then reconstructed using the series of recorded 2DUS images and their geometric information. To avoid distortions and inaccuracies in the final image, the potentiometers must be accurate and precise, and the arms must not flex. Sufficient accuracy can be achieved by keeping individual arms as short as possible and reducing the number of movable joints. However, increased precision and accuracy in the 3DUS data are achieved at the expense of reducing flexibility in scanning and in the size of volume that can be imaged.

There have been a number of successful implementations of the articulated arm approach primarily for echocardiographic measurements of ventricular volume. A number of research groups have reduced the number of movable joints to three to simplify the mechanism and increase its accuracy (5,54–57). Recently, articulated arm mechanisms have become available commercially (FARO Technologies, Inc.) that have sufficient accuracy to be useful for 3DUS imaging, although the volume that can be scanned is limited in size.

Magnetic Field Sensors

Currently, the most popular and successful free-hand positioning approach makes use of a six-degrees-of-freedom magnetic field sensor to measure the ultrasound transducer's position and orientation. This approach, shown in Figure 2-8, uses a transmitter, which produces a spatially varying magnetic field, and a small receiver containing three orthogonal coils to sense the magnetic field strength. By measuring the strength of three components of the local magnetic field, the ultrasound transducer's position and angulation can be determined.

Although magnetic field sensors allow for less constraining geometric tracking, they are susceptible to noise and errors. These devices are sensitive to electromagnetic interference from sources such as CRT monitors, AC power cabling, and ultrasound transducers. In addition, ferrous and highly conductive metals will distort the magnetic field causing errors in the recorded geometric information. All these effects can combine to produce

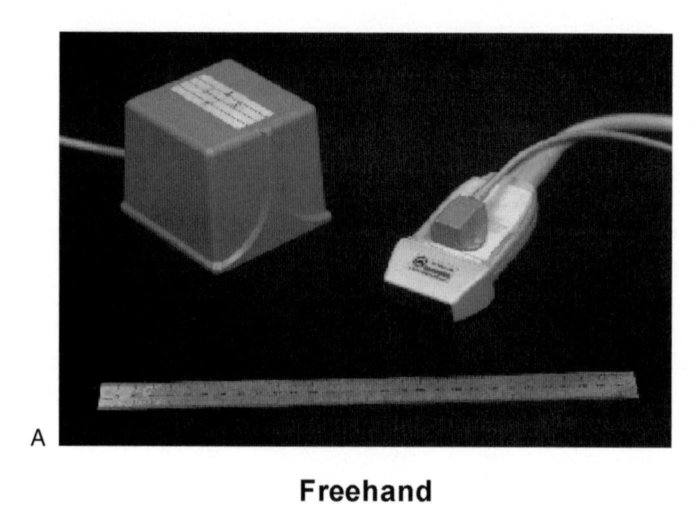

Figure 2-8. A: An ultrasound linear array transducer (Advanced Technology Laboratories, Bothell, WA) and a magnetic localizer (Flock of Birds, Ascension Technologies, Colchester, VT) are shown. (Image courtesy of S. Sherebrin.) **B:** Schematic diagram showing a free-hand scanning approach with sensing of the location of the transducer. The position and orientation must be obtained to allow reconstruction of a 3DUS volume. This information can be obtained using position sensing techniques such as acoustic, articulated arm, and magnetic field. This provides a flexible method of performing free-hand vascular ultrasound scans.

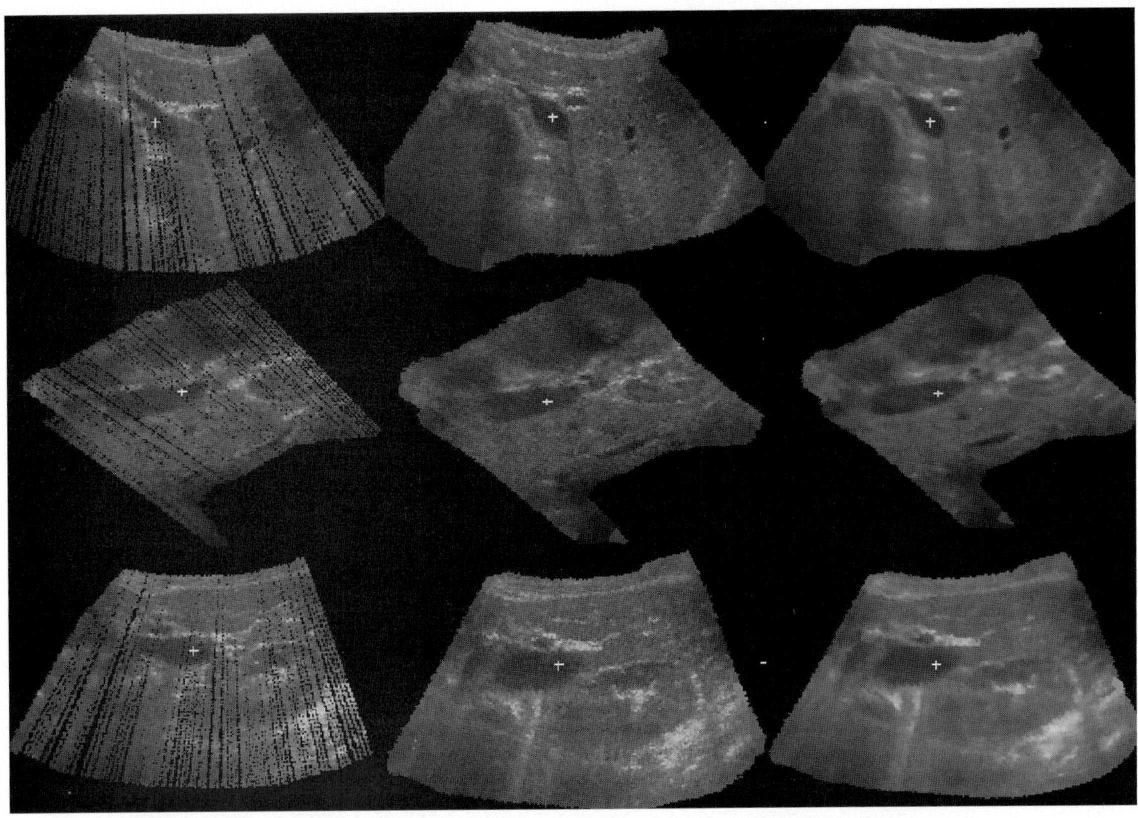

Figure 2-9. A 3DUS free-hand scan with position sensing using an electromagnetic positioning system showing three orthogonal planes (across) through the liver. The first vertical set of images (**A**) is the original acquisition with gaps where no data were acquired. The center vertical set of images (**B**) is the same data after application of a nearest neighbor augmentation algorithm to estimate the value of the missing data. The right vertical set of images (**C**) is following application of a 3-pt cubic (3 × 3 × 3) median filter to the data. The median filter reduces speckle without significantly reducing resolution of the data.

long-range 3DUS data distortions and local misregistration in the 3D reconstruction. In addition, this approach is susceptible to errors in the determination of the location of the moving transducer if the magnetic field sampling rate is not sufficiently high or if the transducer is moved too quickly. This image lag artifact can be overcome with a sampling rate of about 100 Hz or higher. By ensuring that the environment of the scanned volume is free of electrical interference and metals, high-quality 3DUS data can be obtained with the appropriate sampling rate.

Two companies are currently producing magnetic positioning devices of sufficient quality for 3D ultrasound imaging: the Fastrack by Polhemus (Colchester, VT) and Flock of Birds by Ascension Technologies. These devices have been used successfully in echocardiography, obstetrics, and vascular imaging (Figure 2-9) (4,10,11,14,22,58–65).

Image Correlation Techniques

It is well known that the image speckle pattern and features change as the transducer is moved across the patient. The rate at which the image features change is a function of the transducer frequency and beam-forming properties and the patient anatomy. With careful calibration, the change in image features can be measured and used to estimate the relative change in the transducer position (66–68). This principle is being exploited on a commercial scanner to obtain 3DUS data (Figure 2-10). An advantage of this method is that no external position-sensing system is required. On the other hand, extremely accurate calibration of the image feature properties is required to accurately estimate the

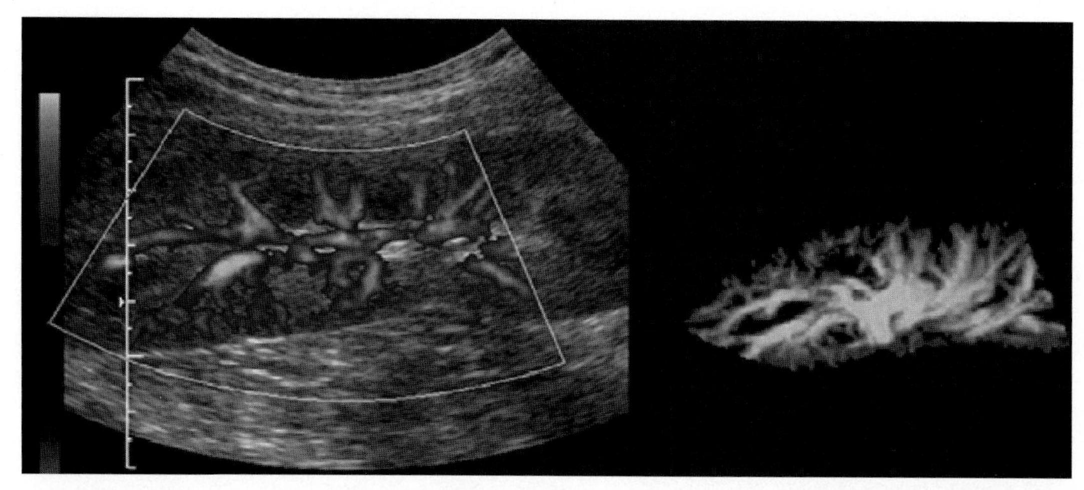

Figure 2-10. 3DUS power Doppler acquisition of a kidney using image-based position sensing to geometrically register 2DUS slices during real-time acquisition. Acquisitions can be performed with a linear, tilt, or rotational technique. The right image is one frame from the 2DUS acquisition. The left frame is from the volume data after segmentation based on the power Doppler signal. (Courtesy of Siemens Ultrasound, Issaquah, WA.) See color version of figure.

movement. Also, reversals in direction are significantly more difficult to calculate. Moreover, error propagation can reduce the overall accuracy across the field of view because even small errors in image-to-image correlation changes can grow to significant levels even over short distances. In the future this method may provide a convenient means of tracking transducer movement while scanning.

FREE-HAND SCANNING WITHOUT POSITION SENSING

Although the sensed free-hand scanning with position sensing allows the transducer to be manipulated in the usual manner, it requires that the position and orientation be measured. Thus the various sensing devices require additional attachment to the transducer, which reduces the convenience at times. An alternative approach is to manipulate the transducer over the patient without any sensing device. Clearly, because no geometric information is provided to the 3D reconstruction algorithm, the transducer must be moved in a regular and steady manner so that the acquired images are obtained with spacing as regular as possible. For example, the transducer may be scanned by hand linearly over the skin while the 2DUS images are recorded at regular temporal intervals. If the operator scans the transducer in a uniform and steady manner, then with the knowledge of the distance scanned, reasonable 3DUS images may be reconstructed (Figure 2-11) (18). Be-

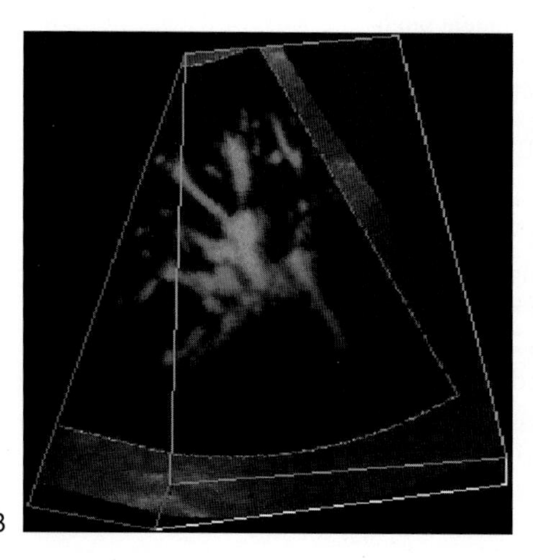

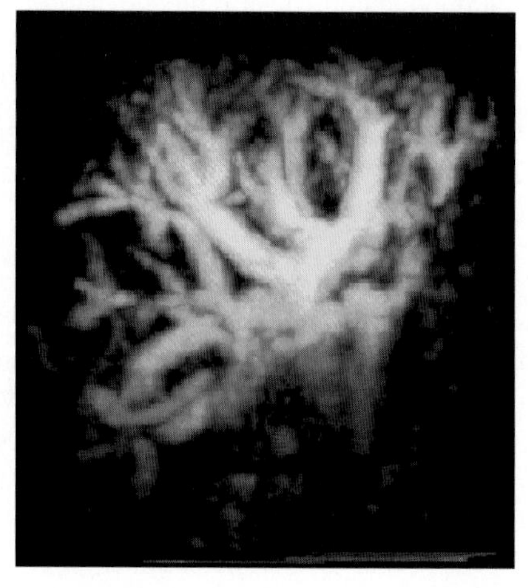

A,B

Figure 2-11. A 3DUS power Doppler image of a volunteer's spleen obtained using the free-hand scanning approach without location sensing: **A** shows the texture mapped 3DUS data; **B** shows a MIP (maximum intensity projection) rendering of the same 3DUS data. Such images may be qualitatively satisfactory but cannot be used for measurements other than those in the plane of acquisition.

cause the image geometry cannot be guaranteed to be correct, 3DUS data obtained using this approach should not be used for measurements of the patient's anatomy.

2D TRANSDUCER ARRAYS

All the 3D imaging techniques described earlier require that a planar beam of ultrasound be swept over the anatomy by the use of a 1D transducer array, which is manipulated either mechanically or free-hand. A better approach would be to keep the transducer stationary but use electronic scanning to sweep the ultrasound beam(s) over the volume. This can be accomplished with the use of 2D transducer arrays that produce an ultrasound beam covering the entire volume. A number of 2D array designs have been described, but the one developed at Duke University for real-time 3D echocardiography is the most advanced (69–76).

In the Duke approach, a transducer composed of a 2D phased array of elements is used to transmit a beam, which is directed to sweep out a pyramidal volume (Figure 2-12). The returned echoes are then processed using multiple banks of electronics to acquire and display multiple planes from the volume in real time. Although the whole 3D volume is not displayed, these planes can be manipulated interactively to allow the user to explore the whole volume under investigation. This approach is being tested for real-time echocardiographic 3D imaging by producing approximately 20 3D views/sec. Although this approach is still in the early stages of development, useful 3DUS data have been produced. However, before this approach becomes practical, a number of problems must be overcome related to the cost and low yields resulting from the manufacture of a large number of small elements, and connecting and assembling a large number of electronic leads. These problems are being overcome by the use of sparse arrays, which have produced useful images. Since the 2D arrays are small, measuring about 2 cm by 2 cm, the volume is relatively small, suitable for cardiac imaging, but not yet for larger organs.

The overall benefits of different 3DUS scanning approaches are shown in Table 2-6.

3DUS VOLUME RECONSTRUCTION

After the 2DUS images have been acquired and their relative position and orientation determined from either a predefined scanning geometry (mechanical scanners) or a free-hand scanning approach, the 3DUS data can be reconstructed. This process refers to the generation of a 3D representation of the anatomy and involves placing each acquired 2DUS image at its correct relative position to all the other images (Figure 2-13). The 3D reconstruction process can be implemented using two distinct methods: feature based and voxel based (Table 2-7).

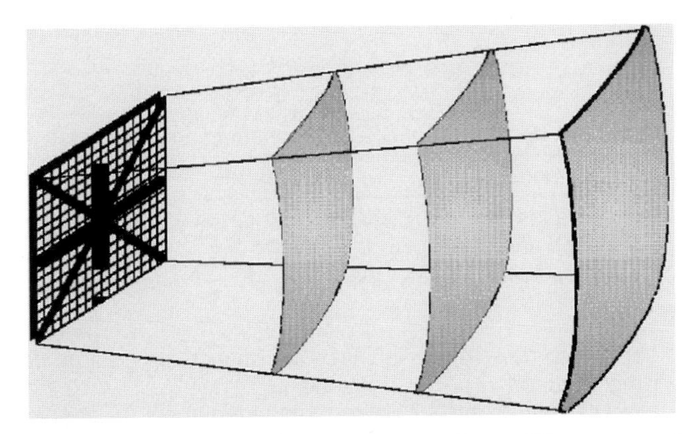

Figure 2-12. Schematic diagram showing a 2D array used in a real-time 3DUS imaging system. Also shown is the pyramidal volume being imaged and ultrasound pulses emanating from the 2D array.

Table 2-6. *Summary of 3DUS scanning approaches*

3D scanning approaches	Advantages	Disadvantages
Mechanical	Accurate geometry Predefined geometry allowing immediate reconstruction	Bulky and heavy mechanisms
Free-hand with position sensing	Easy-to-manipulate magnetic sensor requires only a small attachment	Subject to potential geometric errors Less accurate than mechanical Reconstruction more complicated
Free-hand without sensing	Easy to manipulate No attachments	Potential for distortions and inaccuracies Cannot make measurements
2D arrays	Real-time 3D imaging, ideal for moving structures such as the heart	Relatively small volume images Specialized system required Signal-to-noise and dynamic range not as good as 1D arrays

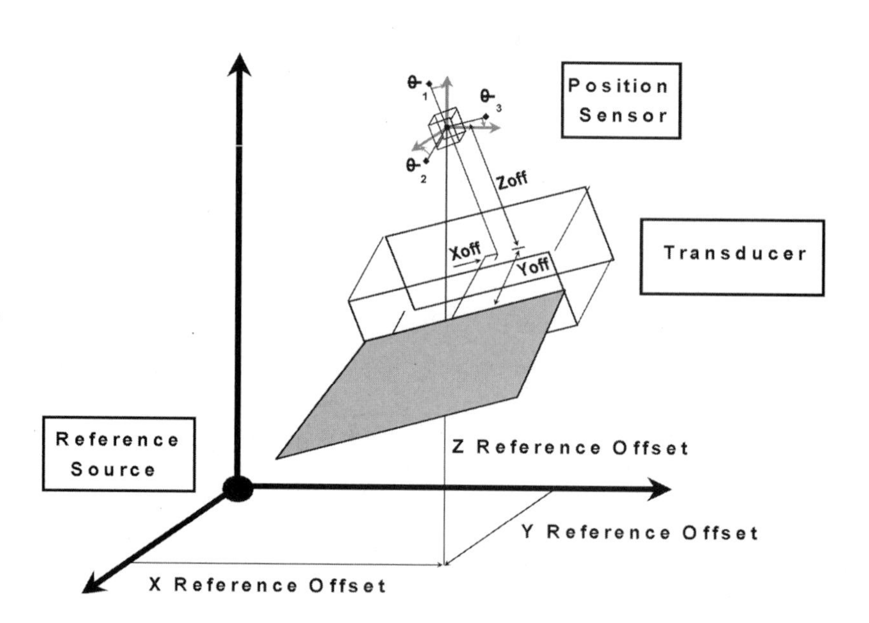

Figure 2-13. Diagram of the primary parameters used in computing the location of an image pixel in a volume referenced to a fixed location in the patient coordinate system. The position sensor provides a relative *x*, *y*, and *z* distance from the source plus three rotation angles (θ_1, θ_2, θ_3) referencing the relative orientation between the source and the sensor. The source field pattern results in two symmetric hemispheres requiring either right- or left-hand coordinate transformations, depending on the relative orientation of the system. Transducer scan plane, image, and position sensor offsets are also necessary to locate an image pixel in the patient coordinate system.

Table 2-7. *Summary of 3DUS reconstruction techniques*

Technique	Methods	Advantages	Disadvantages
Feature-based	Segmentation of surfaces and/or volumes from acquired images	Fast reconstruction Simple display of anatomy	Removes subtle tissue information Boundary description may be erroneous and tedious
Voxel-based	Place each pixel in the acquired 2DUS images into its proper location in 3D	All image information is retained Different rendering techniques are possible	Large data files must be stored and manipulated Requires special algorithms to achieve fast reconstruction

Feature-Based Reconstruction

In feature-based reconstruction, the 2DUS images are first analyzed, and desired features in each image are identified and classified. For example, in echocardiographic or obstetric imaging, the boundaries of the fluid-filled regions and tissues may be outlined manually or using an automated computerized method. The outlined structures can be assigned a value or color and other structures are eliminated or assigned a different value or color, generating solid representations of the desired structures. Alternatively, the set of outlines may be built into a surface description of the structure, which can be displayed in 3D (Figure 2-14). This approach has been used extensively in echocardiographic imaging to identify the boundaries of the ventricles either manually or automatically. From these boundary descriptions of the heart chambers, a 3D surface model is developed that can then be viewed with a computer workstation as well as used to measure the ventricular volume. A similar approach also has been used to reconstruct the vascular lumen in 3D using intravascular ultrasound imaging (IVUS).

The main advantage of this approach is that the process of reducing the 3DUS data of the anatomy to a simple description of boundaries or a few structures substantially reduces the amount of information. This leads to short 3D reconstruction times and to efficient and inexpensive viewing of the 3D information. In addition, this approach artificially increases the contrast between different structures by the manual or automatic classification step, leading to improved appreciation of the 3D anatomy (77). However, this approach also may result in major disadvantages. Because the classification process identifies boundaries and stores only simple descriptions of the anatomy, image information (e.g., subtle

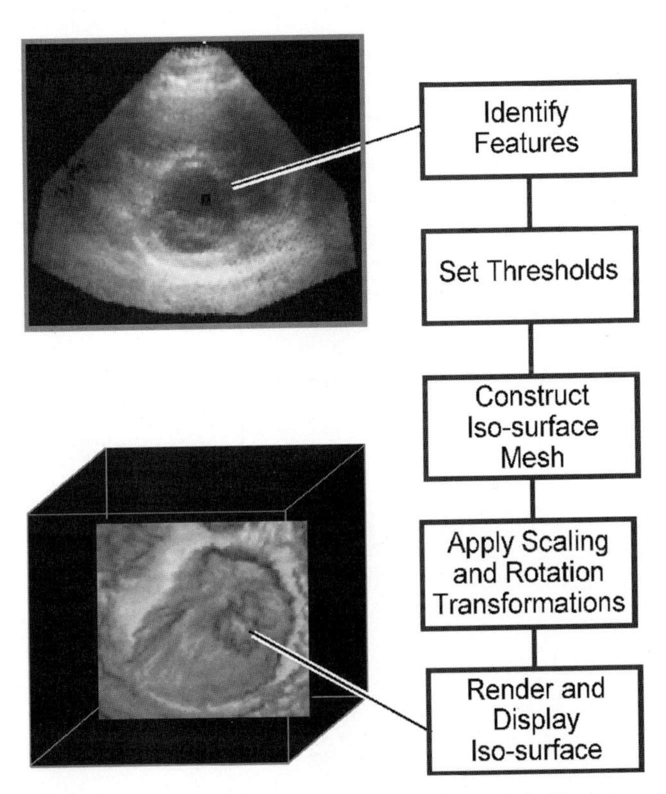

Figure 2-14. A block diagram of the basic reprojection algorithm for performing feature-based reconstruction. A feature is identified and segmented using a threshold, or some other parameter of interest. Those parts of the image and volume data set meeting the segmentation criteria are included and rendered on the display after scaling and coordinate transformations are performed. (Rendered heart image of a prolapsed mitral valve courtesy of TomTec, Inc., Munich, FRG.)

features and tissue texture) is generally lost at the initial phase of the reconstruction. The classification process also artificially accentuates the contrast between different structures, again distorting or misrepresenting subtle image features. In addition, the boundary identification step is tedious and time-consuming if done manually, and it may be erroneous if done automatically by a computer.

To avoid image artifacts and generation of false information, the classification process must be accurate. In 3DUS imaging this generally can be achieved when the image contrast is high, as found between soft tissues and fluid-filled regions, but it is very difficult in situations where the image contrast is small, as found between tumors and tissue, or even different soft tissues. Because of the tedious and potentially erroneous aspects of the boundary identification process, this method for reconstruction of 3DUS images is not used frequently. However, it is still useful in situations where volume measurements of fluid-filled regions are desired.

Voxel-Based Reconstruction

The second and more popular approach to the 3DUS image reconstruction is based on the use of the set of acquired 2DUS images to build a voxel-based volume (i.e., 3D grid of picture elements). This is accomplished by placing each pixel in the acquired 2D image in the correct location in the 3D volume matrix using its relative geometric information (Figure 2-15). If a particular voxel in the volume is not sampled by the scanning process (i.e., no 2D-acquired image intersects it), then the voxel value (color or gray scale) is calculated by interpolation using the values from neighboring voxels in the same acquired image or neighboring images. This approach does not remove any image information but rather preserves all the original information so that the original 2DUS images can always

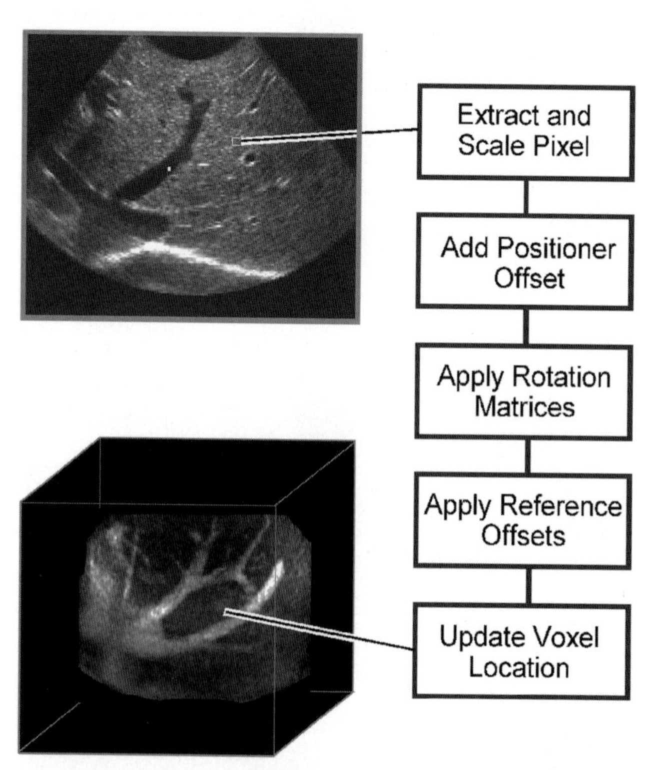

Figure 2-15. A block diagram of the basic reprojection algorithm for performing voxel-based reconstruction. Two-dimensional ultrasound image data are reprojected into a volume data set. Prior to reprojection, scaling and offset calibration data are obtained, yielding an absolute *x,y,z* coordinate for each pixel in the original image.

be generated as well as other new views not found in the original set of acquired images. However, if the volume is not sampled properly, leaving gaps between acquired images with a distance greater than half the resolution, the interpolation process will fill the gap with information that does not represent the true anatomy. In this situation, the resolution in the 3DUS data will be degraded and anatomic information will be missed.

Clearly, the voxel-based reconstruction approach makes no assumptions about the desirability of any ultrasound information and retains all the information present in the original 2DUS images. However, to avoid gaps in the scanned volume, this approach necessitates that large data files be maintained and manipulated in 3D. Typical data files can range from 16 MB to 96 MB, depending on the application. For example, data files of 96 MB have been reported in 3D prostate imaging for cryosurgical guidance (16,40). It is only in the past few years that inexpensive desktop computers could handle these sizes of 3DUS data efficiently.

Because voxel-based reconstruction maintains all the image information, the 3DUS data can be reviewed repeatedly using a variety of different rendering techniques. For example, the operator can view the complete 3DUS data and then decide on the rendering technique that may best display the desired features. The operator may classify different features, measure volumes, segment boundaries, or perform various volume-based rendering operations. If the process has not achieved the desired results, then the operator can return to the original 3DUS data and attempt a different procedure. This approach has been used successfully by several commercial companies: TomTec (Germany), Medison (Korea), Life Imaging Systems (Canada), and Aloka (Japan).

HOW TO DO IT
Practical Clinical Tips for 3DUS Data Acquisition

- Use a sufficient quantity of US gel to cover the area of scanning.
- Obtain a proper access window to the organ under investigation.
- Make sure step size between images is sufficiently small (less than one-half elevational resolution).
- *Mechanical scanning:* hold transducer assembly still during acquisition (4–10 sec).
- *Free-hand scanning:* move transducer at a speed that ensures step size is small enough; remove sources of RF noise and magnetic materials with magnetic field sensors.
- *Free-hand without sensing:* use smooth sweep with uniform velocity, without tilting or twisting.

REFERENCES

1. Baba K, Satoh K, Sakamoto S, Takashi O, Ishii S. Development of an ultrasonic system for three-dimensional reconstruction of the Fetus. *J Perinat Med* 1989;17:19–24.
2. Brinkley JF, Muramatsu SK, McCallum WD, Popp RL. In vitro evaluation of an ultrasonic three-dimensional imaging and volume system. *Ultrason Imaging* 1982;4:126–139.
3. Fenster A, Downey DB. 3-D ultrasound imaging: a review. *IEEE Eng Med Biol* 1996;15:41–51.
4. Ganapathy U, Kaufman A. 3D acquisition and visualization of ultrasound data: visualization in biomedical computing. *SPIE* 1992;1808:535–545.
5. Geiser EA, Ariet M, Conetta DA, Lupkiewicz SM, Christie LG Jr, Conti CR. Dynamic three-dimensional echocardiographic reconstruction of the intact human left ventricle: technique and initial observations in patients. *Am Heart J* 1982;103:1056–1065.
6. Ghosh A, Nanda NC, Maurer G. Three-dimensional reconstruction of echocardiographic images using the rotation method. *Ultrasound Med Biol* 1982;8:655–661.
7. Greenleaf JF, Belohlavek M, Gerber TC, Foley DA, Seward JB. Multidimensional visualization in echocardiography: an introduction. *Mayo Clin Proc* 1993;68:213–219.
8. Kirbach D, Whittingham TA. 3DUS—the Kretztechnik Voluson® Approach. *Eur J Ultrasound* 1994;1:85–89.
9. King DL, Gopal AS, Sapin PM, Schroder KM, Demaria AN. Three-dimensional echocardiography. *Am J Card Imaging* 1993;3:209–220.

10. Nelson TR, Pretorius DH. Three-dimensional ultrasound of fetal surface features. *Ultrasound Obstet Gynecol* 1992;2:166–174.
11. Nelson TR, Pretorius DH. Interactive acquisition, analysis, and visualization of sonographic volume data. *Int J Imaging Sys Technol* 1997;8:26–37.
12. Rankin RN, Fenster A, Downey DB, Munk PL, Levin MF, Vellet AD. Three-dimensional sonographic reconstruction: techniques and diagnostic applications. *AJR* 1993;161:695–702.
13. Belohlavek M, Foley DA, Gerber TC, Kinter TM, Greenleaf JF, Seward JB. Three- and four-dimensional cardiovascular ultrasound imaging: a new era for echocardiography. *Mayo Clin Proc* 1993;68:221–240.
14. Nelson TR, Elvins TT. Visualization of 3D ultrasound data. *IEEE Comput Graph Appl* 1993;(Nov):50–57.
15. Ofili EO, Nanda NC. Three-dimensional and four-dimensional echocardiography. *Ultrasound Med Biol* 1994;20:669–675.
16. Downey DB, Chin JL, Fenster A. Three-dimensional US-guided cryosurgery. *Radiology* 1995;197(P):539.
17. Blankenhorn DH, Chin HP, Strikwerda S, Bamberger J, Hestenes JD. Common carotid artery contours reconstructed in three dimensions from parallel ultrasonic images. *Radiology* 1983;148:533–537.
18. Downey DB, Fenster A. Vascular imaging with a three-dimensional power Doppler system. *AJR* 1995;165: 665–668.
19. Downey DB, Fenster A. Three-dimensional power Doppler detection of prostatic cancer. *AJR* 1995;165: 741.
20. Greenleaf JF. Three-dimensional imaging in ultrasound. *J Med Systems* 1982;6:579–589.
21. Silverman RH, Rondeau MJ, Lizzi FL, Coleman DJ. Three-dimensional high-frequency ultrasonic parameter imaging of anterior segment pathology. *Ophthalmology* 1995;102:837–843.
22. Fenster A, Tong S, Sherebrin S, Downey DB, Rankin RN. Three-dimensional ultrasound imaging. *SPIE Phys Med Imaging* 1995;2432:176–184.
23. Guo Z, Fenster A. Three-dimensional power Doppler imaging: a phantom study to quantify vessel stenosis. *Ultrasound Med Biol* 1996;22:1059–1069.
24. Picot PA, Rickey DW, Mitchell R, Rankin RN, Fenster A. Three-dimensional colour Doppler imaging of the carotid artery. *SPIE Proceedings: Image Capture, Formatting and Display* 1991;1444:206–213.
25. Picot PA, Rickey DW, Mitchell R, Rankin RN, Fenster A. Three-dimensional colour Doppler imaging. *Ultrasound Med Biol* 1993;19:95–104.
26. Pretorius DH, Nelson TR, Jaffe JS. 3-Dimensional sonographic analysis based on color flow Doppler and gray scale image data: a preliminary report. *J Ultrasound Med* 1992;11:225–232.
27. Bamber JC, Eckersley RJ, Hubregtse P, Bush NL, Bell DS, Crawford DC. Data processing for 3-D ultrasound visualization of tumour anatomy and blood flow. *SPIE* 1992;1808:651–663.
28. Carson PL, Li X, Pallister J, Moskalik A, Rubin JM, Fowlkes JB. Approximate quantification of detected fractional blood volume and perfusion from 3-D color flow and Doppler power signal imaging. *1993 Ultrasonics Symposium Proceedings*. Piscataway, NJ: IEEE, 1993:1023–1026.
29. King DL, King DL Jr, Shao MY. Evaluation of *in vitro* measurement accuracy of a three-dimensional ultrasound scanner. *J Ultrasound Med* 1991;10:77–82.
30. Guo Z, Moreau M, Rickey DW, Picot PA, Fenster A. Quantitative investigation of *in vitro* flow using three-dimensional colour Doppler ultrasound. *Ultrasound Med Biol* 1995;21:807–816.
31. Pandian NG, Nanda NC, Schwartz SL, Fan P, Cao Q. Three-dimensional and four-dimensional transesophageal echocardiographic imaging of the heart and aorta in humans using a computed tomographic imaging probe. *Echocardiography* 1992;9:677–687.
32. Ross JJ Jr, D'Adamo AJ, Karalis DG, Chandrasekaran K. Three-dimensional transesophageal echo imaging of the descending thoracic aorta. *Am J Cardiol* 1993;71:1000–1002.
33. Wollschlager H, Zeiher AM, Geibel A, Kasper W, Just H. Transesophageal echo computer tomography of the heart. In: Roelandt JRTC, Sutherland GR, Iliceto S, Linker DT, eds. *Cardiac ultrasound.* London: Churchill Livingstone, 1993:181–185.
34. Delabays A, Pandian NG, Cao QL, et al. Transthoracic real-time three-dimensional echocardiography using a fan-like scanning approach for data acquisition: Methods, Strengths, Problems, and Initial Clinical Experience. *Echocardiography* 1995;12:49–59.
35. Downey DB, Nicolle DA, Fenster A. Three-dimensional orbital ultrasonography. *Can J Ophthalmol* 1995; 30:395–398.
36. Gilja OH, Thune N, Matre K, Hausken T, Odegaard S, Berstad A. *In vitro* evaluation of three-dimensional ultrasonography in volume estimation of abdominal organs. *Ultrasound Med Biol* 1994;20:157–165.
37. Sohn C, Stolz W, Kaufmann M, Bastert G. Die dreidimensionale ultraschalldarstellung benigner und maligner brusttumoren-erste klinische erfahrungen. *Geburtsh u Frauenheilk* 1992;52:520–525.
38. Elliot TL, Downey DB, Tong S, Mclean CA, Fenster A. Accuracy of prostate volume measurements *in vitro* using three-dimensional ultrasound. *Acad Radiol* 1996;3:401–406.
39. Tong S, Downey DB, Cardinal HN, Fenster A. A three-dimensional ultrasound prostate imaging system. *Ultrasound Med Biol* 1996;22:735–746.
40. Chin JL, Downey DB, Onik G, Fenster A. Three-dimensional prostate ultrasound and its application to cryosurgery. *Technique Urol* 1997;2:187–193.
41. Onik GM, Downey DB, Fenster A. Three-dimensional sonographically monitored cryosurgery in a prostate phantom. *J Ultrasound Med* 1996;16:267–270.
42. Belohlavek M, Foley DA, Gerber TC, Greenleaf JF, Seward JB. Three-dimensional ultrasound imaging of the atrial septum: normal and pathologic anatomy. *J Am Coll Cardiol* 1993;22:1673–1678.
43. Martin RW, Bashein G. Measurement of stroke volume with three-dimensional transesophageal ultrasonic scanning: comparison with thermodilution measurement. *Anesthesiology* 1989;70:470–476.
44. Ludomirsky A, Silberbach M, Kenny A, Shiota T, Rice MJ. Superiority of rotational scan reconstruction strategies for transthoracic 3-dimensional real-time echocardiographic studies in pediatric patients with CHD. *J Am Coll Cardiol* 1994;169A (abst).
45. McCann HA, Chandrasekaran K, Hoffman EA, Sinak LJ, Kinter TM, Greenleaf JF. A method for three-dimensional ultrasonic imaging of the heart *in vivo*. *Dynam Cardiovasc Imaging* 1987;1:97–109.

46. Roelandt JRTC, Ten Cate FJ, Vletter WB, Taams MA. Ultrasonic dynamic three-dimensional visualization of the heart with a multiplane transesophageal imaging transducer. *J Am Soc Echocardiogr* 1994;7:217–229.
47. Brinkley JF, McCallum WD, Muramatsu SK, Liu DY. Fetal weight estimation from lengths and volumes found by three-dimensional ultrasonic measurements. *J Ultrasound Med* 1984;3:163–168.
48. King DL, King DL Jr, Shao MYC. Three-dimensional spatial registration and interactive display of position and orientation of real-time ultrasound images. *J Ultrasound Med* 1990;9:525–532.
49. King DL, Harrison MR, King DL Jr, Gopal AS, Kwan OL, Demaria AN. Ultrasound beam orientation during standard two-dimensional imaging: assessment by three-dimensional echocardiography. *J Am Soc Echocardiogr* 1992;5:569–576.
50. Levine RA, Handschumacher MD, Sanfilippo AJ, et al. Three-dimensional echo-cardiographic reconstruction of the mitral valve, with implications for the diagnosis of mitral valve prolapse. *Circulation* 1989;80:589–598.
51. Moritz WE, Pearlman AS, McCabe DH, Medema DH, Ainsworth ME, Boles MS. An ultrasonic technique for imaging the ventricle in three dimensions and calculating its volume. *IEEE Trans Biomed Eng* 1983; BME-30:482–491.
52. Rivera JM, Siu SC, Handschumacher MD, Lethor JP, Guerrero JL. Three-dimensional reconstruction of ventricular septal defects: validation studies and *in vivo* feasibility. *J Am Coll Cardiol* 1994;23:201–208.
53. Weiss JL, Eaton LW, Kallman CH, Maughman WL. Accuracy of volume determination by two-dimensional echocardiography: defining requirements under controlled conditions in the ejecting canine left ventricle. *Circulation* 1983;67:889–895.
54. Geiser EA, Christie LG Jr, Conetta DA, Conti CR, Gossman GS. A mechanical arm for spatial registration of two-dimensional echocardiographic sections. *Cathet Cardiovasc Diagn* 1982;8:89–101.
55. Nikravesh PE, Skorton DJ, Chandran KB, Attarwala YM, Pandian N, Kerber RE. Computerized three-dimensional finite element reconstruction of the left ventricle from cross-sectional echocardiograms. *Ultrason Imaging* 1984;6:48–59.
56. Raichlen JS, Trivedi SS, Herman GT, St. John Sutton MG, Reichek N. Dynamic three-dimensional reconstruction of the left ventricle from two-dimensional echocardiograms. *J Am Coll Cardiol* 1986;8:364–370.
57. Sawada H, Fujii J, Kato K, Onoe M, Kuno Y. Three-dimensional reconstruction of the left ventricle from multiple cross-sectional echocardiograms: value for measuring left ventricular volume. *Br Heart J* 1983; 50:438–442.
58. Detmer PR, Bashein G, Hodges T, et al. 3D ultrasonic image feature localization based on magnetic scanhead tracking: *in vitro* calibration and validation. *Ultrasound Med Biol* 1994;20:923–936.
59. Hodges TC, Detmer PR, Burns DH, Beach KW, Strandness DE Jr. Ultrasonic three-dimensional reconstruction: *in vitro* and *in vivo* volume and area measurement. *Ultrasound Med Biol* 1994;20:719–729.
60. Hughes SW, D'Arcy TJ, Maxwell DJ, et al. Volume estimation from multiplanar 2D ultrasound images using a remote electromagnetic position and orientation sensor. *Ultrasound Med Biol* 1996;22:561–572.
61. Leotta DF, Detmer PR, Martin RW. Performance of a miniature magnetic position sensor for three-dimensional ultrasound imaging. *Ultrasound Med Biol* 1997;24:597–609.
62. Ohbuchi R, Chen D, Fuchs H. Incremental volume reconstruction and rendering for 3D ultrasound imaging. *SPIE* 1992;1808:312–322.
63. Pretorius DH, Nelson TR. Prenatal visualization of cranial sutures and fontanelles with three-dimensional ultrasonography. *J Ultrasound Med* 1994;13:871–876.
64. Raab FH, Blood EB, Steiner TO, Jones HR. Magnetic positioning and orientation tracking system. *IEEE Trans Aerospace Electron Systems* 1979;15:709–717.
65. Riccabona M, Nelson TR, Pretorius DH, Davidson TE. Distance and volume measurement using three-dimensional ultrasonography. *J Ultrasound Med* 1995;14:881–886.
66. Chen JF, Fowlkes JB, Carson PL, Rubin JM. Determination of scan-plane motion using speckle decorrelation: theoretical consideration and initial test. *Int J Imaging Systems Technol 1997* 1997;8:38–44.
67. Tuthill TF, Krücker JF, Fowlkes JB, Carson PL. Automated three-dimensional US frame positioning computed from elevational speckle decorrelation. *Radiology*, 1998;209:515–582.
68. Nelson TR, Davidson TE. Determination of transducer position from backscattered signal intensity. *J Ultrasound Med* 1996;15(3):77.
69. Pearson AC, Pasierski T. Initial clinical experience with a 48 by 48 element biplane transesophageal probe. *Am Heart J* 1991;122:559–568.
70. Shattuck DP, Weinshenker MD, Smith SW, von Ramm OT. Explososcan: a parallel processing technique for high speed ultrasound imaging with linear phased arrays. *J Acoust Soc Am* 1984;75:1273–1282.
71. Smith SW, Pavy HG Jr, von Ramm OT. High-speed ultrasound volumetric imaging system. Part I. Transducer design and beam steering. *IEEE Trans Ultrason Ferroelectr Freq Contr* 1991;38:100–108.
72. Smith SW, Trahey GE, von Ramm OT. Two-dimensional arrays for medical ultrasound. *Ultrason Imaging* 1992;14:213–233.
73. Snyder JE, Kisslo J, von Ramm OT. Real-time orthogonal mode scanning of the heart. I. system design. *J Am Coll Cardiol* 1986;7:1279–1285.
74. Turnbull DH, Foster FS. Beam steering with pulsed two-dimensional transducer arrays. *IEEE Trans Ultrason Ferroelectr Freq Contr* 1991;38:320–333.
75. von Ramm OT, Smith SW. Real time volumetric ultrasound imaging system. *SPIE* 1990;1231:15–22.
76. von Ramm OT, Smith SW, Pavy HG Jr. High-speed ultrasound volumetric imaging system. Part II. Parallel processing and image display. *IEEE Trans Ultrason Ferroelectr Freq Contr* 1991;38:109–115.
77. Fishman EK, Magid D, Ney DR, et al. Three-dimensional imaging. *Radiology* 1991;181:321–337.

3

Visualization and Display Methods

OVERVIEW

This chapter provides an overview of various methods of visualizing three-dimensional ultrasound (3DUS) data. Visualization approaches that include planar slices and surface and volume rendering will be discussed. The use of animation and volume-editing techniques will be considered. Optimization of data presentation for the clinician will be reviewed with clinical examples.

KEY CONCEPTS

- Volume visualization methods project a multidimensional data set onto a two-dimensional image plane to enhance understanding of the structure contained within the volumetric data.
- Medical volume visualization techniques must offer physicians:
  - Understandable data representations.
  - Quick data manipulation.
  - Fast rendering.
  - Involvement with data review.
- Volume ultrasound visualization plays an important role in demonstrating normalcy and reassuring patients.
- Three basic approaches to volume visualization: slice projection, surface fitting, and volume rendering:
  - Planar slices provide interactive review from arbitrary orientations:
    - Closely resembles clinical scanning procedures.
    - Offers projections that may not be available during patient scanning.
    - Does not require intermediate processing or filtering of ultrasound image data.
  - Surface fitting provides for rapid evaluation of the overall surface features of the object:
    - Sensitive to noise in the data.

KEY CONCEPTS, *continued*

- Volume-rendering methods produce high-quality images:
  - Relatively tolerant of noise in the data.
  - Filtering can improve results as long as it does not obscure fine detail.
  - Careful selection of opacity values helps provide an accurate rendition of anatomy structure being studied.
  - Transparency permits viewing surface and subsurface features to better understand spatial relationships.
  - Maximum-intensity projection methods give a clear view of structures such as bones of the skeleton.
  - Stereo viewing can assist in more readily appreciating 3D spatial relationships.

Volume visualization methods project a three-dimensional/four-dimensional (3D/4D) data set onto a two-dimensional (2D) image plane with the goal of gaining an understanding of the structure contained within the volumetric data. Medical volume visualization techniques must offer understandable data representations, quick data manipulation, and fast rendering to be useful to physicians. Physicians should be able to change parameters and see the resulting image in real time. Improved display hardware capability at affordable prices has made interactive visualization possible on workstations used in the medical imaging environment. Optimization of volume visualization algorithms is an important area of study, with an understanding of the fundamental algorithms essential to optimize analysis methods (1–5).

Volume visualization methods for medical data—although available for some time in computed tomography (CT), single-photon emission computed tomography (SPECT), positron emission tomography (PET), and magnetic resonance imaging (MRI) (6)—have not achieved widespread clinical use because of the time required to obtain and process high-resolution image data. Real-time two-dimensional ultrasound (2DUS) imaging intrinsically provides interactive visualization of underlying anatomy while providing flexibility in viewing images from different orientations in real time.

Although real-time 2DUS has enabled physicians to make important contributions to patient management, there are occasions when it is difficult to develop a 3D impression of the patient anatomy, particularly with curved structures, when there is a subtle lesion in an organ, when a mass distorts the normal anatomy, or when there are tortuous vessels or structures not commonly seen. Complex cases often make it difficult for even specialists to understand 3D anatomy based on 2D images. Abnormalities may be difficult to demonstrate with 2D sonography because of the particular planes that must be imaged to develop the entire 3D impression. Integration of views obtained over a region of a patient with volume sonography may permit better visualization in these situations and allow for a more accurate diagnosis (7–9). As a result, ultrasound volume visualization methods must offer interactivity to be competitive with current patient imaging equipment. 3DUS visualization also has an important role in demonstrating normalcy and reassuring patients.

VOLUME VISUALIZATION METHODS

Technological advances continue to improve all areas of 3DUS, but one of the most critical areas of development is in the area of volume visualization (2–5,10). The promise of 3DUS lies in enabling the physician to completely understand patient anatomy and function even after the patient has left the clinic, even across networks, in collaboration with specialists. Interactive visualization methods that clearly show the anatomy of interest and permit interactive review of patient data will be essential for complete clinical acceptance.

Although ultrasound volume data are well suited to visualization (11), the optimal method for physicians to review and interpret patient data has yet to be determined, in part because acoustic data do not represent density and cannot be classified like CT and MRI, since tissues exhibit similar acoustic properties and have relatively lower signal-to-noise characteristics than other types of medical image data. Instead, ultrasound signal intensity provides a differential measure of how the acoustic impedance changes as sound passes through tissues. Image intensity increases at the interface between tissues of two different impedance values. It is possible to transform ultrasound volume data into another parameter domain that more clearly differentiates structures in the data, such as using Doppler-shifted data from regions of moving blood displayed as color-coded images merged with gray-scale image data to differentiate vascular from soft-tissue structures. However, segmentation of 3DUS data remains a challenging goal in general.

The quality of the volume data plays an important part in determining the overall quality of ultrasound visualization. One factor affecting image quality is speckle, which in ultrasound images often exceeds the specular echo intensity. As a result, traditional boundary extraction and segmentation algorithms are difficult to implement. The signal quality of the volume data can be improved by image compounding, which occurs when pixels in multiple planes are reprojected through the same voxel. Image compounding reduces speckle intensity compared with 2DUS and improves segmentation performance (12,13). However, precise image registration is necessary for accurate image compounding, which presumes that the patient does not move, that accurate system calibration has occurred, and that the velocity of sound is constant. Because these criteria can be difficult to meet in clinical settings, partly because of displacement of patient tissues resulting from transducer motion when scanning across tissues, compounding has had limited success in volume imaging up to this time.

3DUS visualization algorithms usually treat volume data as an array of voxels. Although adjacent 2DUS images are generally closely spaced to each other, there may be gaps in the resulting volume due to undersampling, particularly for wedge, rotational, or other nonparallel acquisition patterns. As a result, some form of interpolation becomes necessary to fill in any gaps in the volume between acquired images. This step is important because gaps in the volume are visibly distracting, can introduce rendering artifacts, and can make interpretation of volume data more difficult. Finally, depending on the visualization requirements, 3D median or Gaussian filters may be used to improve the signal-to-noise ratio and reduce speckle prior to application of visualization algorithms (11,14,15). Generally, the same processing is applied to the entire volume uniformly. Although having a regular grid with identical elements has advantages, many medical imaging techniques use a rectilinear grid, where, because of nonisotropic image resolution, the voxels are axis-aligned rectangular prisms rather than cubes. Because the rendering process often needs to resample the volume between grid points, the voxel approach assumes that the area around a grid point is the same value as the grid point. This approach has the advantage of making no assumptions about the behavior of data between grid points (i.e., only known data values are used for generating an image).

A further difficulty with 3DUS visualization is that the volume is generally a dense object; that is, one where the majority of voxels contain nonzero, although not necessarily clinically relevant, information. This makes it difficult to define a clearly identifiable threshold to separate adjacent structures, particularly since different tissues often have similar intensities, with only the interface being visible. Thus far an important visualization requirement has been that the viewing station be interactive to assist the operator to optimize viewing. 3DUS visualization (11) generally uses three basic approaches (3–5,16–18): slice projection, surface fitting, and volume rendering, which are discussed in the following sections (Figure 3-1).

Slice Projection

Slicing methods are one of several interactive techniques available for physician review of patient data (11,19,20). Extraction of a planar image of arbitrary orientation at a particular location in a 3DUS data set utilizes standard coordinate transformations and rotations. This is the most computationally straightforward approach to review data throughout the

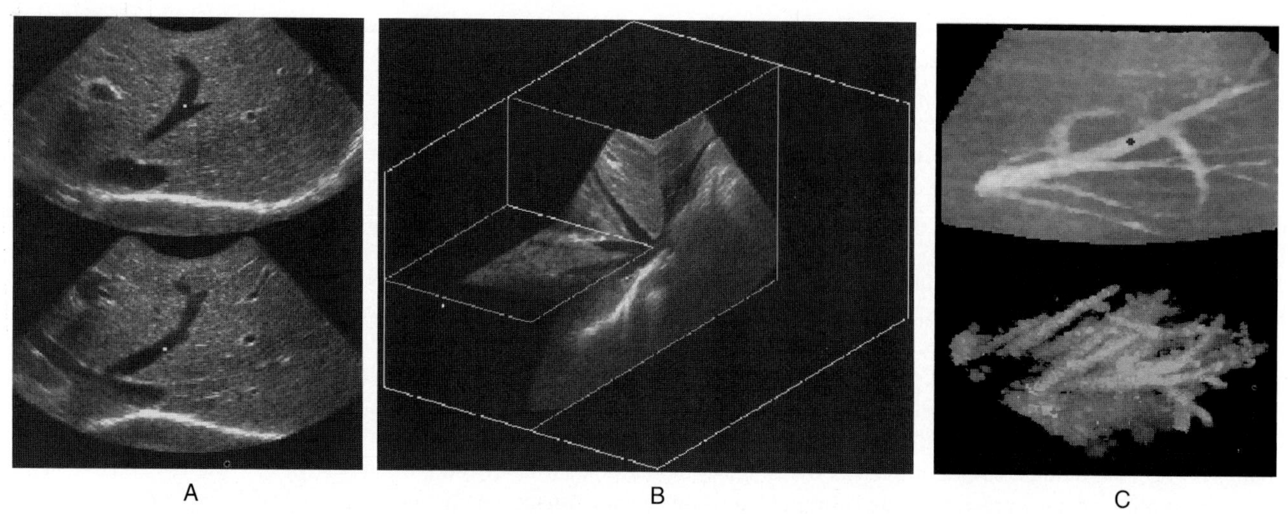

Figure 3-1. Different methods of visualizing 3DUS data. **A:** Two single planes extracted from a liver 3DUS volume at arbitrary orientations and positions. **B:** Application of clipping planes to produce a subcube in the volume. **C:** Application of volume-rendering methods to reveal internal structure of the object; the upper image is inverted minimum-intensity projection; the lower image is maximum-intensity projection after segmenting using the power Doppler signal. See color version of figure.

volume, requiring minimal processing with isotropic data. Interactive display of planar slices offers the physician retrospective evaluation of anatomy, particularly viewing of arbitrary planes perpendicular to the primary exam axis and other orientations not possible during data acquisition. Multiple slices displayed simultaneously can be particularly valuable to assist in understanding patient anatomy. Typically, slicing methods are fully interactive, replicating the scanner operational feel. Multiplane displays are often combined with rendered images to assist in localization of slice-plane position (Figures 3-2, 3-3). Currently, these review approaches are the most common method used to review 3DUS data.

Surface Fitting

Surface-fit algorithms typically fit some type of planar surface primitive (e.g., polygons, patches) to values defined during the segmenting process (21–29). The surface-fit approaches include contour connecting, marching cubes, marching tetrahedra, dividing cubes, and others. Once the surface is defined, interactive display is typically faster than volume-rendering methods because fewer data points are used, because surface-fitting methods only traverse the volume once to extract surfaces, compared with reevaluating the entire volume in volume rendering. After extracting the surfaces, rendering hardware and standard rendering methods can be used to render quickly the surface primitives each time the user changes a viewing or lighting parameter. Changing the surface-fitting threshold value is time-consuming because it requires that all the voxels be revisited to extract a new set of surface primitives. The general approach is shown in Figure 3-4 (30).

Surface-fitting methods can suffer from occasional false-positive and false-negative surface pieces, and incorrect handling of small features and branches in the data. Artifacts can be a serious concern in medicine because they could be incorrectly interpreted by physicians as features in the data. Misinterpretations of artifacts can be avoided by viewing rendered data in combination with planar data rather than in isolation. Although surface fitting provides a good means for visualizing spatial relationships for the entire volume in a readily comprehended manner, small features may be poorly visualized unless a significant number (>500,000) of polygons are used (11,26). Large numbers of polygons degrade display performance to the point where other volume-rendering methods become superior. For these reasons surface-fitting methods are not widely used in 3DUS visualization.

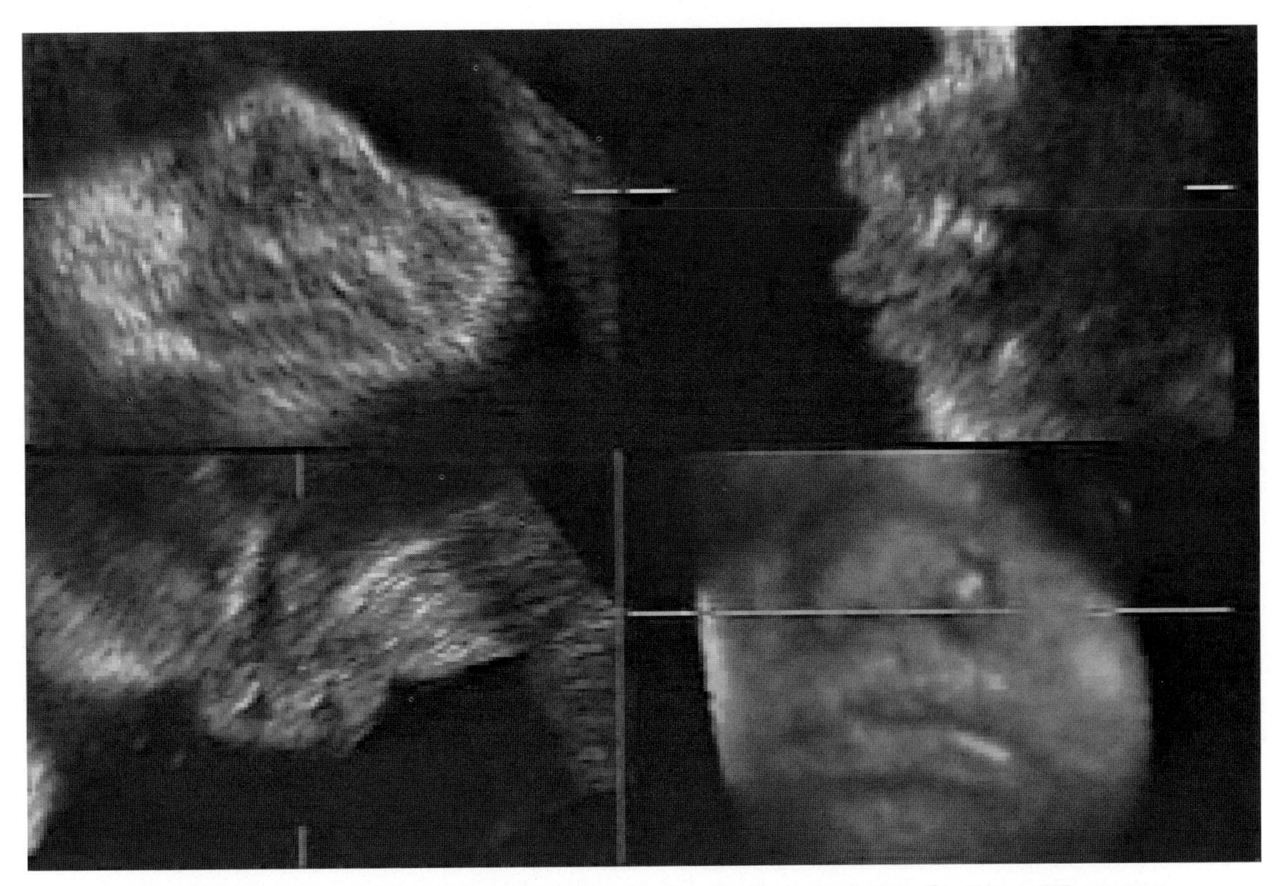

Figure 3-2. Volume data for a 35-week fetal scan using the Kretz-Technique Combison 530 scanner. Three orthogonal slices are extracted simultaneously and reoriented prior to display to put them into a standard anatomic orientation representing the coronal (frontal), sagittal (profile), and axial views. The lower right image is a volume-rendered image of the fetal face with a line showing the level of the axial plane in the rendered image. Combination of slices with volume-rendered data offers a good method of identifying the specific location of a slice within a volume.

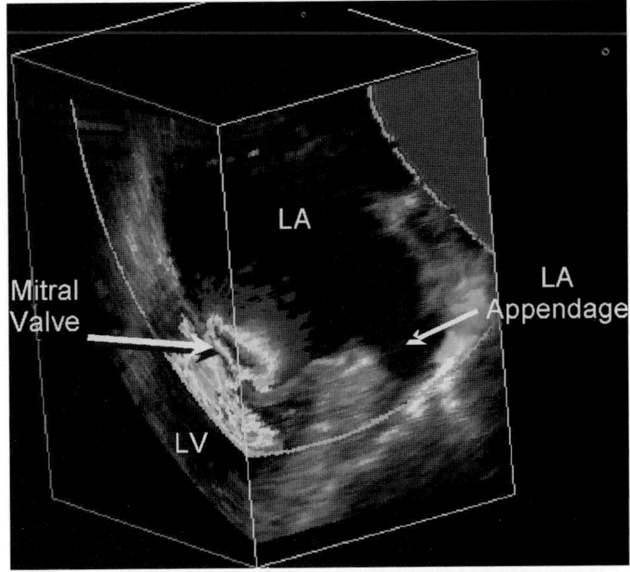

Figure 3-3. 3DUS color Doppler echocardiographic study of a patient with mitral stenosis. Cube view color Doppler data in the image shows some mild mitral regurgitation. **LA,** left atrium; **LV,** left ventricle. (Image courtesy of Dr. D. Boughner.) See color version of figure.

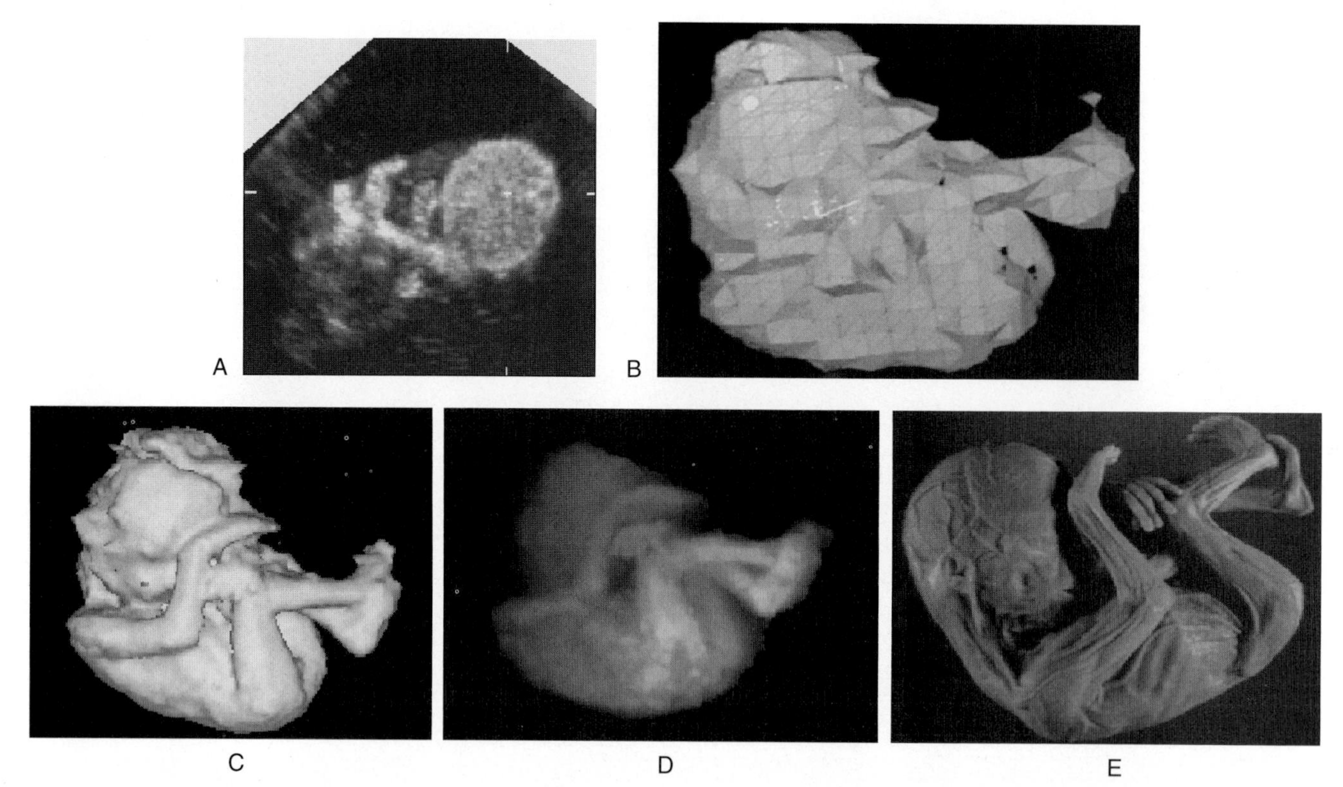

Figure 3-4. A schematic of the process of surface-rendering a volume of data for a fetal specimen. **A:** The 2DUS scan is projected into a volume data set, as has already been described. After determination of the appropriate threshold, a polygonal iso-surface mesh is generated using standard surface-rendering methods. **B:** A low-resolution mesh is shown. Determination of the optimal threshold is an interactive process of adjustment and review of results. **C:** The final high-resolution polygonal mesh may have greater than 250,000 polygons to preserve the fine detail of the data. **D:** A volume-rendered ray-traced image from the same data. **E:** A photograph of the original specimen showing the high fidelity possible with both types of rendering.

Data Classification and Segmentation

Data classification is the most difficult function in volume visualization. Data classification generally involves choosing a threshold for use with a surface-fitting algorithm or choosing color (brightness) and opacity (light attenuation) values to go with each possible range of data values for use with volume-rendering algorithms.

Accurate and automatic segmentation of ultrasound data, essential for high-quality surface fitting, is a particularly difficult problem, because most tissues in the data volume have similar signal characteristics, making it difficult to distinguish one tissue or organ from another based on signal intensity alone (12,31–39). Furthermore, because acoustic signal intensities vary with depth and tissue overburden, some form of image equalization is necessary to obtain a uniform image field, particularly in areas subject to acoustic shadowing.

Segmentation based on signal void greatly simplifies extraction of structural features. Recent work by Baba (7,40) has demonstrated fetal surfaces in near real-time imaging using simple thresholding to identify the fetal surface in the amniotic fluid (Figure 3-5). Blood-flow data—whether from Doppler, contrast, or signal void—may be reprojected and processed in concert with or separately from 3D tissue data (41–62). Figure 3-6 shows a ray-cast image of liver vessels using a simple segmentation based on the signal void and power Doppler signal to segment the liver vessels. Also, using colors that approximate the color of the tissue being studied can heighten both the realism and cognition process, as when stereotypical colors are associated with a certain object (e.g., red for arteries and blue for veins) (Figure 3-7). During rendering, slight changes in opacity values often have

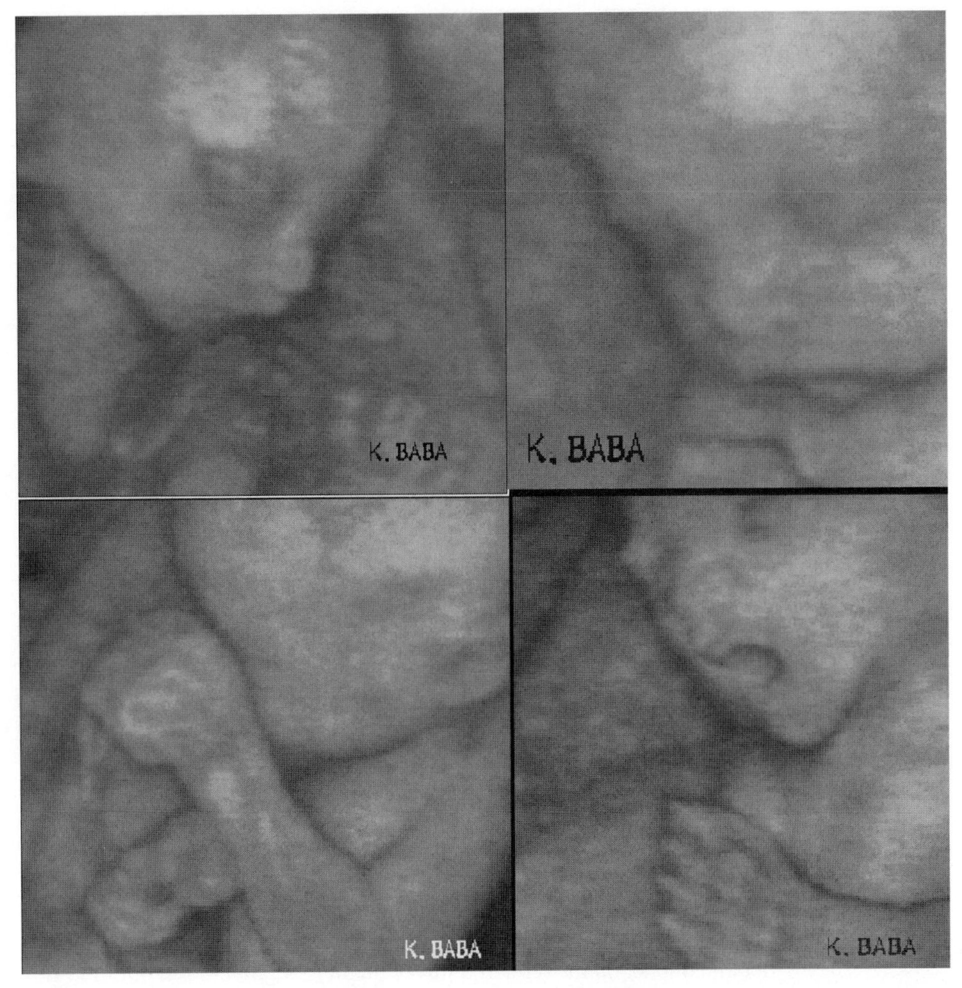

Figure 3-5. Images from a rapid surface-imaging system showing excellent fetal surface details. This system depends on the presence of sufficient amounts of amniotic fluid to produce clear views of fetal anatomy. (Images courtesy of Dr. K. Baba.)

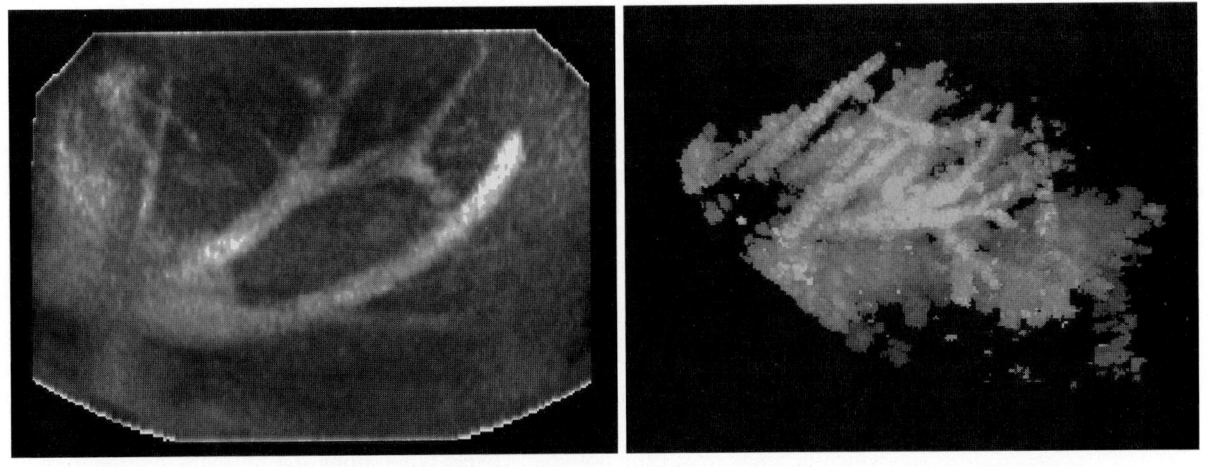

Figure 3-6. Two approaches to segmentation of vessels. The left image uses the inverted minimum-intensity projection of a liver data set to clearly show the 3D structure of the vessels. Minimum-intensity projection is an important method for presenting echo-poor structures embedded in tissue, such as vessels, cysts, and so on, requiring minimal parameter adjustment and can be performed very rapidly. The right image segments the vessels from the gray-scale data based on the power Doppler signal. Volume rendering of the flow signal produces a clear picture of the continuity of the vessel flows within the liver. See color version of figure.

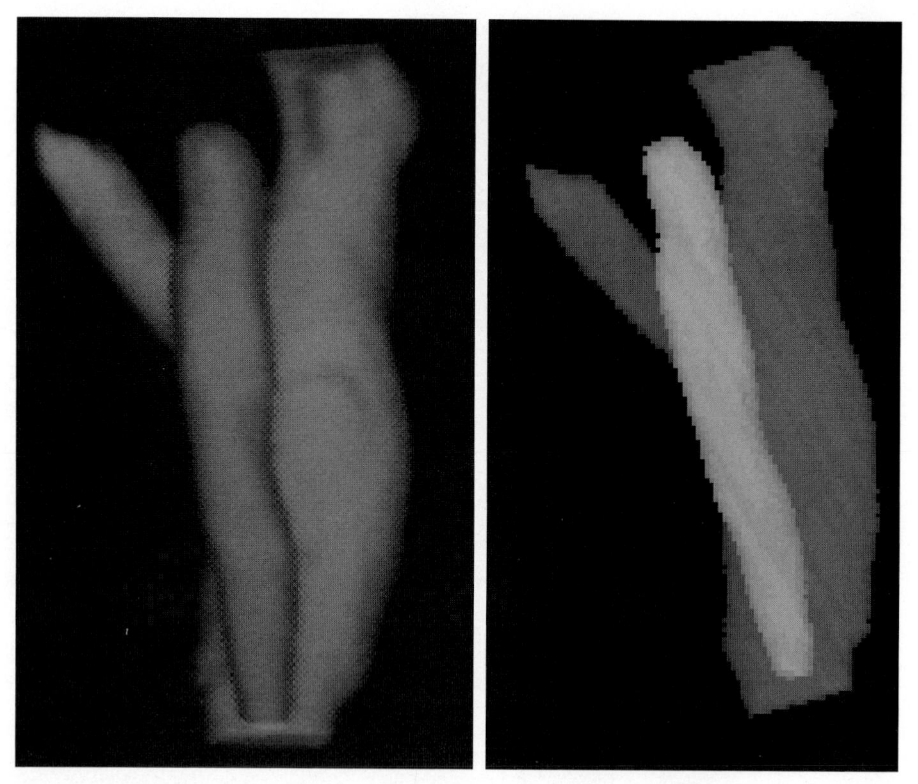

Figure 3-7. Volume data from a velocity Doppler study of the carotid artery. The left image is a surface-rendered image based on a defined threshold for the color velocity data. The right image is a volume-rendered, ray-cast image of the same data. See color version of figure.

a large impact on the rendered image quality. Segmentation of blood flow data is relatively straightforward compared with tissue and organ segmentation, where interfaces are less visible. Ongoing work is showing promising results in segmentation of the prostate based on echo properties (34,63) and spectrum analysis (64). Interactive performance greatly assists segmentation by providing feedback to optimize the image rendering parameters.

Volume Rendering

Volume-rendering methods map individual voxels (V) directly onto the screen without using geometric primitives (18,65–71) but require that the entire data set be sampled each time an image is rendered or re-rendered. In some situations a low-resolution pass is sometimes used to create low-quality images quickly for optimization of viewing position or parameter setting with a high-resolution image produced immediately after values have stopped changing. The most often used volume visualization algorithm for the production of high-quality images is ray casting.

Ray casting uses the color and opacity of each voxel along a ray (R) passing through a volume (72). The trajectory of the ray is defined by the relative viewing orientation of the observer and the volume. Along a particular ray, the intensity of the emerging ray (R_{out}) is determined by the current voxel [$V(k)$] and the intensity of the entering ray (R_{in}), which depends on all the previous voxels lying along the ray path and the current (k^{th}) voxel. Depending on the blending or evaluation function, transparency/opacity ($\alpha(k)$) values R_{out} reflects surface or subsurface features within the volume, (Figure 3-8) with:

$$R_{out} = R_{in}(1 - \alpha(k) + c(k)\alpha(k))$$

where $\alpha(k)$ is the opacity value based on a mapping function relating opacity to the kth voxel brightness and/or color. When $\alpha(k) = 0$, the voxel is transparent; and when $\alpha(k) =$

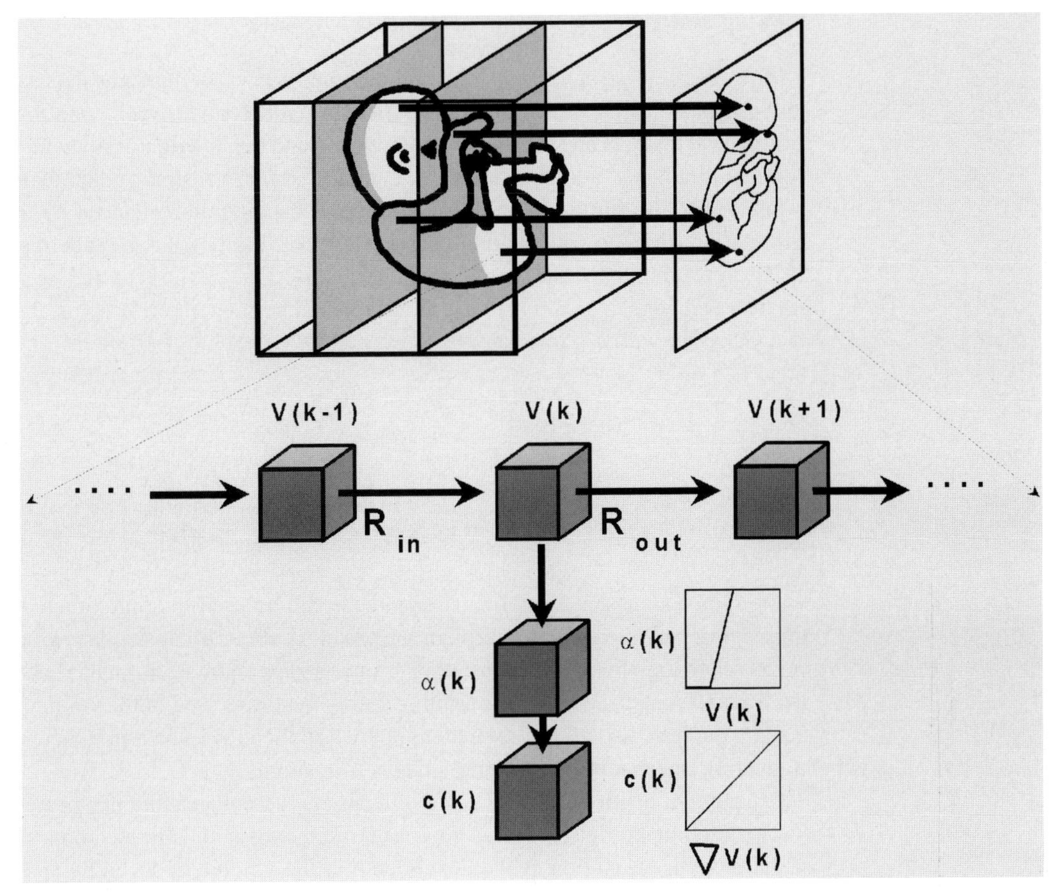

Figure 3-8. A diagram showing the basic concepts of ray casting. In ray casting, the voxel intensity is propagated forward toward the viewing plane along each ray from back to front. Each voxel (V(*k*)) contributes to the final image intensity based on shading ($c(k)$) and transparency ($\alpha(k)$) values. Prototype examples of the $c(k)$ and $\alpha(k)$ mapping functions are shown in the small graphs. Once (V(*k*)) = 1, full opacity has been reached and further passage of the ray into the volume does not contribute to the final image.

1, the voxel is opaque. $c(k)$ is a shade value based on the local gradient or some other modifying parameter for the voxel. The specific rendering result depends on the choice of the mapping functions for $\alpha(k)$ and $c(k)$. The opacities, shades, and colors encountered along the ray are blended, and the final opacity and color of R_{out} is a pixel [P(R)] in the image on the display (18):

$$P(R) = \sum_{k=0}^{K} [c(R,k)\ \alpha(R,k) \prod_{i=k+1}^{K} (1 - \alpha(R,i))]$$

where: R,k is the kth voxel along the Rth ray and

$$^{c}(R,0) = c_{background}\ \alpha(r,0) = 1$$

No shadows or reflections are generated in ray casting, and these useful visual cues must be added to optimize visual presentation (11).

Maximum- and minimum-intensity projection (MIP) methods are one form of ray casting where only the maximum (minimum) voxel value is retained as the ray traverses the data volume. These techniques are quite simple to implement and provide good-quality results for many applications (30,73,74).

Although ray casting is CPU intensive, the images show the entire data set, depending on opacity and intensity values, not just a collection of thin surfaces, as in surface fitting. Ray casting can be parallelized at the pixel level because rays from all the pixels in the image plane can be cast independently.

Viewing and Depth Cue Enhancement

Most medical visualization uses orthographic views so that physicians will not see patient data warped by the perspective transformation. Under such conditions other cues to structural features and depth such as depth-fog and depth-brightness attenuation, shading based on gradients, and animation can enhance comprehension of data features.

Subtle surface features in ultrasound volume data can be enhanced by using gradient information from the volume to highlight interfaces. Many volume-rendering and surface-fitting algorithms use some type of gradient shading. One method for finding the directional gradient at a voxel $V(x,y,z)$ is to use the Mark–Hildreth operator, which is the convolution of a Gaussian with a Laplacian (18):

$$V(x,y,z) = \nabla^2 [V(x,y,z) ** G(x,y,z)]$$

where:

$$\nabla^2 V(x,y,z) = \left(\frac{\partial^2 V}{\partial x^2} + \frac{\partial^2 V}{\partial y^2} + \frac{\partial^2 V}{\partial z^2}\right)$$

where G is a Gaussian function. The gradient can be used to approximate the normal to an imaginary surface passing through the voxel location. Most standard graphics shading models can be applied to shading elements once an approximate normal is known. Ambient, diffuse, and sometimes specular lighting components are used in the volume visualization shading process. As in other computer graphics applications, shading is an important factor in creating understandable volume images of anatomy.

Exponential depth shading is a useful technique to enhance the perspective of depth in a volume-rendered image. In this approach, often used with MIP techniques, the intensity of the voxel is a function of the voxel depth in the volume, reduced by an amount based on an exponential function (73) (Figure 3-9).

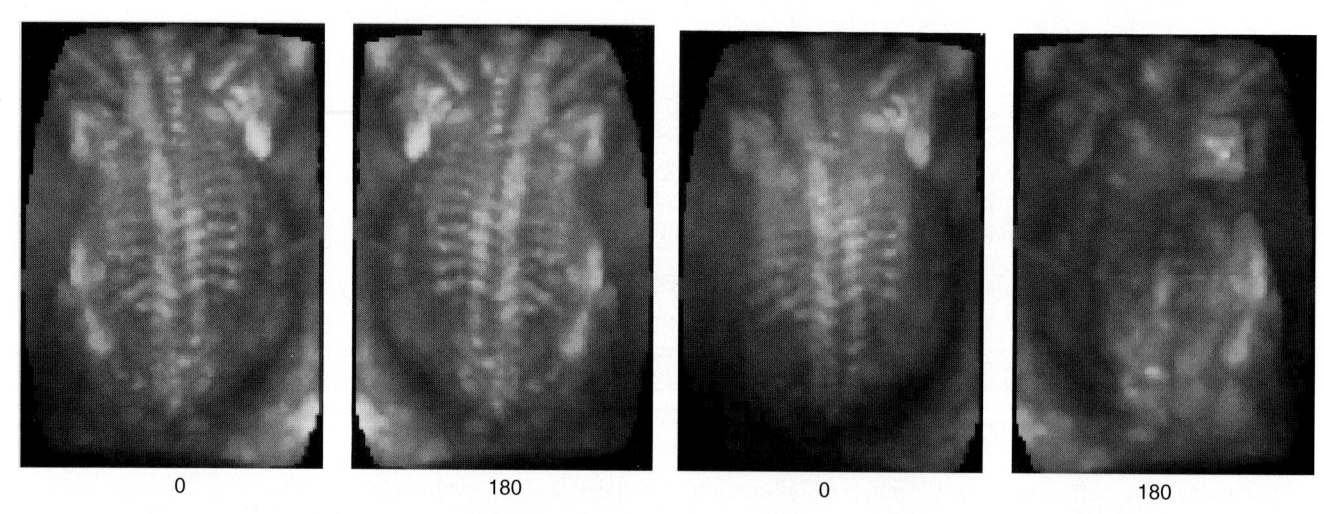

Figure 3-9. 3DUS maximum-intensity projection images comparing the effect of depth shading. All images are produced from the same data set. The two images on the left show front (0-degree) and back (180-degree) views of the 22-week fetal spine. The two left-hand images are produced without a depth-weighting function. The image appears the same from both orientations (0 and 180 degrees), which results in visual aliasing during rotation when the rotating object appears to change direction (or orientation) spontaneously. Objects close to the back appear to be moving in the opposite direction to objects close to the front. With both near and far objects having the same intensity, the viewer is unable to determine the true position and location. Structures nearest the viewer are emphasized through use of an exponential weighting function. The two right-hand images incorporate a depth-weighting function to emphasize structures closest to the viewer. Depth shading eliminates visual aliasing.

$$\text{without depth shading: } P(R) = \max_{k=0}^{K} V(R,k)$$

$$\text{with depth shading: } P(R) = \max_{k=0}^{K} [V(R,k)\ e^{-\beta(k)}]$$

where β is the attenuation factor at the kth location along the Rth ray.

Depth shading reduces visual aliasing during rotation when near and far points of the volume cross, and it greatly assists in clarifying structural relationships. Applications of depth shading with different rendering strategies are shown for the placental vasculature (Figure 3-10) (75), a kidney with polycystic disease demonstrating segmentation based on both the power Doppler signal from the vessels and the signal void from the cysts in the kidney (Figure 3-11), and the fetal face rendered with a variety of different algorithms that emphasize different features (Figure 3-12) (9).

Animation

Animation sequences such as rotation and gated cine-loop review greatly assist volume visualization. Without the animation display of rendered volumetric images offered by real-time processing or precalculation, the physician often has a difficult time extracting 3D information from 2D displays. Motion should be viewed at normal, accelerated, or reduced speed to enhance comprehension. Furthermore, analysis of dynamic function can increase the diagnostic value of many studies (Figure 3-13) (76).

Interactive volume-rendering methods make it possible to follow structural curvature that cannot be viewed in any planar orientation. Rotation of anatomy to a standard presentation facilitates identification of subtle anatomic landmarks by clarifying the relative location of the overlapping structures. Interactive stereo viewing incorporating motion cues makes it possible to separate structures, identify the continuity of structures, and enhance anatomic visualization (77).

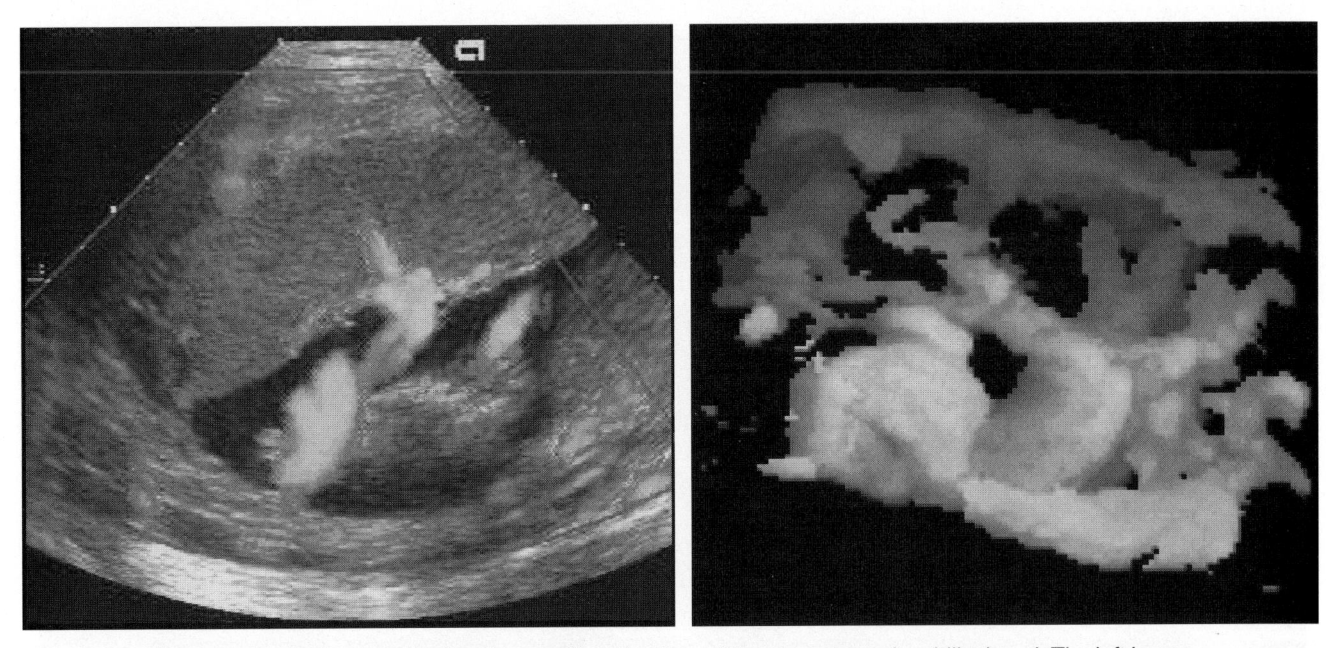

Figure 3-10. Volume rendering of power Doppler data of the placenta and umbilical cord. The left image is a single slice from the 2DUS acquisition. The right image is a volume-rendered image of the vascular anatomy. The gray-scale signal has been removed. The rendering is a modified maximum-intensity method with depth coding. See color version of figure.

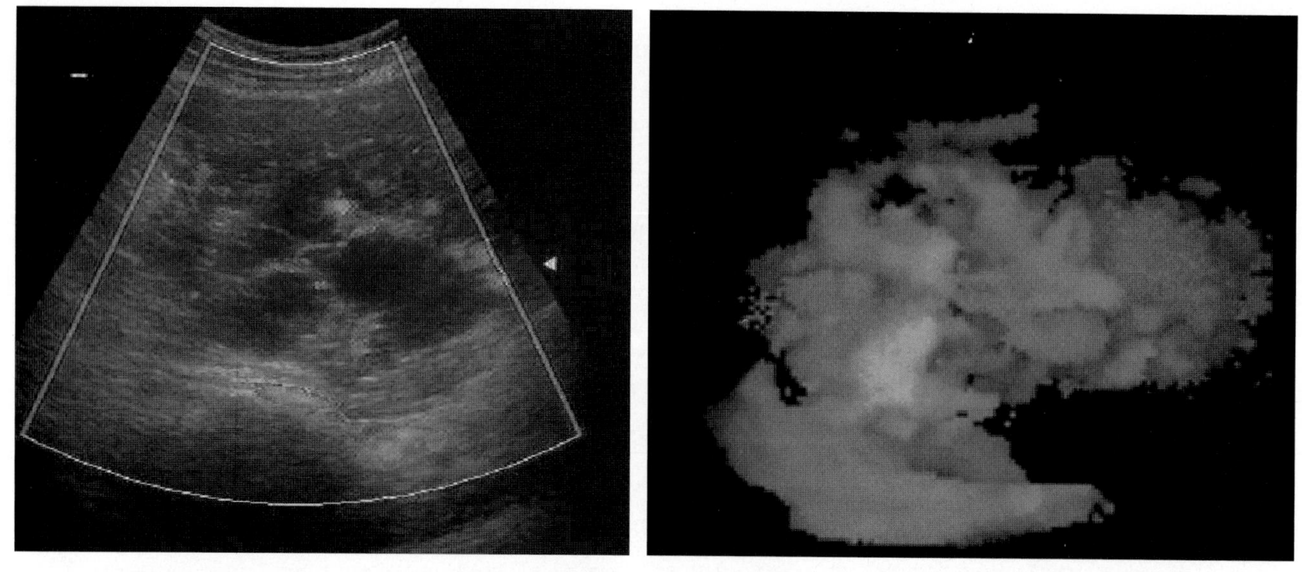

Figure 3-11. Volume rendering of power Doppler data of a polycystic kidney. The left image is a single slice from the 2DUS acquisition. The signal voids represent the cysts in the kidney. The right image is a volume-rendered image segmented using both the power Doppler signal from the vascular anatomy and the signal void from the cysts. Both methods are a modified maximum-intensity method with depth coding. Note that the relative position of the cysts and vessels is clearly shown in the volume-rendered image using two different segmentation methods on the same data. See color version of figure.

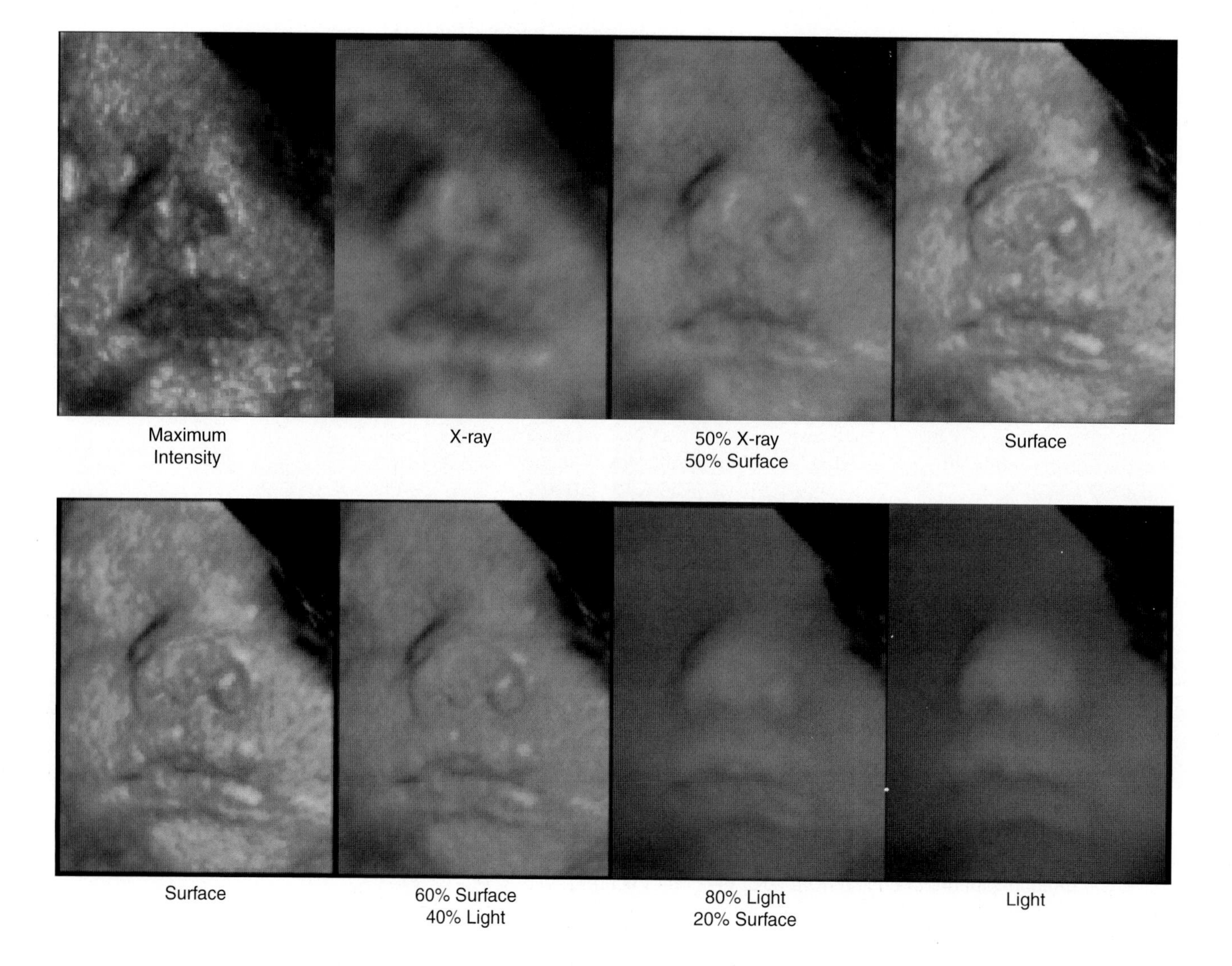

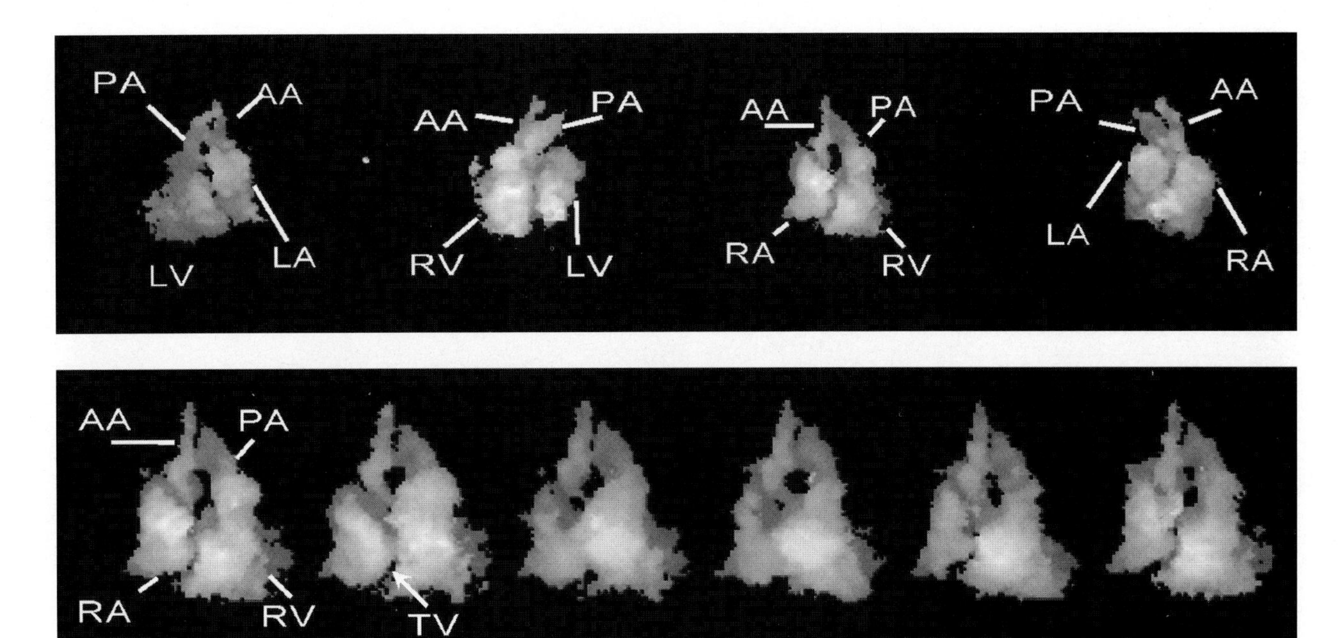

Figure 3-13. Imaging of the fetal heart. Volume-rendered images of the cardiac chambers and vessels. The signal void in the original acquisition has been extracted to show only the blood signal. Chamber and vessel rendering uses a modified maximum-intensity method with depth-coding. The upper panel shows volume-rendered images from several orientations.The lower panel contains a series of volume-rendered images at different points in the cardiac cycle. Note the tricuspid valve (TV), which is clearly seen in each image. The TV is closed in the first left-hand image and proceeds to open during ventricular diastole in the next few frames as blood flows from the atria into the ventricle. Toward the end the TV is closed once again as blood is ejected into the pulmonary artery (PA). Interactive display of cardiac dynamics greatly assists comprehension of cardiac anatomy and function.

Volume Editing Tools

Physicians often need to visualize a 3D structure located within a data volume, which may require editing to remove undesirable echoes to see the important anatomy. Segmentation of ultrasound data is possible when there are very different acoustic interfaces, such as vessels imaged with Doppler methods. Unfortunately, when adjacent organs have similar acoustic properties, segmentation of ultrasound data can be difficult. The Voluson 530D (Medison, Korea) system provides editing boxes that can be positioned around the object of interest to exclude obscuring structures. The boxes can be combined with a threshold control to further eliminate unwanted echoes. Identifying the boundary of an organ or structure in 3D ultrasound data for morphometric analysis also may benefit from the use of stereo displays and 3D input devices.

Another means of removing unwanted echoes is to use interactive editing tools that function in a volume-rendered or virtual reality environment. In this situation the physician

Figure 3-12. Images of a fetal face showing the effect of different types of rendering. Several different methods are used to optimize display of facial details: maximum-intensity projection; x-ray projection, surface rendering, and light rendering. The upper left images show maximum-intensity and x-ray projection images, demonstrating internal structures such as bones. X-ray projection sums all the voxels along a given ray and is of limited value for most ultrasound data. Surface-rendering methods are based on selection of a threshold/gradient to highlight a specific acoustic feature. The next images show a gradation from surface-rendered image using a combination threshold–gradient method that clearly shows surfaces but sometimes produces a "stone-hard" surface to light that demonstrates the primary surface features with a a semitransparent surface "painted" by the original gray values of the volume that preserves high detail and includes original gray-scale values on the 3D surface.

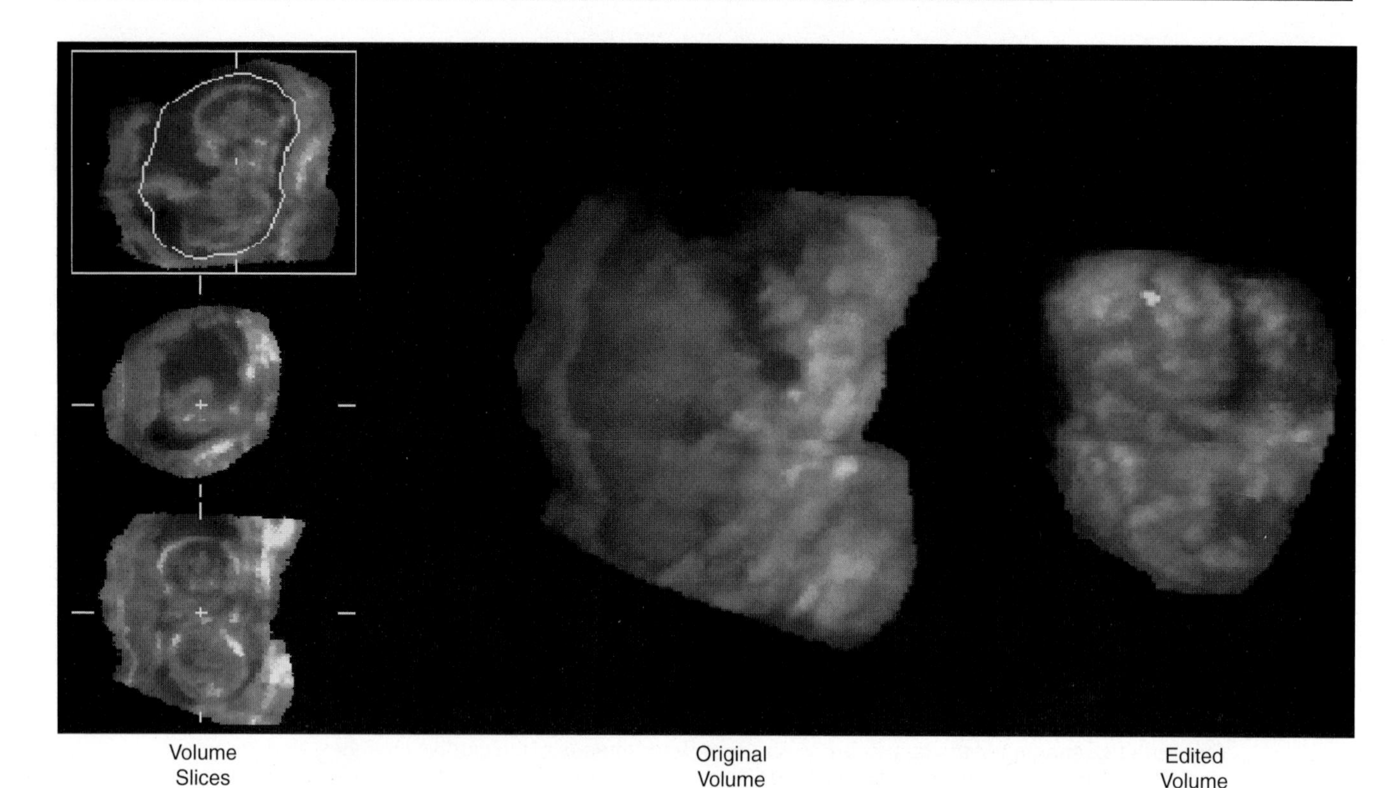

Figure 3-14. Volume editing of an 18-week fetal study with an electronic scalpel. The original volume data contain the uterus, the fetus, and the amniotic fluid. Volume-rendering methods cannot readily differentiate what is important and what is superfluous. Removal of unimportant objects is readily accomplished by editing either the individual slices or the volume directly. The electronic scalpel is designed to remove a few voxels between the viewer and the object of interest by direct application to the volume-rendered data. Alternatively, a single slice can be edited as shown in the upper left-hand image slice with the contour drawn around the fetus. Once the contour has been defined, that contour is applied to the entire volume in that orientation and similarly for the two other orthogonal slices. Direct editing of the volume also is used to refine the extraction of the fetus (or object of interest). Using volume-editing methods makes it possible to completely and accurately extract the fetus from the surrounding tissue in less than 30 sec, compared with 5–10 min for conventional single-slice editing. Furthermore, volume-editing tools make it possible to correct any errors directly on the volume. Editing is conducted from any orientation, permitting rapid optimization of feature extraction.

uses an electronic scalpel to dissect away tissues that are not part of the organ of interest (Figure 3-14) (78). The dissection process is greatly assisted by using volume-rendered data, compared with single-plane data, particularly if some type of stereoscopic display can be used to create the impression of working in a 3D environment. Under these conditions it is possible to remove nonimportant tissue signals and isolate the structure of interest rapidly. Interactive, user-directed erosion techniques where the user removes layers of 3D data with the resulting volume immediately rendered so that the user can analyze and adjust the amount of data removed can be helpful in improving interactivity. If the user has only affected a small area of the volume, then the image is adaptively rendered to save CPU time. Allowing users to separate structures of interest from surrounding diagnostically less important data enhances previously difficult-to-see anatomic features.

INTERACTIVE VIEWING USER INTERFACES

Clinical user acceptance of 3DUS capability will depend to a large extent on the type of user interface provided. Many current techniques for user–display interaction are often awkward, nonintuitive, and unsuited for use in a clinical setting requiring significant effort and commitment to learn. To facilitate usage in a clinical environment it is important to have an intuitive, easy-to-use interface with a rapid learning curve that facilitates physician operation of a system. As a result, human factor ergonomics considerations will become increasingly important with 3DUS equipment to facilitate rapid learning and intuitive interaction with the scanner and volume data by the operator.

The key factors critical to rapid learning and ready clinical user acceptance in 3DUS are (a) a user interface with a clearly labeled set of functions appropriate to the task at hand, (b) rapid response to user commands permitting interactive review, adjustment and manipulation of volume data, (c) intuitive interactivity devices (e.g., trackballs, touch screens), and (d) functions that permit rapid extraction of clinically relevant data (i.e., measurements, segmentation, visualization). This section is intended to discuss some of these important factors.

Task-Based User Interface Functions

A key area for 3DUS acceptance is to use a menu system that provides only those choices relevant to the particular task at hand. Current ultrasound scanner user interfaces often provide all the possible choices simultaneously, even if not relevant to the task, requiring the operator to remember which options (or key combinations) are required to utilize a particular feature. An alternate approach employs simple-to-learn interfaces that incorporate reconfigurable displays coupled with touch screens to match the function choices to the task and are now becoming available on commercial systems.

Rapid Response to User Commands

Once the desired operation is identified and selected, the system must respond quickly to permit interactive review, adjustment, and manipulation of volume data. Interactive review of volume data is essential to clinical use, and high-performance computing systems are becoming more cost-effective, greatly enhancing interactivity. In addition to orienting volume data to the optimal position, it is important to adjust rendering and display parameters to emphasize the structure of interest. This may include changing from a maximum-intensity projection to a ray-cast projection while adjusting the window–level controls. In other situations it may be necessary to segment the volume data to visualize the vessels, which might require thresholding, or adjusting signal levels while viewing the images. In other cases measurement of length, area, and volume might require removing part of the volume interactively while reviewing the remaining structure to verify the accuracy of the removal. In each of these situations the system must rapidly provide feedback to the operator to assist in optimizing the process.

Segmentation is an important part of isolating structures of interest from surrounding tissues for visualization or measurement. Because this is often a tedious and time-consuming process, tools that enhance the task are essential. Such tools may include thresholding, texture analysis, or region growing. They also may include 3D display of the result using stereoscopic displays (Figure 3-15) utilizing depth information to assist in volume operations such as the electronic scalpel discussed earlier. Certainly, alignment of two volume data sets (such as soft tissue and blood flow) from the same patient to create a hybrid image may be more readily performed with stereo displays and fast alignment algorithms.

Intuitive Interactivity Devices

Review of patient data must include the flexibility to rotate, scale, and view objects from perspectives that optimize visualization of the anatomy of interest. Incorporation of touch screens, knobs, trackballs, and 3D positioning systems simplifies operator control of operations, especially if the response is immediate. Touch screen systems do not rely on keyboard or mouse input devices, are intuitive, and provide physicians with an easy-to-use and learn interface to the computer system, greatly improving interactivity. With this approach, learning time to use the system can be significantly reduced. Stereo displays can be used by physicians to integrate displays and interactive devices, obviating the use of the computer keyboard. Instead they interact with volume data in 3D space via a stereo display, in some cases using head-mounted display systems.

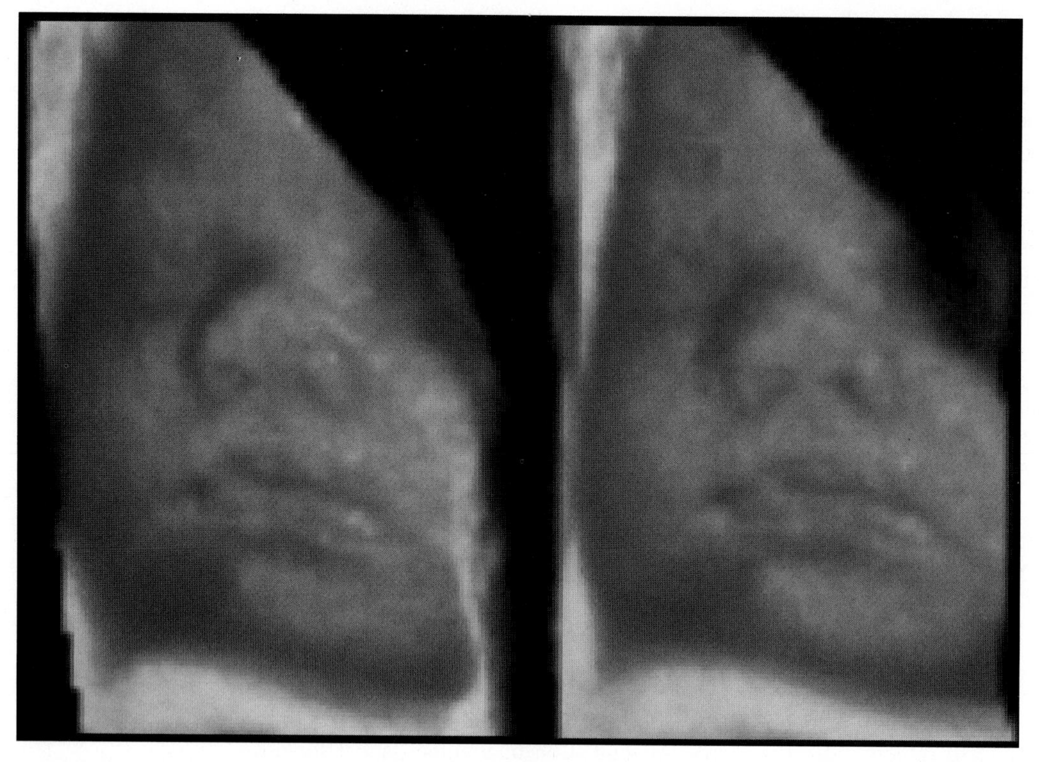

Figure 3-15. 3DUS volume-rendered data of the fetal face presented as a stereo pair. The adjacent images may be viewed in stereo, which provides additional depth cues.

Rapid Extraction of Clinically Relevant Data

One of the most important factors affecting the evaluation of 3DUS data in a clinically useful system is providing clinicians with an interactive means of reviewing patient data and extracting the diagnostically important information from the study (79–82). User interfaces should incorporate interactive data review to permit optimization of viewing orientation and image presentation (83,84). Displays that incorporate slices through the volume and rendered images from the volume, similar to that shown in Figure 3-2, use the volume-rendered image to provide important navigational landmarks to identify the location of the slices. When coupled with a standard viewing orientation, localization on the critical anatomic feature is greatly facilitated. Volume-rendering methods can enhance evaluation of volume data in a readily comprehensible, unambiguous manner by presenting information from the entire volume to the physician in a single image.

SUMMARY

Three-dimensional ultrasound imaging is a new, exciting technology that allows physicians to use ultrasound to view anatomy and pathology as a volume, thereby enhancing comprehension of patient anatomy. Ongoing developments in computers and technology now permit acquisition, analysis, and display of volume data in seconds, facilitating many opportunities for rapid diagnosis and interventional techniques. Commercial and academic interest in 3DUS is mounting, and continued advancements and understanding are expected in the near future.

An important part of viewing 3DUS data is the ability to review patient data interactively. The advantages of 3DUS compared with 2DUS are shown in Table 3-1. The flexibility to rotate, scale, and view objects from perspectives that optimize visualization of the

Table 3-1. *Visualization advantages of 3DUS compared with 2DUS*

- Volume data can be viewed using a standard anatomic orientation
- Planar images
  - Provide simultaneous display of coronal, sagittal, and axial planes
  - Display views difficult or impossible to obtain with 2DUS due to anatomic constraints
- Volume rendering
  - Display entire volume demonstrating continuity of curved structures in a single image
- Interactivity
  - Rotation permits optimizing view of anatomy from multiple orientations
- Guidance of interventional procedures for accurate placement of needles/catheters
- Archived data may be used for subsequent review or teaching after patient has left clinic

anatomy of interest is critical to clinical acceptance, as is physician involvement in optimizing and enhancing visualization tools and in the ongoing evaluation of visualization techniques. There is a continuing need for development of intuitive user interfaces and semiautomatic data classification and segmentation tools, so physicians can quickly learn and use volume visualization systems. Minimizing artifacts from volume visualization algorithms is a matter of diagnostic, ethical, and legal importance where lingering flaws could produce images that lead to an incorrect medical diagnosis. Errors in images could have serious ramifications in surgical procedures that rely on multidimensional data. A standard means for validating algorithms and performance is essential to future visualization efforts. Future volume visualization systems also will need to fuse volume data from different modalities for the same patient.

Volume visualization quality is highly dependent on 2DUS image quality (Table 3-2); poor images result in poor volume data. If the patient moves during the scan, then the scan must be repeated, or some form of correction for motion must be made. Faster affordable hardware is needed to shorten the time between acquisition and display. Some equipment is now becoming available for which the volume data are acquired directly into the rendering engine, permitting direct volume visualization immediately upon completion of the image acquisition, which will greatly enhance clinical acceptance. As experience with 3DUS increases, continued refinement of analysis and visualization software will further assist in making a more accurate diagnosis.

Stereoscopic viewing systems that let physicians utilize their binocular interpretative system to clarify structural relationships (77,85–90) can enhance identification of small structures with greater confidence in less time. Physician involvement in optimizing and enhancing visualization tools is essential as part of the ongoing optimization of visualization techniques.

As newer 3DUS imaging systems become available to acquire volume data directly into the ultrasound scanner, physicians and sonographers will be able to immediately examine, visualize, and interpret the patient's anatomy on-line. Regardless of which viewing technique is used, one key benefit of 3DUS will be that once the patient has been scanned, the original 2DUS image acquisition data may be analyzed for the entire region

Table 3-2. *Optimization of 3DUS visualization*

- Overall image quality is dependent on scanner setup and patient habitus
  - Distance between structure and transducer may lead to one region imaged more clearly than another
- 2DUS image quality may be better in one plane than in others due to elevational resolution
- Rendered image quality is affected by the anatomic orientation and beam pathway
  - Signal dropout from shadowing due to overlying structures can significantly obscure structural detail
  - Compounding data obtained from different orientations can minimize the adverse effects of signal dropout, although elastic deformation of tissues due to pressure from the transducer during scanning can distort and complicate realignment of scans

Table 3-3. *Optimization of 3DUS viewing to achieve optimal clinical utility*

- Planar slices
  - Provide a straightforward method of display for interactive review closely resembling clinical scanning procedures
  - Provide projections that may be unavailable during patient scanning
  - Generally do not require intermediate processing or filtering minimizing delay time to viewing
- Surface fitting
  - Provides for rapid evaluation of the overall surface features of the object
  - Sensitive to noise in the data—bumpy or erratic surface can significantly distort features
  - Noise sensitivity potentially limits use except for specialized applications (e.g., vascular imaging)
- Volume rendering
  - Produces high-quality images that are relatively tolerant of noise
  - Filtering can improve results as long as it does not remove fine detail
  - Careful transparency value selection is essential to produce an accurate rendition of anatomy
  - Viewing surface and subsurface features helps establish global spatial relationships
  - Maximum-intensity methods give a clear view of structures such as bones of the skeleton, although they are often best applied to selected regions of the volume
- Feature localization
  - Combination of planar and volume rendering assists feature identification and localization

or magnified subregions without the need to rescan the patient. The optimum method of 3DUS data review will depend on the clinical application (Table 3-3); however, all approaches will require interactivity to permit the viewer to optimize the display of patient anatomy.

HOW TO DO IT
Practical Clinical Tips

Overall approach to volume visualization after volume is reconstructed:

- View planar slices:
  - Individual planes or multiple orthogonal planes.
  - Reorient to standard anatomic orientation.
  - Identify area of interest.
- View rendered image:
  - Select type of rendering.
    - Ray casting.
    - Maximum/minimum projection.
    - Surface fitting.
  - Threshold/segment data.
  - Select rendering parameters.
- Interactively adjust viewing parameters and orientation.

REFERENCES

1. Brodlie KW, Carpenter KW, et al., eds.. Scientific visualization: techniques and applications. New York: Springer-Verlag, 1992.
2. McCormick B, DeFanti T, Brown M. Visualization in scientific computing. *Comput Graphics* 1987;21(6).
3. Wolff RS. Volume visualization I: basic concepts and applications. *Comput Phys* 1992;6(4):421–461.
4. Wolff RS. Volume visualization II: Ray-tracing of volume data. *Comput Phys* 1992;6(6):692–695.
5. Wolff RS. Volume visualization III: polygonally based volume rendering. *Comput Phys* 1993;7(2):158–161.
6. Fishman EK, Magid D, Ney DR, et al. Three-dimensional imaging. *Radiology* 1991;181:321–337.
7. Baba K, Okai T, Kozuma S, Taketani Y, Mochizuki T, Akahane M. Real-time processable three-dimensional US in obstetrics. *Radiology* 1997;203(2):571–574.

8. Baba K, Jurkovic D. *Three-dimensional ultrasound in obstetrics and gynecology.* New York: Parthenon Press, 1997.
9. Pretorius DH, Nelson TR. Fetal face visualization using three-dimensional ultrasonography. *J Ultrasound Med* 1995;14:349–356.
10. Coatrieux JL, Toumoulin C, Hamon C, Lou L. Future trends in 3D medical imaging. *IEEE Eng Med Biol* 1990;12:33–39.
11. Nelson TR, Elvins TT. Visualization of 3D ultrasound data. *IEEE Comput Graph Applic* 1993;13:50–57.
12. Bashford GR, von Ramm OT. Speckle structure in three dimensions. *J Acoust Soc Am* 1995;98:35–42.
13. Trahey GE, Allison JW, Smith SW, von Ramm OT. A quantitative approach to speckle reduction via frequency compounding. *Ultrason Imaging* 1986;8:151–164.
14. Pratt WK. *Digital image processing.* New York: Wiley, 1991.
15. Russ JC. *The image processing handbook.* Boca Raton, FL: CRC Press, 1992.
16. Kaufman A, Hohne KH, Kruger W, Rosenblum L, Schroder P. Research issues in volume visualization. *IEEE Comput Graph Applic* 1994;14:63–67.
17. Kaufman AE. Volume visualization. *ACM Comput Survey* 1996;28:165–167.
18. Watt A, Watt M. Volume rendering techniques. In: Wegner P, ed. *Advanced animation and rendering techniques: theory and practice.* New York: ACM Press, 1992:297–321.
19. Fenster A, Downey DB. 3-D ultrasound imaging: a review. *IEEE Eng Med Biol* 1996;15:41–51.
20. Jones MW, Min Chen. Fast cutting operations on three-dimensional volume datasets. *Visualization Scientific Computing Proceedings* 1995:1–8.
21. Ekoule AB, Peyrin FC, Odet CL. A triangulation algorithm from arbitrary shaped multiple planar contours. *ACM Trans Graph* 1991;10(2):182–199.
22. Kim JJ, Jeong YC. An efficient volume visualization algorithm for rendering isosurface and ray casted images. *1996 Pacific Graphics Conference Proceedings* 1996:91–105.
23. Levoy M. Display of surfaces from volume data. *IEEE Comput Graph Applic* 1988;8(3):29–37.
24. Levoy M. Volume rendering, a hybrid ray tracer for rendering polygon and volume data. *IEEE Comput Graph Applic* 1990;10:33–40.
25. Lorensen WE, Cline HE. Marching cubes: a high resolution three-dimensional surface construction algorithm. *Comput Graph* 1987;21(7):163–169.
26. Sakas G, Walter S. Extracting surfaces from fuzzy 3D-ultrasound data. *Proceedings SIGGRAPH 95. Comput Graph* 1995:465–474.
27. Sander TS, Zucker SW. Stable surface estimation. *IEEE* 1986:1165–1167.
28. Schroeder W, Lorensen B. 3-D surface contours. *Dr. Dobb's J* 1996;7:26–86.
29. Udupa J, Hung HM, Chuang KS. Surface and volume rendering in 3D imaging: a comparison. *J Digit Imaging* 1991;4:159–168.
30. Nelson TR, Pretorius DH. Three-dimensional ultrasound of fetal surface features. *Ultrasound Obstet Gynecol* 1992;2:166–174.
31. Bamber JC, Eckersley RJ, Hubregtse P, et al. Data processing for 3-D ultrasound visualization of tumour anatomy and blood flow. *SPIE* 1992;1808:651–663.
32. Bashford GR, von Ramm OT. Ultrasound three-dimensional velocity measurements by feature tracking. *IEEE Trans Ultrason Ferroelectr Freq Contr* 1996;43:376–384.
33. Bovik AC. On detecting edges in speckle imagery. *IEEE Trans Acoust Speech Signal Processing* 1988; 36(10):1618–1627.
34. Chen CH, Lee JY, Yang WH, Chang CM, Sun YN. Segmentation and reconstruction of prostate from transrectal ultrasound images. *Biomed Eng Applic Basis Communications* 1995;8:287–292.
35. Cootes TF, Hill A, Taylor CJ, Haslem J. The use of active shape models for locating structures in medical images. *Image Vision Comput* 1994;12:276–285.
36. Fine D, Perring MA, Herbetko J, et al. Three-dimensional ultrasound imaging of the gallbladder and dilated biliary tree: reconstruction from real-time B-scans. *Br J Radiol* 1991;64:1056–1057.
37. Richard WD, Keen CG. Automated texture-based segmentation of ultrasound images of the prostate. *Comput Med Imaging Graph* 1996;20:131–140.
38. Sakas G, Schreyer L-A, Grimm M. Visualization of 3D ultrasonic data. *Proceedings Visualization '94* (Cat. No. 94CH35707) 1994;CP42:369–373.
39. Sakas G, Schreyer L-A, Grimm M. Preprocessing and volume rendering of 3D ultrasonic data. *IEEE Comput Graph Applic* 1995;15:47–54.
40. Baba K, Okai T, Kozuma S. Real-time processable three-dimensional fetal ultrasound [letter]. *Lancet* 1996; 348:1307.
41. Ashton EA, Phillips D, Parker KJ. Automated extraction of the LV from 3D cardiac ultrasound scans. *Proceedings of the International Society for Optical Engineering.* 1996;2727:423–429.
42. Blankenhorn DH, Chin HP, Strikwerda S, Bamberger J, Hestenes JD. Common carotid artery contours reconstructed in three dimensions from parallel ultrasonic images. Work in progress. *Radiology* 1983;148: 533–537.
43. Bruining N, von Birgelen C, Di Mario C, et al. Dynamic three-dimensional reconstruction of ICUS images based on an EGG-gated pull-back device. *Comput Cardiol 1995* 1995:633–636.
44. Carson PL, Adler DD, Fowlkes JB, Harnist K, Rubin J. Enhanced color flow imaging of breast cancer vasculature: Continuous wave Doppler, three-dimensional display. *J Ultrasound Med* 1992;11:377–385.
45. Cavaye DM, Tabbara MR, Kopchok GE, Laas TE, White RA. Three-dimensional vascular ultrasound imaging. *Am Surg* 1991;57:751–755.
46. Downey DB, Chin JL, Fenster A. Three-dimensional US-guided cryosurgery. *Radiology* 1995;197(P):539.
47. Downey DB, Fenster A. Three-dimensional power Doppler detection of prostatic cancer. *AJR* 1995;165: 741.
48. Downey DB, Fenster A. Vascular imaging with a three-dimensional power Doppler system. *AJR* 1995;165: 665–668.

49. Downey DB, Nicolle DA, Fenster A. Three-dimensional orbital ultrasonography. *Can J Ophthalmol* 1995; 30:395–398.
50. Ehricke H-H, Donner K, Koller W, Strasser W. Visualization of vasculature from volume data. *Comput Graph* 1994;18:395–406.
51. Ferrara KW, Zagar B, Sokil-Melgar J, Algazi VR. High resolution 3D color flow mapping: applied to the assessment of breast vasculature. *Ultrasound Med Biol* 1996;22:293–304.
52. Guo Z, Moreau M, Rickey DW, Picot PA, Fenster A. Quantitative investigation of *in vitro* flow using three-dimensional colour Doppler ultrasound. *Ultrasound Med Biol* 1995;21:807–816.
53. Guo Z, Fenster A. Three-dimensional power Doppler imaging: a phantom study to quantify vessel stenosis. *Ultrasound Med Biol* 1996;22:1059–1069.
54. Hashimoto H, Shen Y, Takeuchi Y, Yoshitome E. Ultrasound 3-dimensional image processing using power Doppler image. *1995 IEEE ultrasonics symposium proceedings: an international symposium.* IEEE 1995; 2:1423–1426.
55. Kitney RI, Moura L, Straughan K. 3-D visualization of arterial structures using ultrasound and voxel modelling. *Int J Cardiac Imag* 1989;4:135–143.
56. Klein H-M, Günther RW, Verlande M, et al. 3D-surface reconstruction of intravascular ultrasound images using personal computer hardware and a motorized catheter control. *Cardiovasc Intervent Radiol* 1992;15: 97–101.
57. Picot PA, Rickey DW, Mitchell R, Rankin RN, Fenster A. Three-dimensional colour Doppler imaging. *Ultrasound Med Biol* 1993;19:95–104.
58. Pretorius DH, Nelson TR, Jaffe JS. 3-Dimensional sonographic analysis based on color flow Doppler and gray scale image data: a preliminary report. *J Ultrasound Med* 1992;11:225–232.
59. Riccabona M, Nelson TR, Pretorius DH. Three-dimensional ultrasound: accuracy of distance and volume measurements. *Ultrasound Obstet Gynecol* 1996;7:429–434.
60. Riccabona M, Nelson TR, Pretorius DH, Davidson TE. Three-dimensional sonographic measurement of bladder volume. *J Ultrasound Med* 1996;15(9):627–632.
61. Ritchie CJ, Edwards WS, Mack LA, Cyr DR, Kim Y. Three-dimensional ultrasonic angiography using power-mode Doppler. *Ultrasound Med Biol* 1996;22:277–286.
62. Selzer RH, Lee PL, Lai JY, et al. Computer-generated 3D ultrasound images of the carotid artery. Proceedings (Cat. No. 88CH2733–4), Washington, DC. *Comput Cardiol* 1989:21–26.
63. Strasser H, Janetschek G, Reissigl A, Bartsch G. Prostate zones in three-dimensional transrectal ultrasound. *Urology* 1996;47:485–490.
64. Feleppa EJ, Fair WR, Kalisz A, et al. Spectrum analysis and three-dimensional imaging for prostate evaluation. *Molec Urol* 1997;1:109–116.
65. Cabral B, Cam N, Foran J. Accelerated volume rendering and tomographic reconstruction using texture mapping hardware. Proceedings of 1994 symposium on volume visualization. *IEEE* 1995;131:91–98.
66. Cohen MF, Painter J, Mehta M, Kwan-Liu M. Volume seedlings. Proceedings of the 1992 symposium on interactive 3D graphics. ACM Press, 1992:139–145.
67. Kajiya JT, Von Herzen BP. Ray tracing volume densities. *Comput Graph* 1984;10(3):165–173.
68. Levoy M. Efficient ray tracing of volume data. *ACM Trans Graph* 1990;9:245–261.
69. Sabella P. A rendering algorithm for visualizing 3D scalar fields. *Comput Graph* 1988;22(4):51–57.
70. Sarti A, Lamberti C, Erbacci G, Pini R. Volume rendering for 3-D echocardiography visualization. Proceedings. (Cat. No. 93CH3384–5) *Comput Cardiol* 1993:209–212.
71. Steen E, Olstad B. Volume rendering of 3D medical ultrasound data using direct feature mapping. *IEEE Trans Med Imaging* 1994;13:517–525.
72. Drebin RA, Carpenter L, Hanrahan P. Volume rendering. *Comput Graph* 1998;22(4):65–71.
73. Nelson TR, Pretorius DH. Visualization of the fetal thoracic skeleton with three-dimensional sonography: a preliminary report. *AJR* 1995;164:1485–1488.
74. Pretorius DH, Nelson TR. Prenatal visualization of cranial sutures and fontanelles with three-dimensional ultrasonography. *J Ultrasound Med* 1994;13:871–876.
75. Pretorius DH, Nelson TR, Baergen RN, Pai E, Cantrell C. Imaging of placental vasculature using three-dimensional ultrasound and color power Doppler. UOG, 1998;12:45–49.
76. Schwartz SL, Cao Q-L, Azevedo J, Pandian NG. Simulation of intraoperative visualization of cardiac structures and study of dynamic surgical anatomy with real-time three-dimensional echocardiography. *Am J Cardiol* 1994;73:501–507.
77. Adelson SJ, Hansen CD. Fast stereoscopic images with ray-traced volume rendering. Proceedings of 1994 symposium on volume visualization, 1995;125:3–9.
78. Nelson TR, Davidson TE, Pretorius DH. Interactive electronic scalpel for extraction of organs from 3DUS data. *Radiology* 1995;197(P):191.
79. Brady ML, Higgins WE, Ramaswamy K, Srinivasan R. Interactive navigation inside 3D radiological images. Proceedings of 1995 biomedical visualization. *IEEE* 1995:33–40, 84.
80. Kaufman AE, Sobierajski LM, Avila RS, Taosong HE. Navigation and animation in a volume visualization system. *Models Techn Comput Anim* 1993:64–74.
81. Robb RA, Barillot C. Interactive 3-D image display and analysis. *SPIE* 1988;939:173–202.
82. Robb RA, Barillot C. Interactive display and analysis of 3-D medical images. *IEEE Trans Med Imaging* 1989;8:217–226.
83. Bajura M, Fuchs H, Ohbuchi R. Merging virtual objects with the real world: seeing ultrasound imagery within the patient. *Comput Graph* 1992;26:203–210.
84. Fuchs H, Levoy M, Pizer SM. Interactive visualization of 3D medical data. *IEEE Comput* 1989;(Aug): 46–51.
85. Herman GT, Vose WF, Gomori JM, Gefter WB. Stereoscopic computed three-dimensional surface displays. *RadioGraphics* 1995;5(6):825–852.
86. Hernandez A, Basset O, Dautraix I, et al. Stereoscopic visualization of 3D ultrasonic data for the diagnosis

improvement of breast tumors. 1995 IEEE ultrasonics symposium proceedings: an international symposium (Cat. No. 95CH35844) *IEEE* 1995;2:1435–1438.

87. Hernandez A, Basset O, Dautraix I, Magnin IE. Acquisition and stereoscopic visualization of three-dimensional ultrasonic breast data. *IEEE Trans Ultrason Ferroelectr Freq Contr* 1996;43:576–580.
88. Howry DH, Posakony G, Cushman R, Holmes JH. Three-dimensional and stereoscopic observation of body structures by ultrasound. *J Appl Physiol* 1956;9:304–306.
89. Lateiner JS, Rubio S II. An efficient environment for stereoscopic volume visualization. *Intel Eng Systems Artificial Neural Networks* 1994;4:761–766.
90. Martin RW, Legget M, McDonald J, et al. Stereographic viewing of 3D ultrasound images: a novelty or a tool? *1995 IEEE ultrasonics symposium proceedings: an international symposium.* IEEE 1995;2:1431–1434.

4

Quantitative Analysis Methods

OVERVIEW

This chapter will provide an overview of the key concepts in making quantitative measurements using three-dimensional ultrasound (3DUS) data as well as provide a discussion of sources or errors that affect these measurements.

KEY CONCEPTS

- Comparison of two-dimensional ultrasound (2DUS) and 3DUS volume measurement methods:
  - 2DUS measurements:
    - Assume a geometric model.
    - Uncertainty in key measurements may lead to inaccuracies and variabilities.
    - Difficult to ensure reproducibility of region being measured for serial studies.
  - 3DUS measurements:
    - Measure the geometry of the organ.
    - Utilize complete organ structure.
    - Accommodate irregularity in the structure.
    - Structure can be viewed in optimum orientation.
- 3DUS quantitative measurements can be both accurate and reproducible:
  - Distance—optimum plane can be selected for measurement.
  - Area—optimum cross-section can be selected for area measurement.
  - Volume—structure volume can be measured accurately using: manual planimetry and automated algorithms.
- Measurement accuracy and precision depends on slice spacing, resolution, and organ complexity:
  - *In vitro*—approximately 2% to 5%.

KEY CONCEPTS, *continued*

- *In vivo*—approximately 5% to 10%, depending on segmentation.
- Inter- and intra-observer variability for 3DUS less than 2DUS by factor of 2–3.

- Sources of measurement error primarily due to resolution, slice spacing, and improper calibration:
  - Mechanical scanning:
    - Measurement accuracy depends on image slice distance and angle spacing.
  - Free-hand scanning with position sensing:
    - Measurement accuracy depends on calibration.
  - Free-hand scanning without position sensing.
    - Measurements not reliable or accurate.

The role of quantitative measurements in ultrasound imaging has progressed with advances in ultrasound technology. In the early days, A-mode ultrasound was used to measure the location of the midline of the brain and the thickness of various tissue layers. With the ongoing developments and improvements, B-mode and color Doppler ultrasound have developed into important methods for measuring anatomic structures and blood flow.

An important part of the patient imaging procedure is to provide data of sufficient quality to measure the length, area, or volume of organs and follow their temporal changes. Although visual assessment is valuable, quantitative data provide a more accurate basis for decision making and comparison against previous studies or reference data bases. Three-dimensional ultrasound (3DUS) images are ideally suited to volume measurement, and results have been reported in a number of areas (1–18).

3DUS ROLE IN MEASUREMENTS

In most applications, measuring the volume of an organ or lesion is the goal, rather than measuring its diameter or perimeter. However, conventional ultrasound imaging provides only a two-dimensional (2D) image, from which the three-dimensional (3D) structure must be estimated. Over the years, techniques have been developed that provide useful estimates of organ volume from 2D ultrasonography, and many studies have been published on the accuracy of these techniques for various applications, such as the estimation of gestational age, the effects of growth retardation, or prostate volume. Although these techniques have proven to be invaluable, the estimation of the volume of an organ or anatomic structure suffers from a number of deficiencies (Table 4-1). These deficiencies may lead to two problems with quantitative measurements of the patient's anatomy: accuracy and reproducibility.

Accuracy

The accuracy of a measurement technique reflects its ability to measure the truth (e.g., how close the measured volume is to the actual organ volume). Because each measurement

Table 4-1. *Deficiencies of organ volume measurement strategies*

- Some "normal" geometric model of the organ that only approximates the organ shape must be assumed so that an estimate of the organ volume can be calculated from its "key" dimensions in a few 2D cross-sections
- The measured estimate of a "key" dimension may be inaccurate because of the difficulty in orienting the ultrasound probe so that the 2D image plane is correctly located within the organ
- The optimal location of the 2D image plane is sometimes inaccessible because of the restrictions imposed on the orientation of the ultrasound probe by the patient's anatomy
- It is difficult to orient the ultrasound probe so that the 2D image plane is placed at exactly the same location as in an earlier exam, in serial monitoring of organ volume, resulting in the inability to follow small changes in organ volume reliably

is subject to some uncertainty, the accuracy is estimated from the mean value of multiple measurements of the same structure. The variation of these measurements about the mean provides information about the precision, but the value of the mean itself determines the accuracy.

Reproducibility

The reproducibility of a measurement technique reflects its consistency (i.e., the mean variation between different measurements). This variation may be due to a number of factors, as discussed earlier, and may result in a volume estimate that may have high accuracy, averaged over many measurements, but that may also vary greatly among different measurements.

Figure 4-1 illustrates one source of the variation in measurements. Figure 4-1A shows an idealized image of a high-contrast edge; Figure 4-1B, a high-resolution image of the edge; and Figure 4-1C, a low-resolution image with the corresponding gray-scale profiles.

If different observers were asked to estimate the edge location in each of the high- and low-resolution images, some would locate it nearer the dark region in each image (underestimating the dark area and overestimating the bright area), some nearer the bright region (vice versa), and some near the midpoint (yielding unbiased estimates of both areas). This may result in two scenarios with differing accuracy and variability.

High Accuracy and Different Variability

The average of each set (high- and low-resolution images) of estimates is equally near the midpoint of each image, with no systematic bias toward either the darker or the brighter regions. Then, on the average, both the high-resolution (see Figure 4-1B) and the low-resolution (see Figure 4-1C) images provide equally accurate estimates of the edge location. However, the variation between location estimates by different observers is of course smaller for the high-resolution image than for the low-resolution image, so that the high-resolution image still yields a more precise estimate of the edge location.

Different Accuracy and Variability

The observers may identify the edge as the area that is close to the dark region. Therefore the average of each set of estimates is biased toward the dark region in each image (e.g., toward a point in the edge profile where the image intensity is higher than the mean dark

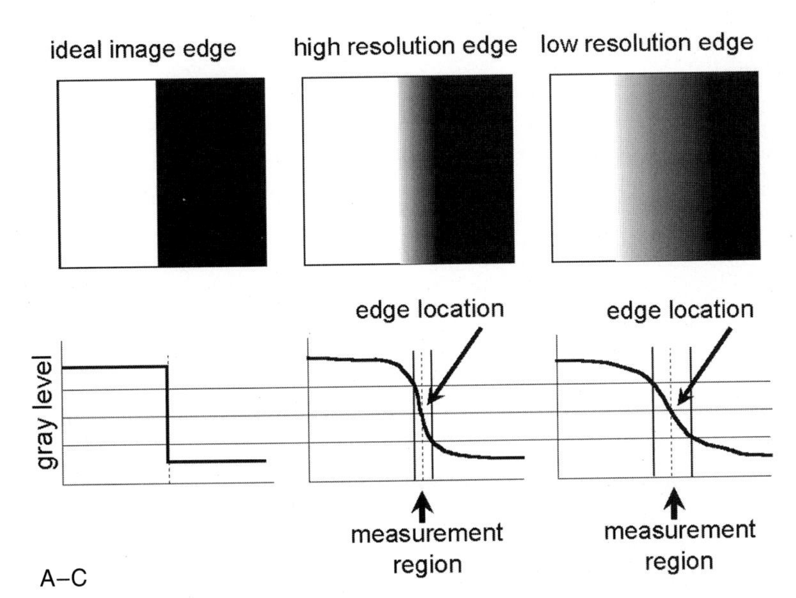

Figure 4-1. Schematic diagram showing **(A)** an idealized image of a high-contrast edge, **(B)** the image of the idealized edge obtained with a high-resolution imaging system, and **(C)** the image of the idealized edge obtained with a low-resolution imaging system. The gray-scale profiles are shown below with the location of the edge as well as the region over which the measurement location often is taken.

value by an amount equal to the noise level). Because image intensity increases more sharply (has a higher gradient) across the edge in the high-resolution image than in the low-resolution image, the selected edge locations are, on the average, closer to the midpoint (the true edge location) in the high-resolution image than in the low-resolution image. Thus the high-resolution image has higher accuracy (unlike the previous scenario, which assumes no bias). Moreover, for the same reason, if the observers tend to select an edge location within a given intensity range, this will correspond to a narrower band of image locations in the high-resolution image than in the low-resolution image, and result in less variation in the selected edge location between observers. Thus the high-resolution image again has higher precision.

These scenarios illustrate the fact that the accuracy of volume estimation (and, indeed, any measurement technique) depends on both system performance and observer performance. Because 3DUS imaging is a recent development, with commercial systems either having been introduced only recently or still being experimentally evaluated, experience in quantitative measurements is only now being gained, and techniques have not yet been standardized. These factors make both system and observer performance important.

3DUS MEASUREMENTS

Because the complete representation of an organ is available, it is possible to measure distance, area (including surface area), and volume from the 3DUS image. Although distance and area may be correctly measured from a conventional 2DUS image, a 3DUS image provides the ability to select the optimal view of the anatomy for performing the measurement.

Distance

Because ultrasound images are tomographic, measurement of length within the 2D image always has been possible. The overall accuracy and precision of the measurement depends on the frequency of the transducer, field of view, and limiting resolution of the scanner. Accuracy is generally possible to a few percent with measurement precision of ± 0.1 mm.

Distance measurement requires that the operator select two points. Because the 3D image is available, these points need not lie on the same original acquired image, but can be located on two different acquired images, which have been geometrically located relative to each other through the 3D reconstruction process (Figure 4-2). Clearly, for any such distance measurement to be accurate, the reconstruction process must be accurate.

Cross-sectional Area

Because 3DUS data can be sliced in any orientation (see Figure 4-2), any desired cross-section of the organ can be obtained. For measurement of cross-sectional area, the optimal slice can be found using a multiplanar reformatting technique and the desired area can then be outlined either manually using planimetric techniques (2,19) or automatically. By counting the pixels in the outlined region and multiplying by the pixel area, the area of the region is measured.

Volume

Historically, multiple regions from different planes could be used to estimate volumes; however, direct measurement of volumes is impossible with 2D methods because the scanner typically does not provide a means of connecting adjacent tomographic slices. However, the algorithms that are being used to estimate organ volume are based on assumptions of ellipsoidal geometry (20–22). Most volume measurements made using conventional 2D ultrasound methods generally are accurate to within $\pm 5\%$ if the organs are

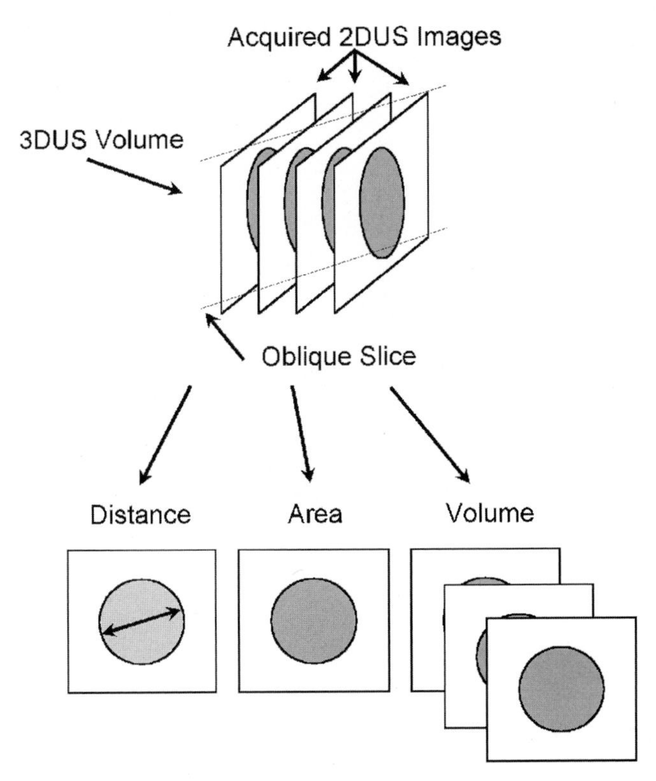

Figure 4-2. Schematic diagram showing that distance, area, and volume measurements can be made. Because the image of the organ is available in the 3D image, measurements can be made at any orientation through the organ.

regularly shaped (i.e., spherical) but are only accurate to within ±20% when they are irregularly shaped.

With a 3DUS image, the volume of an organ may be measured by either manual or automated techniques. However, automated techniques are still under development in many laboratories and are not yet being used routinely. These techniques will be discussed in the section entitled ''Volume Segmentation.'' In the manual technique, called *manual planimetry*, the multiplanar reformatting technique is used to slice the 3D image into a series of uniformly spaced parallel 2D images. For each 2D image, the cross-sectional area of the organ is manually outlined on the computer screen using a mouse, trackball, or other pointing device (Figure 4-3). The areas on these slices are then summed and multiplied by the interslice distance to obtain an estimate of the organ volume (see Figure 4-2).

This approach permits measurement of volume for regular, irregular, and disconnected objects with an accuracy of better than 5%. For small objects in which machine resolution is a significant factor, both the relative (normalized by the volume) accuracy and precision will decrease accordingly. In general, the improved measurement accuracy afforded by volume-sonographic methods makes possible accurate quantitative measurement of heart chambers, vessel dimensions, and organ volumes.

APPLICATIONS OF 3DUS VOLUME MEASUREMENT

Many investigators have reported on the improved accuracy, precision, and reproducibility in volume estimation provided by 3DUS. The earliest reports appeared with the developments in 3D echocardiography. More recently, the capabilities of 3DUS volume estimation have been reported for a wide range of applications, which will be discussed in later chapters. In this section we review some of the more recent reports, to provide information describing the levels of accuracy that can be achieved.

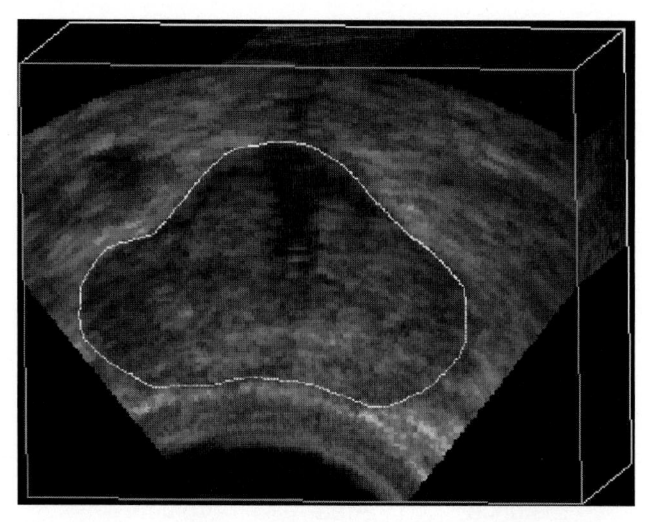

Figure 4-3. A 3DUS image of a prostate showing that the cross-sectional area can be measured by outlining the prostate in a "slice" through the prostate. The perimeter of the prostate is outlined in each slice and areas from all the slice regions are summed to calculate the volume.

In Vitro Studies

A number of *in vitro* studies have demonstrated that the error of volume measurements using 3DUS can be less than 5% (2,5,7,12–14,19). In studies using water-filled balloons, Riccabona et al. (12) reported a mean error of 2.2% for volumes ranging from 23 ml to 2400 ml, and Tong et al. (19) reported a mean error of 0.9% for volumes in the smaller range of 23–66 ml. Tong et al. (19) also reported an error of less than 1.5% in distance measurements of a string phantom.

In a study using cadaveric specimens, imaged in a water bath, Elliot et al. (2) found mean absolute and relative errors of 0.36 ± 1.17 ml and 1.6% for prostates ranging from 23 to 100 ml in volume. In an *in vitro* study using fetal livers, Hughes et al. (23) found that the accuracy of volume estimates ranged from 5.3% to 3.1%, depending on the scanning method. In a study reporting on the accuracy of distance measurements obtained with a 3DUS imaging system using a mechanical fan scanning technique, Tong et al. (19) reported that the error was less than 1.5% in measurements of distance. These studies clearly demonstrate that, using a properly calibrated 3DUS imaging system under ideal laboratory conditions, the accuracy of 3D volume estimation is very good.

In Vivo Studies

Volume measurements also show good results in a variety of organ systems, including the thyroid (24) and spleen (25). Brunner (26) showed an increased accuracy in follicular volume measurements. Chang et al. (27) has shown good volume measurement results for fetal heart volumes across a range of gestational ages. Liess et al. (28) demonstrated volume measurement errors of approximately 6% in liver lesions. Wolf et al. (29) have shown that 3DUS not only is more accurate than 2DUS and comparable to three-dimensional computed tomography (3DCT) for measurement of liver volume but also is faster than 3DCT. Riccabona et al. (13) showed good accuracy between voided urine volumes and measured bladder volumes prevoid (Figure 4-4). Prostate volume measurements also have been shown to be accurate (2,30). Hughes et al. (8) measured a variety of organ volumes and found that the errors ranged from 2% to 6%. Favre et al. (3,4) reported on studies in which they measured fetal arm and thigh circumference using 3DUS for fetal weight estimation. They obtained the most accurate results among macrosomic fetuses with a mean error of −1.4%. Lee et al. (31), however, measured thigh and abdomen

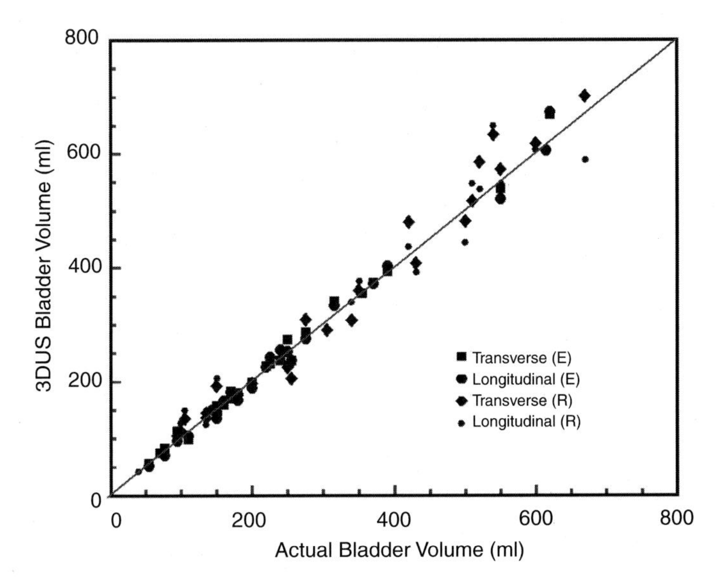

Figure 4-4. Plot comparing actual volume (voided urine) with the 3DUS measured volume for the urinary bladder showing excellent agreement between methods. (Adapted from Ref. 13.)

volume for birth weight prediction and compared the results with 2D ultrasound and found that the mean systematic error using 3DUS was $-0.03\% \pm 6.1\%$ while the results for 2D ultrasound were $-0.60\% \pm 8.8\%$.

Interslice Distance Selection

Because volume estimation requires that the 3D image of the organ be sliced into individual parallel slices, the question arises as to what the maximum interslice distance is that still provides accurate volume estimates. If this distance is small so that numerous slices intersect the organ, the volume will certainly be estimated accurately, but manually outlining the organ in all slices is tedious. On the other hand, if the interslice distance is too large, so that only a few slices intersect the organ, then the labor will be reduced, but so will the accuracy.

In general, the more irregular or complex a structure, the smaller the interslice distance that must be used to estimate its volume accurately. However, using an interslice distance that is less than the spacing of the original 2D images used to reconstruct the 3D image (i.e., less than the elevational resolution) will not result in increased accuracy. Because of the varying sizes and geometries of different organs, it is difficult to generalize the choice of interslice distance. However, in their prostate volume estimation study, Elliot et al. (2) investigated the effect of varying the interslice distance over a range of 1–15 mm in three different planes (transaxial, sagittal, and coronal) for a prostate of 25.5 ml volume. The results showed that volume estimation remained accurate in all three viewing planes for interslice distances of up to 8 mm. Interslice distances of more than 8 mm resulted in errors of 4% to 8% (Figure 4-5). Other reports addressing this issue have recommended interslice distances ranging from 2 to 5 mm for measuring prostate volume using 2DUS and serial planimetry (17,32).

Intra- and Interobserver Variability

Volume estimation via 2DUS techniques is often based on assuming an idealized organ geometry, such as an ellipsoidal shape for the prostate (17,20,33). In this case the volume is estimated by measuring the prostate's height (H), width (W), and length (L), in two

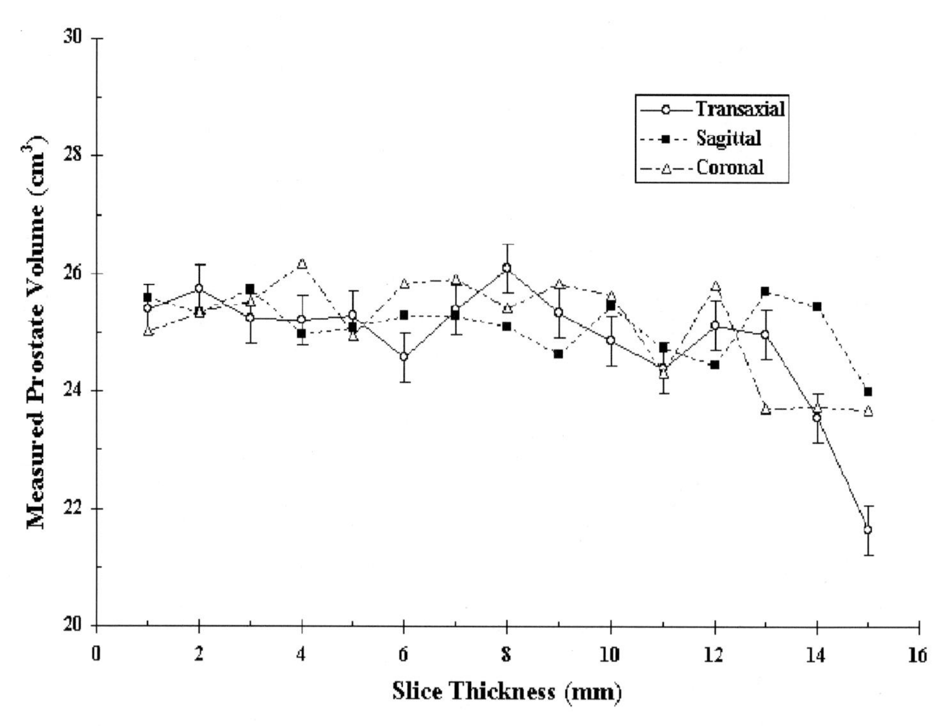

Figure 4-5. Plot of the prostate volume measurement obtained by manual planimetry for interslice distances ranging from 1 to 15 mm. The error bars indicate a 1 SD of volume estimation by planimetry (i.e., 1.7%). The plot shows results for planimetry carried out in three viewing planes: transaxial, sagittal, and coronal.

orthogonal (transaxial and sagittal) 2D images, and calculating the volume of the corresponding ellipsoid. This is called the HWL method that, along with similar methods used for other organs, has several deficiencies (Table 4-2). Estimating organ volume from a 3DUS image overcomes these difficulties, since the entire organ volume can be measured, thereby potentially reducing both the intra- and interobserver variabilities.

Chang et al. (27) have reported on the reproducibility of 3DUS versus the HWL method in the assessment of fetal liver volume. This report demonstrated that both the intra- and interobserver reproducibilities are improved with 3DUS. Volume measurements had an intraobserver standard deviation of 8.46 cm^3 for the HWL method compared with 2.15 ml for 3DUS, and an interobserver standard deviation of 6.60 cm^3 for the HWL method, but only 2.02 cm^3 for 3DUS.

Similar results comparing HWL and 3DUS methods have been obtained in prostate volume measurements made from 3D transrectal ultrasound (19). They showed that volume estimates had an intraobserver variability of 21.9% for the HWL method versus 15.5% for 3DUS, and an interobserver variability of 11.4% for the HWL method versus 5.1% for 3DUS (19).

Table 4-2. *Deficiencies in geometric methods for estimating volume*

- Many organs (e.g., prostate) are generally nonellipsoidal, so that the choice of which three chords in the 2D images are to be selected to measure H, W, and L is not obvious, and subject to observer preference, resulting in high interobserver variability
- Even for a single observer, the choice of chords is still arbitrary, leading to high intraobserver variability
- The placement of the thin 2D image planes within the organ is variable and arbitrary, increasing further the variability of the selected chord locations and hence the intra- and interobserver variabilites

SOURCES OF MEASUREMENT ERROR

Three-dimensional ultrasound image data provide unique advantages for measurement of distance, area, and volume. With the exception of systems using 2D arrays, volume data are reconstructed from multiple 2DUS images using their relative positions and orientations derived from geometric parameters associated with the 2DUS image scan. Thus any errors in the values of the parameters used for the 3D image reconstruction will lead to distortions and hence to measurement errors. Lack of border definition from acoustic shadowing or poor transducer contact also will pose a problem for accurately identifying structural details in 3DUS measurements similar to those encountered in 2DUS.

Mechanical Scanners: Linear

In mechanical linear scanning, the acquired 2DUS images are constrained to be parallel. Therefore two geometric parameters must be known accurately to produce accurate measurements: the distance (d) between the acquired 2DUS images and the tilt angle (θ) between the acquired 2DUS images in addition to the scanning direction (Figure 4-6). Consider a plane parallel to the scanning direction that cuts across n acquired 2DUS image planes, which are seen edge-on in Figure 4-6B. Suppose that parallelogram $WXYZ$ in this orientation has sides of lengths $D = |WX| = |YZ|$ and height $H = |XY| = |ZW|$, then the area (A) is given by

$$A = D\,H \cos\theta$$

The volume is the summation of the cross-sectional regions from multiple slices parallel to each other multiplied by the interslice separation (d) that is in the elevational direction and orthogonal to the 2DUS image plane and known accurately.

The measurement error can be estimated by observing that errors due to a distance error (Δd) in d and/or an angle error ($\Delta\theta$) in θ will result in a volume error estimate (ΔV) in an arbitrary volume (V) with relative volume error ($\Delta V/V$) that will be the same as the area error estimate (ΔA) in an arbitrary area (A) with relative area error ($\Delta A/A$) and is given by

$$\frac{\Delta A}{A} = \frac{\Delta V}{V} \cong \left(1 + \frac{\Delta d}{d}\right)(1 + \Delta\theta \tan\theta)\cos\Delta\theta - 1$$

Thus when the tilt angle is known exactly the percentage error in volume or area will essentially equal that in d. As a result, to ensure a volume error of less than 5% with an acquired image spacing of $d = 1$ mm, the systematic error in the distance between acquired images, Δd, must be less than 0.05 mm.

Mechanical Scanners: Tilting

Tong et al. (34) have performed an analysis of the linear, area, and volume measurement errors for a tilting transducer, which produces a radial fan of 2DUS image planes (Figure 4-7). In this geometry, the ultrasound transducer is tilted through a scan angle θ about its face by mechanically rotating the probe about its axis, while a series of 2DUS images are digitized at N equally spaced angular intervals (θ). The digitized region in each 2DUS image is depicted in Figure 4-7. Thus the 3DUS region used in the measurement is an annular sector of a cylinder, subtending an angle θ about the rotation axis, of height $H = |XY| = |WZ|$, from an inner radius R_0 to an outer radius $R_0 + H$, where R_0 is the distance from the rotation axis to the top of the digitized region. Thus to reconstruct and calculate the 3DUS volume correctly from the digitized 2DUS images, two geometric parameters R_0 and θ must be known accurately.

Consider a plane perpendicular to the rotational axis that cuts across the radial fan of acquired 2D images seen edge-on, as in Figure 4-7B. Suppose that the annular region

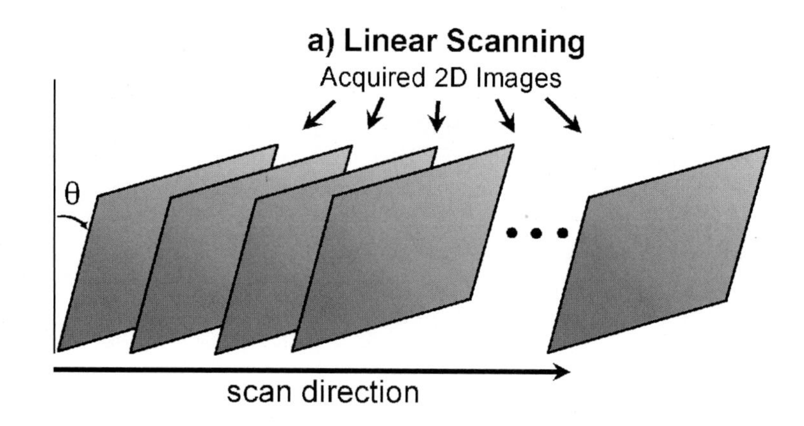

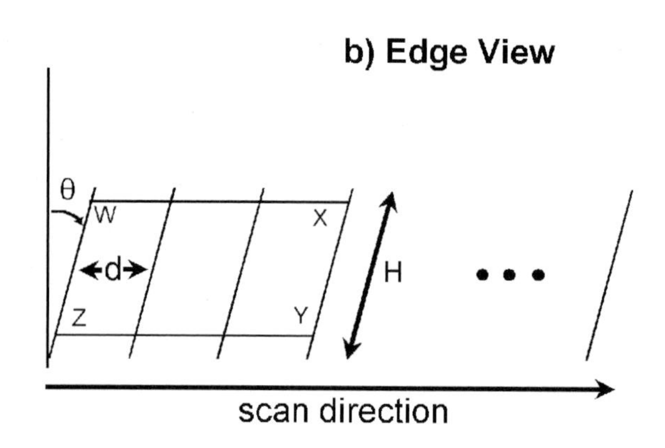

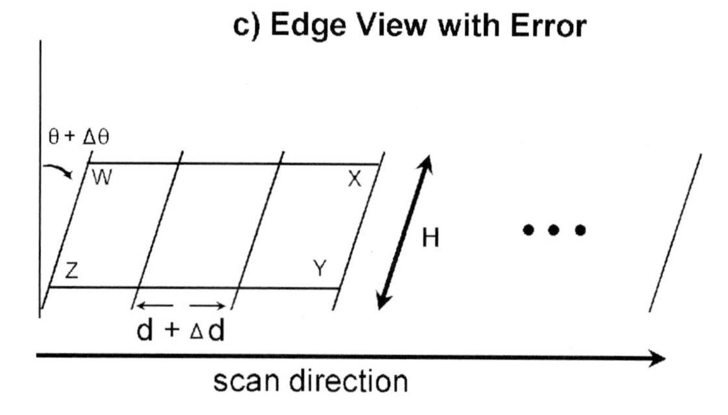

Figure 4-6. Schematic diagram of (**A**) a linear 3DUS scanning system showing multiple 2DUS images used in the 3D reconstruction; (**B**) the correct relative position and orientation of the 2DUS images are used in the reconstruction; and (**C**) errors in both the orientation and the distance between the acquired 2DUS images comprising the 3D volume.

WXYZ in this plane cuts across n images, subtending an angle $\theta = n/(N - 1)\theta$ about the axis and that the region inner radius $R_1 = R_0$ and outer radius $R_2 = R_0 + H$ corresponding to distances from the top of each acquired image are known accurately. Then the area A of *WXYZ* is:

$$A = \left(R_2^2 - R_1^2\right)\frac{\theta}{2}$$

The volume can be estimated by multiplying the area of the cross-sectional regions in multiple image slices by the interslice separation that varies with the radius of rotation, is in the elevational direction, and is orthogonal to the 2D image plane and known accurately. Measurement errors due to an error (ΔR_0) in R_0 or ($\Delta\theta_0$) in θ_0 will result in a relative volume error $\Delta V/V$ in an arbitrary volume V will be the same as the relative area error $\Delta A/A$ in an arbitrary area A in the same way as for a linear scanner and is given by:

$$\frac{\Delta V}{V} = \frac{\Delta A}{A} = \left(1 + \frac{\Delta R}{R}\right)\left(1 + \frac{\Delta\theta}{\theta}\right) - 1$$

$$= \frac{\Delta R}{R} + \frac{\Delta\theta}{\theta} + \frac{\Delta R}{R}\frac{\Delta\theta}{\theta}.$$

where $R = (R_2 + R_1)/2$ is the mean radius of the measured volume from the axis of rotation.

Thus if the axis location is known exactly, the percentage error in the measurement will equal the error in θ. As a result, to ensure a volume error of less than 5% for a scanning angle of $\theta = 80$ degrees, $\Delta\theta$ must be less than 4 degrees, so that, for a scan

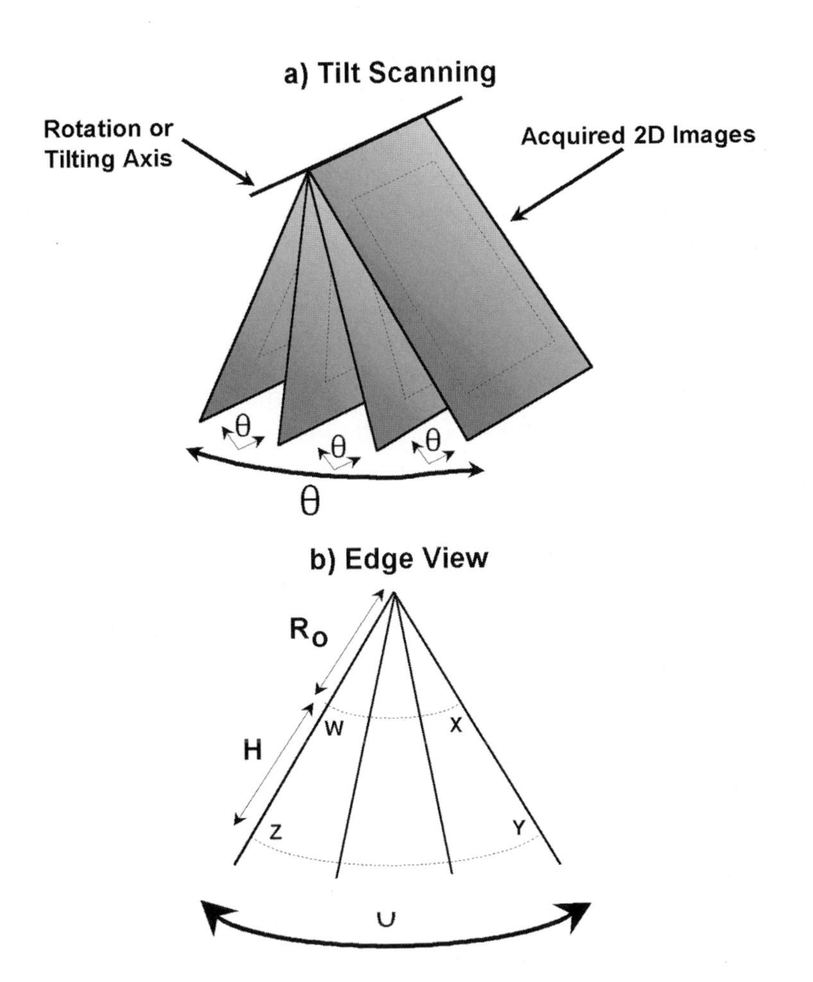

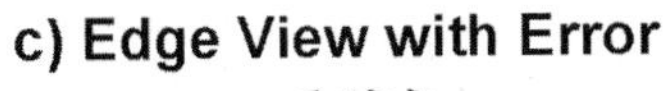

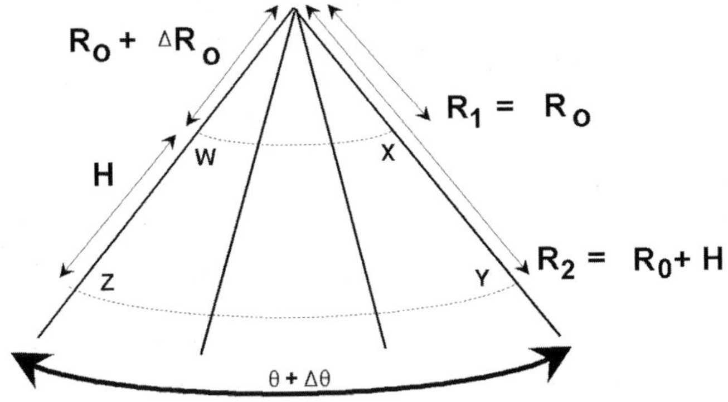

Figure 4-7. Schematic diagram of (**A**) a tilting 3DUS scanning system showing the set of 2DUS images used in the 3D reconstruction; (**B**) the correct relative position and orientation of the 2DUS images used in the reconstruction; and (**C**) errors in the angle between the images, as well as in the distance between the axis of orientation and the digitized region of interest.

containing 100 2D images, the angular accuracy must be within 0.04 degree. Similarly, if the scan angle is known exactly, the percentage error in volume will equal that in R. Thus to ensure a volume error of less than 5% for $R = 10$ mm requires that ΔR be less than 0.5 mm. For a more detailed analysis of the shape, distance, area, and volume distortions that arise due to these errors, a calibration method for virtually eliminating them, and an exact method for calculating the mean radius R of a volume, the reader is referred to the report by Tong et al. (34).

Free-hand Scanners

Because free-hand scanning systems do not constrain the movement of the ultrasound transducer, the position and orientation of each digitized image must be accurately determined individually. This information can be obtained using a variety of techniques, such as mechanical arms (35), spark gaps (9), optical sensors (36), and magnetic sensors (8,11,37–39).

To obtain high-quality 3DUS measurements, both random and systematic errors must be avoided. However, the most important requirement for measurement accuracy is system calibration. This is due to the fact that although random errors are important with regard to image quality, they tend to average out over the distances and volumes to be measured, so that systematic errors do not lead to measurement errors.

The issues involved in calibrating free-hand scanners have been addressed by a number of investigators. State et al. (36) describe a technique that uses a small bead and cross-wire calibration method. King et al. (9), using a multiple pin approach, reported a standard deviation of 0.7 to 1.3 mm for locating a point with an acoustic positioning sensor. Detmer et al. (37) investigated in detail the accuracy of a magnetic positioning device (Flock of Birds, Ascension Technology Corp., Burlington, VT) and used a cross-wire calibration method to demonstrate that the root-mean-square (rms) error at a tissue depth of 60 mm ranged from 2.1 to 3.5 mm, depending on the distance from the stationary transmitter to the small receiver mounted on the ultrasound transducer. Leotta et al. (39), reporting on two calibration methods for magnetic positioning devices, found an rms calibration error of about 2.5 mm. Subsequently, Leotta et al. (39) examined an upgraded version of the device. They found an rms calibration error ranging from 0.5 to 2.5 mm, depending on tissue depth, and found that the separations between beads arranged in a grid pattern could be measured with an error of only 0.06 ± 0.68 mm.

VOLUME SEGMENTATION

Measurement of volumes by manual planimetry is tedious and time-consuming. A better approach is to use a computerized method that recognizes structures automatically, requiring minimal user involvement. In the field of image processing, this process is known as segmentation (40). Although humans can easily distinguish separate image regions, such as organs or tumors, using a wide variety of image cues (e.g., texture, brightness, shape, and location), it is extremely difficult to develop a computer algorithm that can use these and other image attributes to perform the same task. Consequently, no general computerized segmentation approach for medical images has yet been developed. However, over the past 20 years, many methods have been developed and applied to 3D computed tomography and 3D magnetic resonance images, with varying degrees of success (41,42).

Table 4-3. *Ultrasound image artifacts affecting boundary contour detection*

- Shadowing (e.g., by calcifications in organs, arterial plaques, bony structures, or gas), which causes distal structures to be obscured
- Speckle, which causes the boundary or surface of an object to be discontinuous and not smooth
- Attenuation of the sound, which causes signal drop-off in deep structures and results in distal boundaries becoming less distinct or sometimes invisible

Table 4-4. *Boundary contour image segmentation algorithm*

- The operator identifies the structure to be segmented
- The operator provides an initial boundary contour or seed point(s)
- The algorithm filters the image to reduce noise and speckle
- The algorithm filters the image to enhance edges
- The algorithm identifies the edge contour either as disconnected segments or as a continuous contour
- The algorithm connects the contour segments (if necessary) and optimizes the fit of the contour to the structure
- The operator inspects the contour and corrects errors for final result

Segmentation techniques have been applied to a number of 3DUS imaging applications. Examples range from simple thresholding of the image to using sophisticated boundary detection techniques in cardiology to identify the ventricles (43–52) for intravascular 3DUS (53–57) and the segmentation of other structures (58–62). In 3DUS, successful segmentation approaches have been based on the definition and recognition of organ boundary contours. However, reliable boundary contour identification requires that sufficient contrast be present in the image, whereas ultrasound images (both 2D and 3D) uniquely suffer from the following image artifacts (Table 4-3). In combination, these artifacts make the fully automatic and general boundary segmentation of ultrasound images virtually impossible. Only simple structures with distinct image contrast can be automatically segmented, such as endocardial contours (63,64).

Most algorithms for ultrasound image segmentation use a semi-automatic approach, in which the operator provides an initial estimate of the boundary contour or can edit the computer-generated contour to correct for obvious errors. In this way, the operator can use specific knowledge and imaging experience to produce a better final result. Adopting this approach, the segmentation algorithm can then include several steps (Table 4-4).

Some reports of ultrasound image segmentation in the literature include the successful application of edge enhancement techniques for the determination of the prostate boundary (32) and the application of active contours to the surfaces of the heart (65–67). Figure 4-8 shows an example of a segmentation approach to measure volume in the fetal heart.

Another approach to segmentation is to use the ultrasound machine to select the structures of interest. For example, Doppler imaging techniques can provide images in which the blood flowing in the vasculature is displayed in color and the stationary anatomy is displayed as a gray-scale image. If this color image is digitized, the computer can remove the gray-scale portion of the image, leaving only an image of the Doppler signal corresponding to the vasculature. This approach has been successfully used by a number of

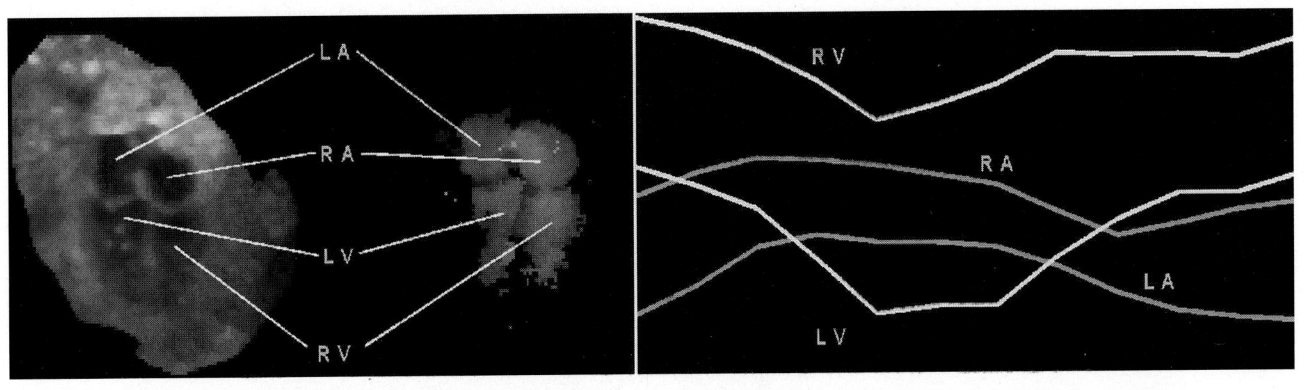

Figure 4-8. Imaging of the fetal heart and chambers showing volume-rendered images of the cardiac chambers and vessels. The signal void in the original acquisition has been used to segment the data to show only the blood signal. Chamber and vessel rendering display uses a modified maximum-intensity method with depth-coding. Once segmented, the chambers are isolated and individual volumes are calculated throughout the cardiac cycle.

investigators to segment the carotid arteries (68–71), the breast vasculature (72–74), the prostate vasculature (75), the kidney (76), and vascular phantoms (77,78).

HOW TO DO IT
Practical Clinical Tips for Measurements

- Be sure that the 3DUS system is properly calibrated.
- Be sure that there are no motion artifacts.
- Be sure that the resolution is optimized in all 3DUS data directions:
  - *Distance:* Place measurement cursor at the midpoint of contrast transition (not the bottom or top).
  - *Area:* For planimetric tracing of the boundary, be sure that the steps along the boundary are sufficiently small to follow the contour accurately.
  - *Area:* For automated techniques, be sure to verify and edit the contour if it is not correct to accommodate shadowing artifacts.
  - *Volume:* For measurements using areas from multiple parallel slices, be sure that the interslice distance is sufficiently small.
  - *Area and volume:* In manual planimetry, verify that the edge has been outlined by viewing it also from orthogonal planes.

REFERENCES

1. Ariet M, Geiser EA, Lupkiewicz SM, Conetta DA, Conti CR. Evaluation of a three-dimensional reconstruction to compute left ventricular volume and mass. *Am J Cardiol* 1984;54:415–420.
2. Elliot TL, Downey DB, Tong S, Mclean CA, Fenster A. Accuracy of prostate volume measurements *in vitro* using three-dimensional ultrasound. *Acad Radiol* 1996;3:401–406.
3. Favre R, Nisand G, Bettahar K, Grange G, Nisand I. Measurement of limb circumferences with three-dimensional ultrasound for fetal weight estimation. *Ultrasound Obstet Gynecol* 1993;3:176–179.
4. Favre R, Bader AM, Nisand G. Prospective study on fetal weight estimation using limb circumferences obtained by three-dimensional ultrasound. *Ultrasound Obstet Gynecol* 1995;6:140–144.
5. Gilja OH, Thune N, Matre K, Hausken T, Odegaard S, Berstad A. *In vitro* evaluation of three-dimensional ultrasonography in volume estimation of abdominal organs. *Ultrasound Med Biol* 1994;20:157–165.
6. Gilja OH, Smievoll AI, Thune N, et al. *In vivo* comparison of 3D ultrasonography and magnetic resonance imaging in volume estimation of human kidneys. *Ultrasound Med Biol* 1995;21:25–32.
7. Hodges TC, Detmer PR, Burns DH, Beach KW, Strandness DE Jr. Ultrasonic three-dimensional reconstruction: *in vitro* and *in vivo* volume and area measurement. *Ultrasound Med Biol* 1994;20:719–729.
8. Hughes SW, D'Arcy TJ, Maxwell DJ, Saunders JE, Sheppard RJ. Volume estimation from multiplanar 2D ultrasound images using a remote electromagnetic position and orientation sensor. *Ultrasound Med Biol* 1996;22:561–572.
9. King DL, King DLJ, Shao MY. Evaluation of *in vitro* measurement accuracy of a three-dimensional ultrasound scanner. *J Ultrasound Med* 1991;10:77–82.
10. Moritz WE, Pearlman AS, McCabe DH, et al. An ultrasonic technique for imaging the ventricle in three dimensions and calculating its volume. *IEEE Trans Biomed Eng* 1993;30:482–491.
11. Nelson TR, Pretorius DH. 3-dimensional ultrasound volume measurement. *Med Phys* 1993;201:927.
12. Riccabona M, Nelson TR, Pretorius DH, Davidson TE. Distance and volume measurement using three-dimensional ultrasonography. *J Ultrasound Med* 1995;14:881–886.
13. Riccabona M, Nelson TR, Pretorius DH, Davidson TE. Three-dimensional sonographic measurement of bladder volume. *J Ultrasound Med* 1996;15:627–632.
14. Riccabona M, Nelson TR, Pretorius DH. Three-dimensional ultrasound: accuracy of distance and volume measurements. *Ultrasound Obstet Gynecol* 1996;7:429–434.
15. Siu SC, Rivera JM, Guerrero JL, et al. Three-dimensional echocardiography. *In vivo* validation for left ventricular volume and function. *Circulation* 1993;88:1715–1723.
16. Sivan E, Chan L, Uerpairojkit B, Chu GP, Reece EA. Growth of the fetal forehead and normative dimensions developed by three-dimensional ultrasonographic technology. *J Ultrasound Med* 1997;16:401–406.
17. Terris MK, Stamey TA. Determination of prostate volume by transrectal ultrasound. *J Urol* 1991;145: 984–987.
18. Weiss JL, Eaten LW, Kallman CH, Maughman WL. Accuracy of volume determination by two-dimensional echocardiography: defining requirements under controlled conditions in the ejecting canine left ventricle. *Circulation* 1983;67:889–895.
19. Tong S, Downey DB, Cardinal HN, Fenster A. A three-dimensional ultrasound prostate imaging system. *Ultrasound Med Biol* 1996;22:735–746.
20. Brinkley JF, Muramatsu SK, McCallum WD, Popp RL. *In vitro* evaluation of an ultrasonic three-dimensional imaging and volume system. *Utrason Imaging* 1982;4:126–139.

21. Brinkley JF, McCallum WD, Muramatsu SK, Liu DY. Fetal weight estimation from lengths and volumes found by three-dimensional ultrasonic measurements. *J Ultrasound Med* 1984;3:163–168.
22. Chan H. Noninvasive bladder volume measurement. *J Neurosci Nurs* 1993;25:309–312.
23. Hughes SW, D'Arcy TJ, Maxwell DJ, JE Saunders. *In vitro* estimation of foetal volume using ultrasound, x-ray, computed tomography and magnetic resonance imaging. *Physiol Meas* 1997;18:401–410.
24. Chanoine JP, Toppet V, Lagasse R, et al. Determination of thyroid volume by ultrasound from the neonatal period to late adolescence. *Eur J Pediatr* 1991;150:395–399.
25. De Odorico I, Spaulding KA, Pretorius DH, Lev-Toaff AS, Bailey TB, Nelson TR. Normal splenic volumes estimated using three-dimensional ultrasound. *JUM* 1999 (*in press*).
26. Brunner M, Obruca A, Bauer P, Feichtinger W. Clinical application of volume estimation based on three-dimensional ultrasonography. *Ultrasound Obstet Gynecol* 1995;6:358–361.
27. Chang FM, Hsu KF, Ko HC, et al. Fetal heart volume assessment by three-dimensional ultrasound. *Ultrasound Obstet Gynecol* 1997;9:942–948.
28. Liess H, Roth C, Umgelter A, Zoller WG. Improvements in volumetric quantification of circumscribed hepatic lesions by three-dimensional sonography. *Z Gastroenterol* 1994;32:488–492.
29. Wolf GK, Lang H, Prokop M, Schreiber M, Zoller WG. Volume measurements of localized hepatic lesions using three-dimensional sonography in comparison with three-dimensional computed tomography. *Eur J Med Res* 1998;3:157–164.
30. Sehgal CM, Broderick GA, Whittington R, Gorniak RJ, Arger PH. Three-dimensional US and volumetric assessment of the prostate. *Radiology* 1994;192:274–278.
31. Lee W, Comstock CH, Kirk JS, et al. Birthweight prediction by three-dimensional ultrasonographic volumes of the fetal thigh and abdomen. *J Ultrasound Med* 1997;16:799–805.
32. Aarnink RG, Giesen RJB, de la Rosette J, Huynen AL, Debruyne FMJ, Wijkstra H. Planimetric volumetry of the prostate: how accurate is it? *Phys Meas* 1995;16:141–150.
33. Stone NN, Ray PS, Smith JA, et al. Ultrasound determination of prostate volume: comparison of transrectal (ellipsoid v planimetry) and suprapubic methods. *J Endourol* 1991;5:252–254.
34. Tong S, Cardinal HN, Downey DB, Fenster A. Analysis of linear, area, and volume distortion in 3D ultrasound imaging. *Ultrasound Med Biol* 1998;24:355–373.
35. Ohbuchi R, Chen D, Fuchs H. Incremental volume reconstruction and rendering for 3D ultrasound imaging. *SPIE* 1992;1808:312–322.
36. State A, Livingston MA, Garrett WF, et al. Technologies for augmented reality systems: realizing ultrasound-guided needle biopsies. Proceedings of SIGGRAPH '96. *Comput Graph* 1996;439–446.
37. Detmer PR, Bashein G, Hodges T, et al. 3D ultrasonic image feature localization based on magnetic scanhead tracking: *in vitro* calibration and validation. *Ultrasound Med Biol* 1994;20:923–936.
38. Fenster A, Downey DB. 3-D ultrasound imaging: a review. *IEEE Eng Med Biol* 1996;15:41–51.
39. Leotta DF, Detmer PR, Martin RW. Performance of a miniature magnetic position sensor for three-dimensional ultrasound imaging. *Ultrasound Med Biol* 1997;24:597–609.
40. Robb RA. *Three-dimensional biomedical imaging: principles and practice.* New York: VCH, 1995.
41. Coatrieux JL, Toumoulin C, Hamon C, Lou L. Future trends in 3D medical imaging. *IEEE Eng Med Biol* 1990;12:33–39.
42. Fishman EK, Magid D, Ney DR, et al. Three-dimensional imaging. *Radiology* 1991;181:321–337.
43. Ashton EA, Phillips D, Parker KJ. Automated extraction of the LV from 3D cardiac ultrasound scans. *Proceedings of the International Society for Optical Engineering* 1996;2727:423–429.
44. Chu CH, Delp EJ, Buda AJ. Detecting left ventricular endocardial and epicardial boundaries by digital two-dimensional echocardiography. *IEEE Trans Med Imaging* 1988;7:81–90.
45. Fazzalari NL, Goldblatt E, Adams APS. A composite three-dimensional echocardiographic technique for left ventricular volume estimation in children: comparison with angiography and established echocardiographic methods. *J Clin Ultrasound* 1986;14:663–674.
46. Klingler JW Jr, Vaughan CL, Fraker TD Jr, Andrews LT. Segmentation of echocardiographic images using mathematical morphology. *IEEE Trans Biomed Eng* 1988;35:925–934.
47. Kuwahara M, Eiho S, Asada N. Left ventricular image processing of 2-D echocardiograms and 3-D reconstruction of the left ventricle. *Front Med Biol Eng* 1988;1:9–17.
48. Martin RW, Bashein G. Measurement of stroke volume with three-dimensional transesophageal ultrasonic scanning: comparison with thermodilution measurement. *Anesthesiology* 1989;70:470–476.
49. McPherson DD, Skorton DJ, Kodiyalam S, et al. Finite element analysis of myocardial diastolic function using three-dimensional echocardiographic reconstructions: application of a new method for study of acute ischemia in dogs. *Circ Res* 1987;60:674–682.
50. Osakada G, Hashimoto S, Murachi T, et al. Pathophysiological analysis of cardiac function by computer processing of echocardiograms. *Rinsho Byori* 1990;38:1114–1118.
51. Skorton DJ, Collins SM. Quantitation in echocardiography. *Cardiovasc Intervent Radiol* 1987;10:316–331.
52. Zoghbi WA, Buckey JC, Massey MA, Blomqvist CG. Determination of left ventricular volumes with use of a new nongeometric echocardiographic method: clinical validation and potential application. *J Am Coll Cardiol* 1990;15:610–617.
53. Bruining N, von Birgelen C, Di Mario C, et al. Dynamic three-dimensional reconstruction of ICUS images based on an EGG-gated pull-back device. *Comput Cardiol* 1995;1995:633–636.
54. Cavaye DM, Tabbara MR, Kopchok GE, Laas TE, White RA. Three-dimensional vascular ultrasound imaging. *Am Surg* 1991;57:751–755.
55. Cavaye DM, Diethrich EB, Santiago OJ, et al. Intravascular ultrasound imaging: an essential component of angioplasty assessment and vascular stent deployment. *Int Angiol* 1993;1993;12:214–220.
56. Cavaye DM, White RA. Intravascular ultrasound imaging: development and clinical applications. *Int Angiol* 1993;12:245–255.
57. Klein H-M, Günther RW, Verlande M, et al. 3D-surface reconstruction of intravascular ultrasound images using personal computer hardware and a motorized catheter control. *Cardiovasc Intervent Radiol* 1992;15: 97–101.

58. Baba K, Okai T, Kozuma S, Taketani Y, Mochizuki T, Akahane M. Real-time processable three-dimensional US in obstetrics. *Radiology* 1997;203:571–574.
59. Baba K, Jurkovic D. *Three-dimensional ultrasound in obstetrics and gynecology.* New York: Parthenon Press, 1997.
60. Blankenhorn DH, Chin HP, Strikwerda S, Bamberger J. Hestenes JD. Work in progress: common carotid artery contours reconstructed in three dimensions from parallel ultrasonic images. *Radiology* 1983;148: 533–537.
61. Ehricke H-H, Donner K, Koller W, Strasser W. Visualization of vasculature from volume data. *Comput Graph* 1994;18:395–406.
62. Kitney RI, Moura L, Straughan K. 3-D visualization of arterial structures using ultrasound and Voxel modelling. *Int J Cardiac Imag* 1989;1989;4:135–143.
63. Lohregt S, Viergever MA. A discrete dynamic contour model. *IEEE Trans Med Imaging* 1995;14:12–24.
64. Nelson TR, Pretorius DH, Sklansky M, Hagen-Ansert S. Three-dimensional echocardiographic evaluation of fetal heart anatomy and function: acquisition, analysis, and display. *J Ultrasound Med* 1996;15:1–9.
65. Chakraborty A, Staib LH, Duncan JS. An integrated approach to boundary finding in medical images. *IEEE Workshop on Biomedical Image Analysis* 1994;(June):13–22.
66. Chalana V, Costa W, Kim Y. Integrating region growing and edge detection using regularization. *SPIE* 1995;2434:262–271.
67. Neveu M, Faudot D, Derdouri B. Recovery of 3D deformable models from echocardiographic images. *SPIE* 1994;2299:367–376.
68. Fenster A, Tong S, Sherebrin S, Downey DB, Rankin RN. Three-dimensional ultrasound imaging. *SPIE Phys Med Imaging* 1995;2432:176–184.
69. Picot PA, Rickey DW, Mitchell R, Rankin RN, Fenster A. Three-dimensional colour Doppler imaging. *Ultrasound Med Biol* 1993;19:95–104.
70. Pretorius DH, Nelson TR, Jaffe JS. 3-Dimensional sonographic analysis based on color flow Doppler and gray scale image data: a preliminary report. *J Ultrasound Med* 1992;11:225–232.
71. Ritchie CJ, Edwards WS, Mack LA, Cyr DR, Kim Y. Three-dimensional ultrasonic angiography using power-mode Doppler. *Ultrasound Med Biol* 1996;22:277–286.
72. Carson PL, Adler DD, Fowlkes JB, Harnist K, Rubin J. Enhanced color flow imaging of breast cancer vasculature: continuous wave Doppler three-dimensional display. *J Ultrasound Med* 1992;11:377–385.
73. Carson PL, Li X, Pallister J, Moskalik A, Rubin JM, Fowlkes JB. Approximate quantification of detected fractional blood volume and perfusion from 3-D color flow and Doppler power signal imaging. *1993 ultrasonics symposium proceedings.* IEEE 1993:1023–1026.
74. Ferrara KW, Zagar B, Sokil-Melgar J, Algazi VR. High resolution 3D color flow mapping: applied to the assessment of breast vasculature. *Ultrasound Med Biol* 1996;22:293–304.
75. Downey DB, Chin JL, Fenster A. Three-dimensional US-guided cryosurgery. *Radiology* 1995;197(P):539.
76. Downey DB, Fenster A. Vascular imaging with a three-dimensional power Doppler system. *AJR* 1995;165: 665–668.
77. Guo Z, Moreau M, Rickey DW, Picot PA, Fenster A. Quantitative investigation of *in vitro* flow using three-dimensional colour Doppler ultrasound. *Ultrasound Med Biol* 1995;21:807–816.
78. Guo Z, Fenster A. Three-dimensional power Doppler imaging: a phantom study to quantify vessel stenosis. *Ultrasound Med Biol* 1996;22:1059–1069.

PART III

Clinical Applications

5

Approach to Clinical Scanning

OVERVIEW

This chapter provides an overview of the approach to clinical three-dimensional ultrasound (3DUS) scanning. Although the specific method varies slightly, depending on the equipment used, the overall principles are similar.

KEY CONCEPTS AND HOW TO DO IT

Practical Clinical Tips for Clinical 3DUS Scanning

- Acquisition of volume:
  - Select the transducer for the organ of interest.
  - Scan the organ to localize the region-of-interest and optimize the orientation.
  - Have the patient breath-hold and not move during volume acquisition.
  - Acquire a volume of data.
- Review of volume:
  - Review 2DUS acquisition slices during acquisition for quality and lack of motion.
  - Review volume data for quality and freedom from artifacts or distortions.
  - Save data volume.
  - Acquire next volume if desired.
  - After all volumes are acquired, dismiss patient unless reviewing study with patient.
  - Review volume data by reslicing and rendering as appropriate.
  - Select optimized images for archiving and diagnostic evaluation.
  - Bookmark optimized images from above.
  - Archive images with volumes.
  - Send study to physician for interpretation.

CLINICAL SCANNING IN A NUTSHELL

The general process of acquiring volume data from a patient is at present based on a strategy similar to that for routine conventional two-dimensional ultrasound (2DUS) scanning. As faster scanners become available that offer the capability of updating complete volumes in less than a second, it is expected that the overall approach to volume scanning will change and become more interactive. Generally, there are two primary parts to clinical three-dimensional ultrasound (3DUS) scanning: (a) scanning of the patient and obtaining adequate, complete volumes from which to make a diagnosis; (b) review of volume data with various types of visualization tools to extract or enhance the diagnostically important information for the physician.

ACQUISITION OF VOLUME DATA

The first step in obtaining a volume data set is to select the appropriate transducer for the organ of interest (Figure 5-1). After adjustment of the general imaging parameters and scanner setup, scan over the organ of interest as per a standard 2DUS scan. Next, determine which orientation is anatomically most important by sweeping back and forth. In general, it is best to acquire the volume from the plane that is most important anatomically, such as the transverse or longitudinal view, because the plane of acquisition always will provide the best resolution in the volume. Currently, the optimal plane of acquisition for specific organs is under evaluation. Generally, a region of interest for the volume acquisition is identified during the 2DUS scan. To obtain optimal volume acquisition quality the patient must remain stationary. This is best accomplished by having the patient hold his or her breath and not move during the acquisition, which is generally less than 20 sec. If motion does occur during the volume acquisition, then the volume may be discarded and reacquired. The overall assumption is that the only movement is due to changes in the transducer position or orientation. Obviously, in the case of cardiac motion, additional processing will be necessary to compensate for motion.

Acquisition of a data volume is accomplished by sweeping through the volume of interest at sweep speed sufficiently slow to fill the volume without gaps yet fast enough to avoid patient motion or oversampling of the volume (Figure 5-2). The best results occur if the sweep is steady, smooth, and continuous using either a linear, rocking motion or a rotating motion. Compounding volume data through multiple sweeps to build up a complete volume, similar to static B-scan methods, is possible, although slight tissue distortions resulting from transducer pressure and patient motion generally preclude this approach.

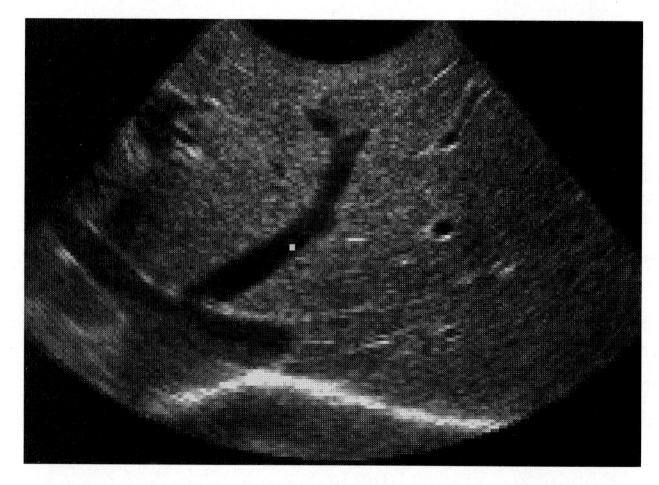

Figure 5-1. Acquisition of 3DUS data. **(a)** Select the transducer for the organ of interest. **(b)** 2DUS prescan the organ to determine the optimum scan parameters, acoustic window, and localize the anatomy of interest.

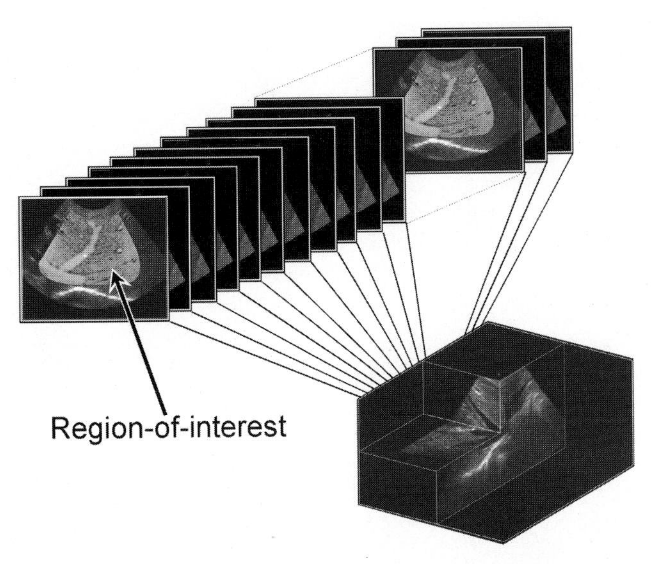

Figure 5-2. Acquisition of 3DUS data. **(a)** Identify region of interest for volume acquisition. **(b)** Acquire 3DUS data volume in linear, tilt, or rotational pattern covering the field of view desired for scan. **(c)** Review 2DUS slices during acquisition to verify image quality and lack of motion. **(d)** Have the patient breath-hold and not move during volume acquisition.

During volume acquisition it is important to monitor the images as they are collected to make sure that the desired organ anatomy is in the field of view, that there is adequate image quality, that the number of images is adequate, that any obvious abnormalities are seen, and that there is no movement.

After acquisition the data should be quickly reviewed in the plane of acquisition (Figure 5-3). After the volume is available for review it is best to reorient the data into a standard anatomic orientation using rotation and verify that the acquisition is adequate as described earlier. If the acquisition is satisfactory then it should be saved to disk or the appropriate archive. Otherwise it should be deleted and another volume acquired. At this point one of two things can happen. Either 2DUS scanning can resume to identify another volume to acquire or a review of the acquired volume(s) can begin. One advantage of volume imaging is that the review can occur after the patient has left the scanning room. The optimal strategy will depend on the experience of the sonographer/physician and his or her confidence as to whether a complete volume has been acquired from which to make a diagnostic assessment. If additional volumes are desired, then the previous scanning approach is repeated until the examination is completed. Generally, at least two volumes are collected for ''routine'' evaluation and four to eight volumes for ''abnormal'' findings. In some cases it may be advantageous to acquire volume data for the same location from different orientations to ensure that no detail is lost. Ultimately, the need for multiple orientations may be reduced as a better understanding of scanner performance and limitations is obtained.

REVIEW AND ANALYSIS OF VOLUME DATA

Depending on the type of study, it is desirable to perform some type of data visualization with the patient present. Certainly visualization of the fetal face or developmental abnormality can be valuable for increasing the patient's awareness of the situation. Generally, only one or two volumes are reviewed or rendered while the patient waits. The other volumes are rendered after the patient has left the clinic. In clinical practice it is important for rendering to occur rapidly, so that it does not delay the clinical schedule.

Typically, multiple volumes will be acquired during the scanning phase of the study.

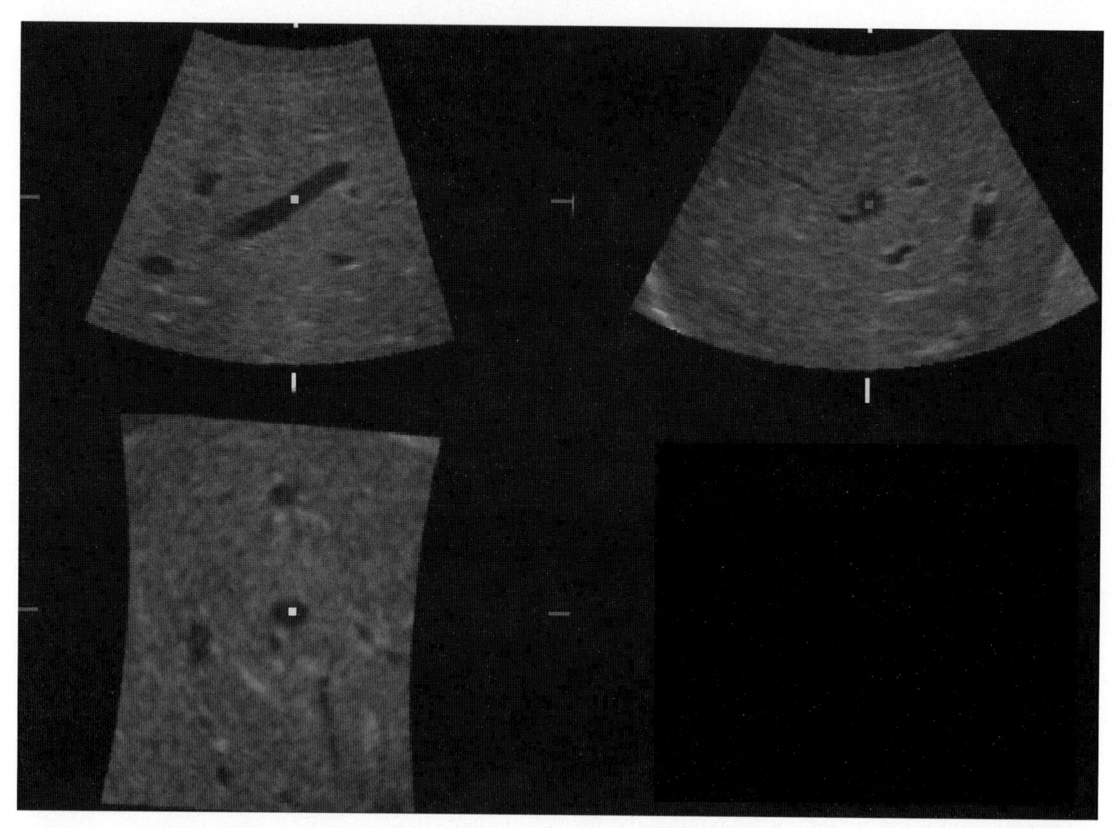

Figure 5-3. Acquisition of 3DUS data. **(a)** Review acquired 3DUS data for study adequacy, image quality, and freedom from artifacts or distortions. **(b)** Reorient data to standard position. **(c)** Save 3DUS data volume. **(d)** Acquire next 3DUS volume if desired. **(e)** After all volumes are acquired, dismiss patient unless reviewing study with patient.

The physician or sonographer needs to review these volumes and determine the clinically "important" volumes to be used for diagnostic analysis by the physician. Generally, all the volumes will not be completely analyzed. The type of review and analysis will depend on the type of study. Volume data analysis can be divided into three general areas.

First, the volume data are resliced, performing a "virtual" scan of the organ. Single or multiple orthogonal slices may be displayed with the sonographer/physician interactively scrolling or panning through the volume. Reorientation of the slice during this process is essential to assist in optimization of the viewing position and to take full advantage of volume data to obtain viewing orientations not possible during the scanning process.

Second, the volume is rendered so as to visualize the overall anatomy of the organ (Figure 5-4). Typically, this may include the fetal face or organ vascularity. Rendering is a valuable complement to reslicing because it provides a more complete view of the anatomic relationships, something that generally is difficult to obtain from 2DUS slices. Rendering is a significantly more sophisticated process than reslicing, and the algorithms are more computationally intensive. As a result, scanner performance will be a significant factor in its success or practicality. Ideally, rendering will be interactive, just as reslicing is. Additionally, selection of the proper rendering algorithm and thresholds will depend on operator experience, type of study, and patient image quality. For example, some studies will be very dependent on rendering (e.g., fetal face); others will not require rendering (e.g., uterine endometrium when evaluating for developmental anomalies such as septate or bicornuate uterus). Combination of rendering with resliced data or orthogonal slices is a valuable means of orienting the operator to the location of the slice within the volume (Figure 5-5).

Third, volume data permit measurements of length, area, and volume that are more

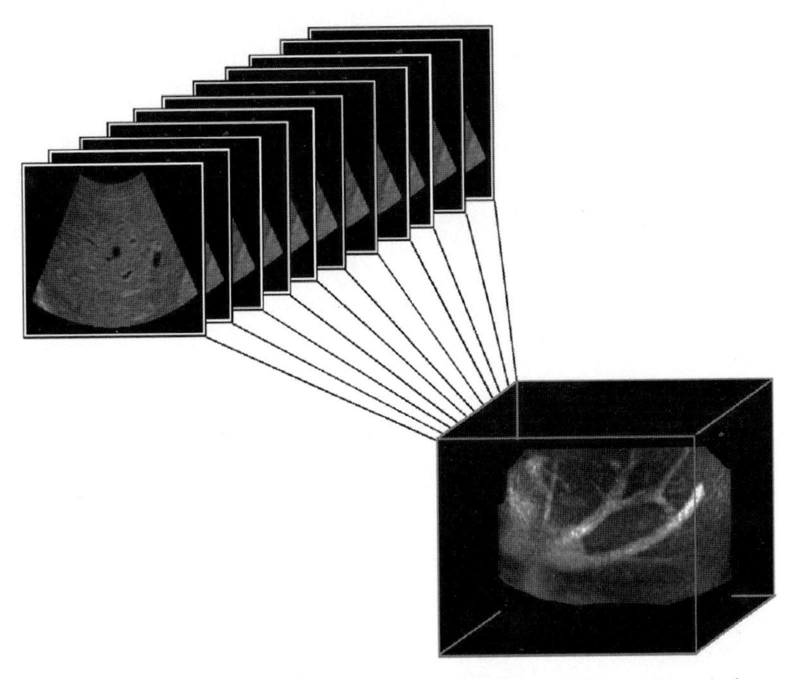

Figure 5-4. Review of 3DUS data. **(a)** Review volume data by reslicing and rendering as appropriate. **(b)** Select rendering method to optimize anatomy of interest (e.g. bones, blood vessels, soft tissues). **(c)** Edit volume data to extract structure or organ of interest.

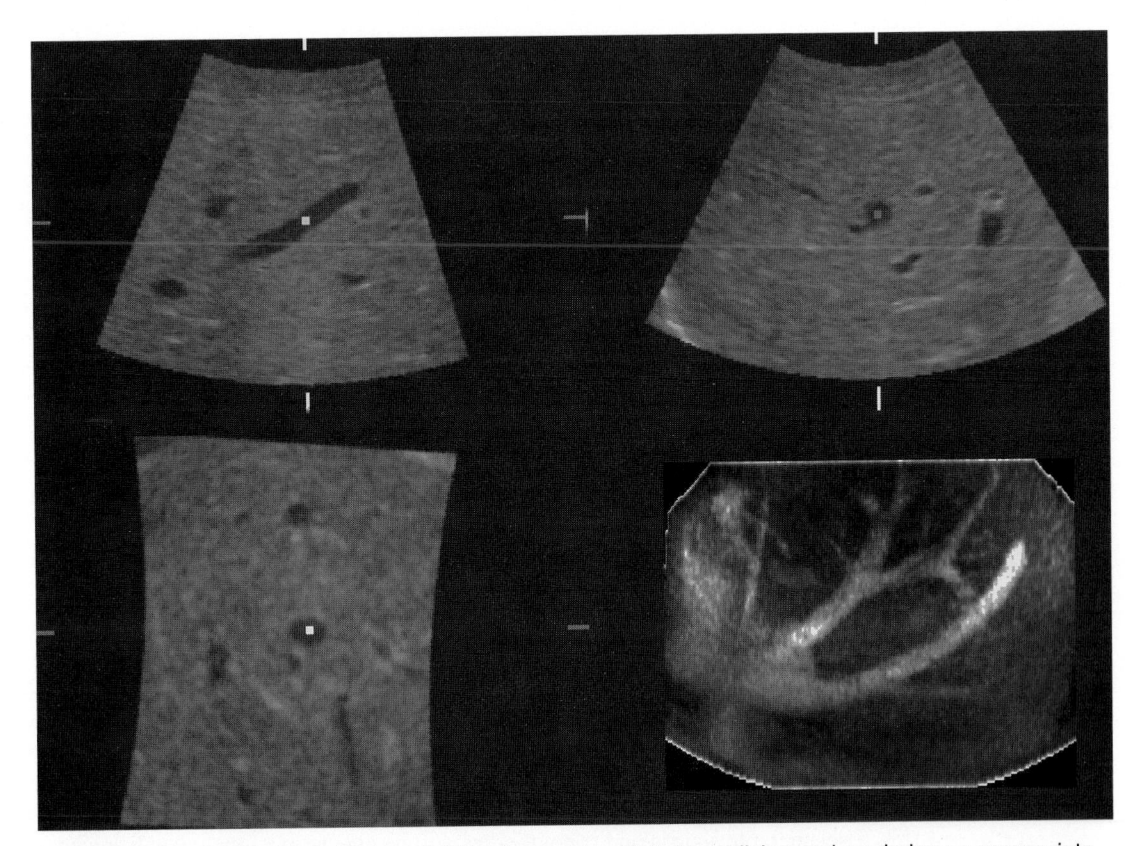

Figure 5-5. Review of 3DUS data. **(a)** Review volume data by reslicing and rendering as appropriate. **(b)** Evaluate combination of planar views with rendered view to coordinate localization and identification of anatomy. **(c)** Select optimized images for archiving and diagnostic evaluation.

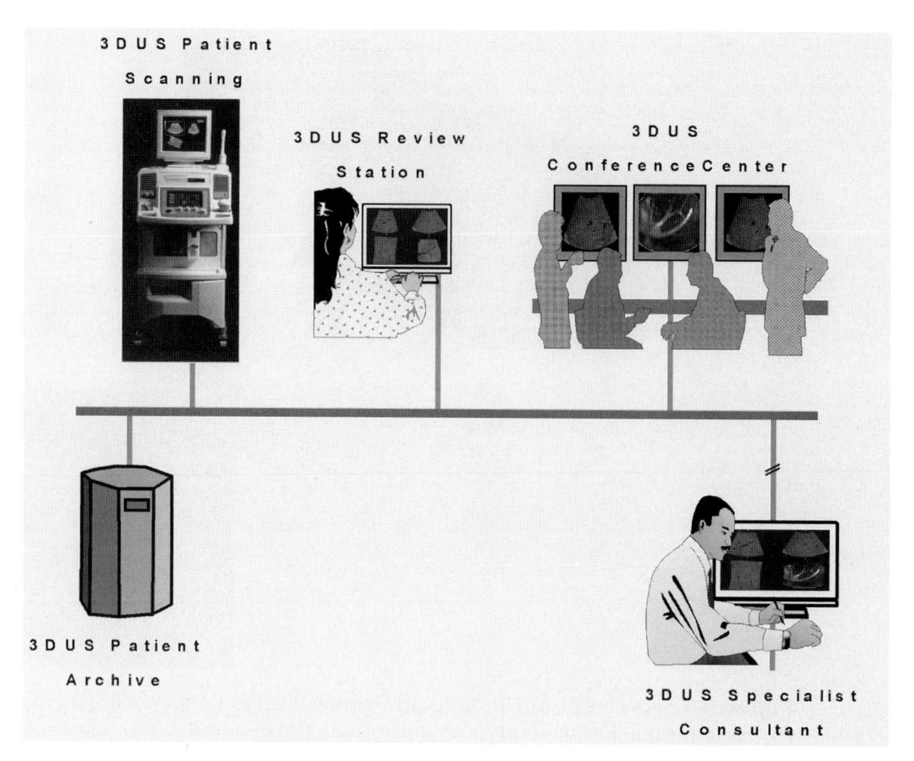

Figure 5-6. Review of 3DUS data. **(a)** Archive images with volumes. **(b)** Send study to 3DUS specialist physician or clinician for interpretation via electronic network. (Voluson 530 scanner image courtesy of Medison, Korea.)

accurate than routine 2DUS scans (see Chapter 4). Measurement tools are needed to assist with rapid outline and identification of objects to be measured.

After the volume has been analyzed, the selected images, including rendered images, should be archived for future reference (Figure 5-6). Currently, archival storage is via film, but in the near future electronic archives will be possible. Typically, resliced images are stored in different ways, such as the orthogonal planar images with rendered image if present, a single planar image, or single rendered image, or perhaps three rendered images in successive degrees of rotation. A valuable capability is to bookmark particular slices or rendered images, with all critical system settings, so that they can be reproduced should additional diagnostic questions arise. Bookmarking is different from saving specific images because it permits re-creation of the image and modification of the parameters without starting the analysis from the beginning.

After the analysis and selection of images is complete, the data should be reviewed by the physician, including the slices, rendered images, and original acquisition data. If the study is prereviewed by the sonographer, then specific volumes can be marked by the sonographer as the most valuable for physician to review. Also, in some cases physicians will find it beneficial to review volumes off-line at a workstation.

SUMMARY

Clinical scanning methods with 3DUS are similar to conventional 2DUS approaches. Depending on the specific transducer/position sensor configuration used, some adjustments in selecting the volume (or organ) of interest may be necessary. The most critical aspect of clinical scanning is to obtain the best possible 2DUS image quality. As newer 3DUS imaging systems become available to acquire volume data directly into the ultrasound scanner, physicians and sonographers will be able to examine, visualize, and interpret the patient's anatomy on-line immediately. Regardless of which viewing technique is used, one key benefit of 3DUS will be that once the patient has been scanned, the original 2DUS image acquisition data can be analyzed for the entire region or magnified subregions without the need to rescan the patient.

Optimization of ultrasound visualization depends on several factors. First, it is important to optimize overall image quality for scanner capability and patient habitus. The distance between the structure and the transducer may lead to one side of the anatomy being imaged more clearly than the other side. Second, 2DUS images should be acquired in the plane that best shows the anatomy; one plane may be better than others because of elevational resolution (thickness) of the 2DUS scan plane. Third, rendered image quality is affected by the anatomic orientation and beam pathway. Signal dropout from shadowing due to overlying structures can significantly obscure structural detail. Compounding data obtained from different orientations can minimize the adverse effects of signal dropout, although elastic deformation of tissues due to pressure from the transducer during scanning can distort and complicate realignment of scans.

The overall quality of the 3DUS study is highly dependent on 2DUS image quality; poor images result in poor volume data. If the patient moves during the scan, then the scan must be repeated or some form of correction for motion must be made. It is important to recognize that faster, affordable hardware is needed to shorten the time between acquisition and display. Some scanners are now becoming available for which the volume data are acquired directly into the rendering computer, permitting direct volume visualization immediately upon completion of the image acquisition. Near real-time viewing of the volume data will greatly enhance clinical acceptance.

Clinical applications of 3DUS are increasing, and continued refinement of analysis and visualization software will assist further in making a more accurate diagnosis. Clinical studies that are affecting patient diagnosis and management will be reviewed in the following chapters.

6

Obstetrics

OVERVIEW

This chapter reviews applications of three-dimensional ultrasound (3DUS) in obstetric imaging that show promise in providing information not readily available with two-dimensional ultrasound (2DUS) regarding normal and abnormal pregnancy and fetal development. Rendering the fetal face and skeleton suggests that 3DUS offers a novel perspective for viewing the fetus and when coordinated with simultaneous multiplanar display enhances comprehension of volume data. Reorientation to a standard anatomic presentation makes structures more recognizable than with 2DUS and allows rapid identification of normal and abnormal anatomy, which is a significant advantage.

KEY CONCEPTS

- Advantages of 3DUS compared with 2DUS in obstetrics:
  - Improved comprehension of fetal anatomy by families.
  - Improved maternal/fetal bonding.
  - Improved identification of suspected or detected anomalies not seen with 2DUS using orientations and planes unobtainable with 2DUS.
  - Improved recognition of anomalies by less experienced physicians using multiplanar and volume rendered images.
  - More accurate identification of the extent and size of anomalies due to anatomic constraints or fetal position.
  - Retrospective review or consultation with specialists/clinicians/fellows if an anomaly is subtle or difficult to assess or after patient has finished examination.
- Applications in obstetrics:
  - First trimester:
    - Sac size may predict prognosis.

KEY CONCEPTS, *continued*

- Embryo size may be measured more accurately.
- More accurate anomaly detection.
- Anomaly detection:
  - Face, central nervous system, extremities/skeletal, heart, lung, genitourinary.
- Fetal growth assessment.

3DUS COLOR/POWER DOPPLER IMAGING

The fetus is a challenging and difficult object to evaluate (1), in part because of its size and movement. Conventional ultrasound, and more recently, 3DUS have shown significant progress in evaluating the fetus. Early work demonstrated the feasibility of 3DUS fetal imaging (2–4). As equipment has become more sophisticated, the quality and detail possible with 3DUS methods have become impressive. Initial reports have shown that 3DUS can improve diagnostic assessment (5–8). Early reports focused on demonstrating fetal anatomy such as the fetal surface (4), skeleton (9,10), and face (Figure 6-1) (11), and proved that 3DUS permitted viewing the fetus in a more global fashion than 2DUS.

ADVANTAGES OF 3DUS COMPARED WITH 2DUS IN OBSTETRICS

Three-dimensional ultrasound has several advantages in obstetrics compared with 2DUS that may have a significant impact on obstetric management during pregnancy and at delivery, and on families' decisions about managing pregnancy. As with any new diagnostic tool the efficacy must be studied to assess whether there is any impact on medical management, and this is a difficult task. Thus far 3DUS studies have shown that there is much promise in evaluating fetal anomalies, but it is difficult to quantify how 3DUS adds to obstetric management. The basic advantages of 3DUS in obstetrics are shown in Table 6-1.

Improved Comprehension of Fetal Anatomy by Families

Fetal anatomy often is very difficult for families to understand. By using plane orientations and rendered images unobtainable with 2DUS we can use 3DUS to show fetal anatomy in an optimal position, making it more comprehensible. In cases with developmental anomalies such as cleft lip and with more subtle underlying anomalies such as cleft palate, volume data can be reviewed using a standard orientation and can be shown to patients after they are sitting up and are looking at the monitor. Viewing 3DUS rendered images of fetal anatomy and fetuses with anomalies may impact families in making management decisions about the pregnancy (Figure 6-2).

Table 6-1. *Advantages of 3DUS compared to 2DUS in obstetrics*

- Improved comprehension of fetal anatomy by families
- Improved maternal/fetal bonding
- Improved identification of suspected or detected anomalies not seen with 2DUS using orientations and planes unobtainable with 2DUS
- Improved recognition of anomalies by less experienced physicians using multiplanar and volume-rendered images
- More accurate identification of the extent and size of anomalies resulting from anatomic constraints or fetal position
- Retrospective review or consultation with specialists/clinicians/fellows if an anomaly is subtle or difficult to assess or after patient has finished examination

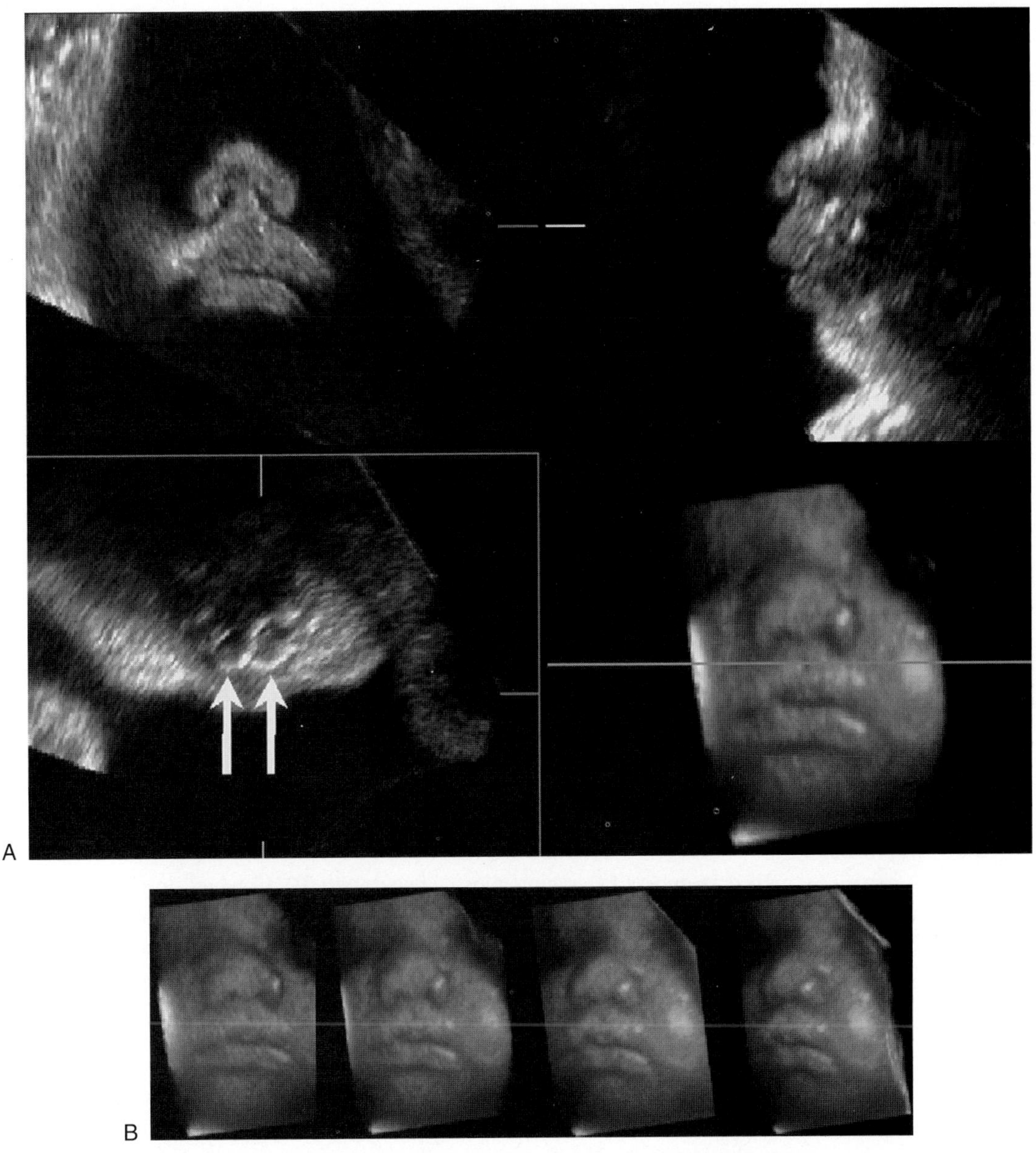

Figure 6-1. Normal face in a 35-week-gestational-age fetus. **A:** Multiplanar and volume-rendered images of the fetal face. The upper-left-hand image shows a coronal view through the nose and lips; the upper-right-hand image depicts a sagittal or profile view of the face. The lower-left-hand image shows an axial plane through the anterior alveolar ridge of the primary palate. The lower-right-hand image shows a volume-rendered image of the face using surface and light techniques. A *line,* shown crossing the upper lip of the fetus, identifies the level of the axial plane in the lower-left-hand box. *Arrows* point to two tooth buds in the anterior alveolar ridge. **B:** Images from multiple rotations of the fetal face. Generally, a volume is rotated with a knob on the work panel or in a cine loop to provide the impression of three dimensionality. (With permission from Ref. 45.)

Improved Maternal/Fetal Bonding

Surface visualization of the normal fetus and multiplanar evaluation of anatomic structures in a standard orientation (e.g., frontal face and profile face) are easily recognized. Steiner et al. (12) have pointed out that for many patients a 2D scan is an abstract image, whereas with 3DUS, the features of the child are readily apparent, regardless of whether the fetus is normal or abnormal. As a result, the potential exists for families to bond more effectively to their fetuses with 3DUS than with 2DUS. 3DUS images provide pregnant women with more security and increased insight into the psychological aspects of their

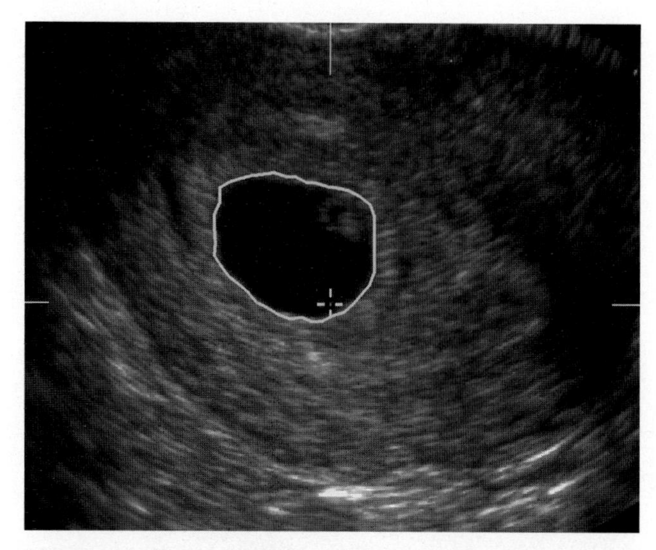

Figure 6-2. Gestational sac volume. Volume imaging of the gestational sac may assist in predicting viable outcome of the pregnancy. The gestational sac volume is measured by using a tractball to draw a region of interest around the gestational sac on multiple parallel planes. The gestational sac volume in this 5-week-gestational-age pregnancy was 2.3 ml.

pregnancy (13). Improved bonding between the mother and fetus offered by 3DUS could assist patients to stop smoking (14) and discontinue other abusive behaviors.

Identification of Anomalies Not Seen with 2DUS

To date, most of the anomalies detected with 3DUS which were not seen with 2DUS have involved the fetal face (micrognathia, cleft lip, mid-face hypoplasia, orbital hypoplasia, facial dysmorphia, cranial ossification defects, dysplastic ear) (5,6,8,15) but other anomalies are now being reported (spinal defects, hypoplastic scapula, choroid plexus cyst (5,14). Some of these have been identified from volume-rendered images and others from image planes unavailable with 2DUS. 3DUS benefits the primary care physician as well as the tertiary care physician in ''seeing'' anomalies that are not immediately apparent on 2DUS because of the more comprehensible nature of the display in a standard anatomic orientation. For example, scoliosis of the fetal spine was easily diagnosed by a tertiary-care physician with 2DUS, yet it was much more apparent to a less experienced perinatologist when viewed from a rotating rendered 3DUS image (16).

More Accurate Identification of the Extent and Size of Anomalies

The extent and size of anomalies can be more accurately identified with 3DUS than with 2DUS (5,6,17). A primary application has been in the evaluation of many anomalies, including neural tube defects, cystic hygroma, and hydrocephalus. 3DUS offers an advantage by being able to obtain orientations and planes not possible with 2DUS. Also, even though such anomalies as cleft lip/palate, clubfoot, hand abnormalities, neural tube defect, and amniotic band syndrome may be detected or suspected on 2DUS, they are more comprehensible with 3DUS (6,18). As a result, 3DUS should bring added confidence to the identification of anomalies, particularly subtle anomalies, and improve patient management.

Retrospective Review or Consultation with Specialists/Clinicians/Fellows

Archived volume data with identified or suspected fetal anomalies may be reviewed with other physicians, either on the 3DUS machine or at a workstation, after the patient has finished the examination. Being able to share 3DUS data with colleagues across a network is advantageous in obtaining second opinions in difficult or subtle cases, in sharing information with clinical colleagues, and in teaching doctors in training without the patient present. An important advantage of 3DUS data is that the volume may be reviewed many times millimeter by millimeter, simulating real-time scanning, which is more effective than reviewing a 2DUS videotape.

EVALUATION OF THE EMBRYO

First-Trimester Evaluation

First-trimester ultrasound examinations are generally performed with transvaginal transducers. This often leads to difficulty in being able to obtain the exact desired plane of viewing because of the limitation of movement of the transvaginal probe. 3DUS provides the technology to arbitrarily examine any plane in a volume acquired with the transvaginal probe.

The volume of the gestational sac has been investigated as a possible predictor of pregnancy outcome (see Figure 6-2). Steiner and co-workers (19) hypothesized that the gestational sac volume (GSV) would reflect the function of the early uteroplacental unit. They evaluated 38 pregnancies between 5 and 11 weeks gestation, 31 of which had a normal outcome and seven of which had an abnormal outcome. An abnormal outcome included pregnancies with blighted ovum or embryonic demise. Their results showed that the gestational sac volume was significantly correlated to gestational age ($r = 0.74$, $p < 0.001$). In looking at the seven abnormal pregnancies, two out of two with a twin sac had a GSV within one SD of the mean, and three out of five cases of missed abortions or blighted ovums had a GSV > 2 SD of the mean. The volume measurements also had a high interobserver correlation ($r = 0.99$, $p < 0.001$), suggesting that GSV volume has potential to predict pregnancy outcome from the first trimester (19).

Feichtinger has suggested that first-trimester fetal biometry can be used accurately to assess gestational age by rotating the image of the scanned embryo into an optimal position (20). It is also possible to measure the yolk sac accurately. Anatomy of the embryo can be assessed in the late first trimester in some embryos, including the limb buds, the stomach, and the bladder (Figure 6-3). The level of the cord insertion into the abdomen

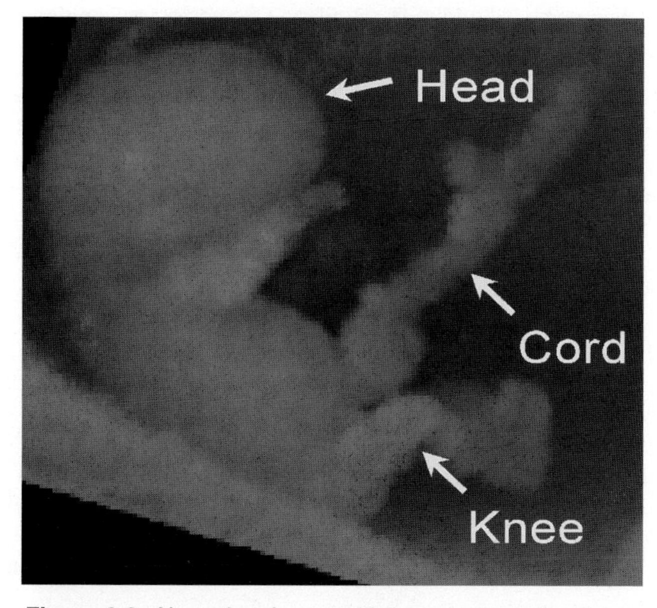

Figure 6-3. Normal embryo at 10.5 weeks gestational age. The head, cord, and knee are identified. (Image courtesy of Dr. Benoit.)

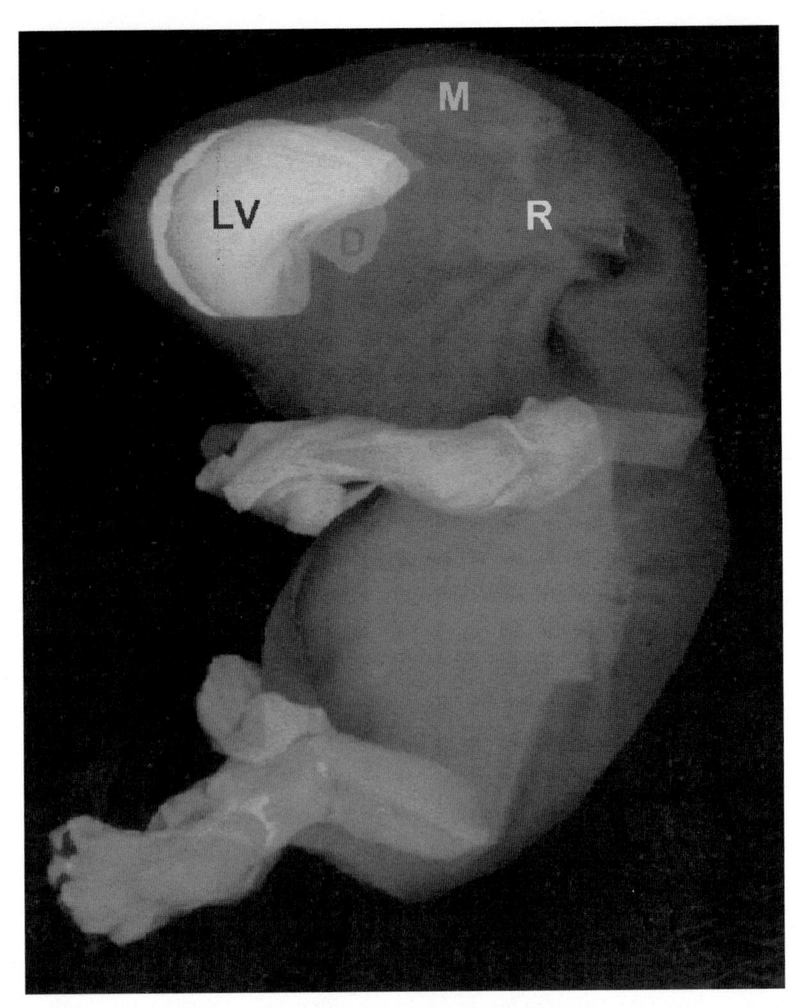

Figure 6-4. Embryo at 9 weeks, 6 days gestational age (29 mm crown-rump length). Surface rendering of the intraventricular cavities is displayed, which may assist in understanding embryologic development. LV, lateral ventricles; D, diencephalon; M, mesencephalon; R, rhombencephalon. (With permission from Dr. H. Blaas.) See color version of figure.

can be seen and the head can be rotated for a biparietal diameter measurement. We predict that the embryo can be aligned in a standard orientation to measure the nuchal region accurately for assessment of a thickened fold or nuchal bleb. The entire volume of the gestational sac may be evaluated in twins, allowing for both embryos to be seen in the same ''scan.''

Feichtinger has also suggested that 3D imaging should be better than conventional transvaginal sonography in detecting the echogenic material or thickened endometrium in cases of probable complete spontaneous abortion that may lead to the necessity of a curettage (20).

Blaas et al. used 3DUS to evaluate the development of the brain cavities in three human embryos *in vivo* at 7, 9, and 10 weeks gestation (21,22). The cavities of both hemispheres, the isthmus rhombencephali, and the third ventricle were seen at 7 weeks. The development of these cavities was shown to progress and overlap as the embryos got older. The connections between the cavities were well seen (Figure 6-4). The actual volumes of these cavities have been calculated from the 3D volumes. Embryologists have long studied the embryo *in vitro*, but cavities often collapse, and it is extremely valuable to study them *in vivo*.

EVALUATION OF THE FETUS

3DUS has shown itself to provide valuable information regarding fetal development, particularly detection of congenital anomalies. It also is important to appreciate that 3DUS

Table 6-2. *Fetal facial malformations identified with 3DUS*

Facial abnormality identified (3DUS advantageous [~70%])	
Cleft lip/palate	Micrognathia
Facial dysmorphia/dysplastic ear	Midface hypoplasia
Hypotelorism	Holoprosencephaly

Adapted from Refs. 6, 18, and 22.

can reassure families by excluding anomalies, particularly in families at increased risk for specific anomalies (4,6,8). Assessment of fetal development benefits from confirming normalcy and accurately identifying anomalies when they are present. To date, the largest series of congenital anomalies studied with both 2DUS and 3DUS was performed by Merz et al. (6). They reported on 204 patients with anomalies and found that 3DUS was advantageous in demonstrating fetal defects in 62% (127/204), was equivalent in 36% (73/204), and was disadvantageous in 2% (4/204). 3DUS was disadvantageous in four cases with cardiac defects due to motion artifacts during data acquisition (6). Richards et al. have reported similar results where 103 anomalies in 63 patients were studied (14,23). 3DUS was advantageous in 51% (53 anomalies), equivalent in 45% (46 anomalies), and disadvantageous in 4% (four anomalies).

Overall, 3DUS images were helpful in detecting a broad range of anomalies, including facial anomalies (Table 6-2); CNS anomalies (Table 6-3); skeletal anomalies (Table 6-4); thorax, abdominal, and cardiac anomalies (Table 6-5); genitourinary anomalies (Table 6-6); and others (Table 6-7). The advantages of 3DUS arise from planar images and volume-rendered images derived from volumetric data.

3DUS was disadvantageous in four cases with cardiac anomalies (6,23) and two cases with complex anomalies (limb–body–wall complex and twin fetus with multiple anomalies) (14,23). The volume data were less helpful with 3DUS than with 2DUS because of motion of the heart during the acquisition in the cardiac cases. In the twin case, one position of the second twin prevented optimal volume acquisition, and in the limb–body–wall complex case, gross anatomic disruption made localization of normal landmarks difficult in the relatively small volume acquisitions.

In another series including both normal (242) and abnormal (216) fetuses Merz et al. found that 3DUS provided a significant diagnostic gain in 64% (294/458) of cases (7). The orthogonal planar display accounted for a diagnostic gain of 46%, whereas a combined 3DUS display of orthogonal planar and volume-rendered views yielded a diagnostic gain of 71.5% (233/326).

Baba et al. and Hata et al. have recently reported on a new 3DUS technique, real-time processible 3DUS (Aloka, Japan) that permits rapid acquisition (4 sec) and nearly immediate rendering (24–26). Fetal images produced with this technique are remarkably photographic, are of high quality (Figure 6-5), and require less expensive computers than other 3DUS equipment currently available. Because the viewing direction is limited to that of the probe, the most desirable direction may not always be available. However, good-quality images of the fetal surface are readily obtainable with sufficient amniotic fluid. High-quality images may be more difficult to obtain in fetuses of less than 24 weeks gestation because skin tissues are thin and their acoustic impedance is virtually the same as that of water. At present, volumetric measurements are impossible because the rendered image is computed on the fly and the volume data are not saved (24). In general, this type

Table 6-3. *Fetal CNS abnormalities identified with 3DUS*

CNS abnormality identified (3DUS advantageous [~70%])	
Anencephaly	Hydrocephalus/holoprosencephaly/hydra
Choroid plexus cyst	Intracranial tumor/teratoma of the face
Dandy–Walker malformation	Microcephalus
Encephalocele	Sacrococcygeal teratoma
Holoprosencephaly	Spina bifida/myelomeningocele

Adapted from Refs. 6, 18, and 23.

Table 6-4. *Fetal skeletal malformations identified 3DUS*

Skeletal abnormality identified (3DUS advantageous [~75%])	
Arthrogryposis multiplex congenital	Polydactyly
Clenched hands	Rocker bottom feet
Club foot	Scoliosis
Contractures	Segmentation defect
Hypoplastic scapulae	Short limb(s)
Leg mass	Short ribs
Osteochondrodysplasia	Single bone forearms

Table 6-5. *Fetal thorax/abdomen/heart malformations identified with 3DUS*

Thorax abnormality identified (3DUS advantageous [~40%])
Diaphragmatic hernia
Hydrothorax
Abdominal abnormality identified (3DUS advantageous [~77%])
Bowel obstruction
Gastroschisis
Intra-abdominal cyst
Omphalocele
Wall defect (bands)
Cardiac abnormality identified (nongated) (3DUS advantageous [~0%])
AV canal
Heart defect/tumor
Transposition

Adapted from Refs 6, 18, and 22.

Table 6-6. *Fetal genitourinary malformations identified with 3DUS*

Genitourinary abnormality identified (3DUS advantageous [~43%])	
Ambiguous genitalia	Hypospadias
Bladder exstrophy	Posterior urethral valves
Bipartite scrotum	Potter's syndrome
Cloacal dysgenesis	Potter I
Dysplastic kidneys	Potter IIa
Hydrocele	Prune belly syndrome
Hydronephrosis	Wilms' tumor

Adapted from Refs. 6, 18, and 22.

Table 6-7. *Miscellaneous fetal malformations identified with 3DUS*

Abnormality identified (3DUS advantageous [~25%])	
Abdominal pregnancy	Lymphangiomatosis
Conjoined twins	Nonimmune hydrops fetalis
Cystic hygroma	Umbilical cord cyst
Hygroma colli	

Adapted from Refs. 6, 18, and 22.

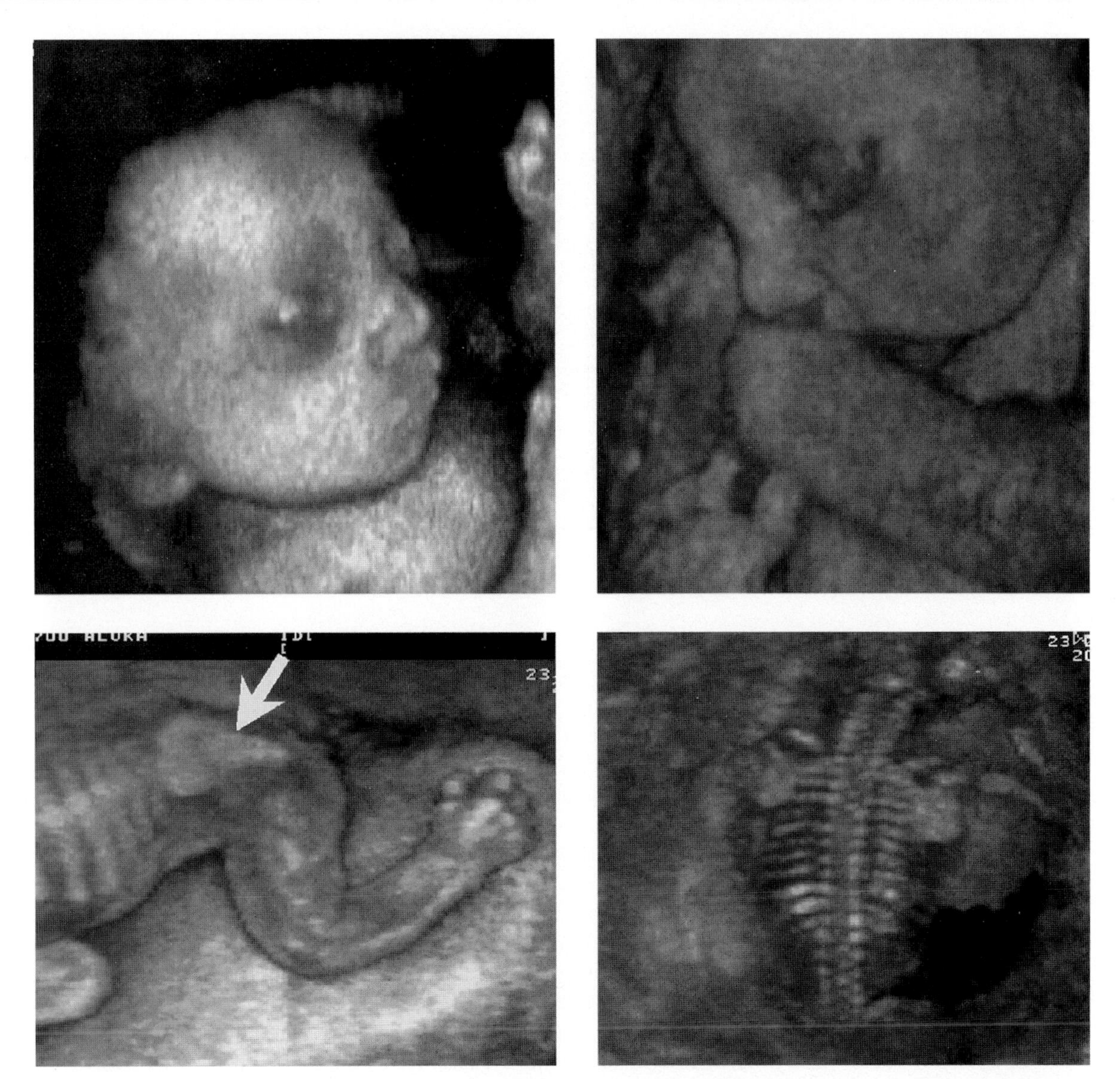

Figure 6-5. Volume rendering of fetuses using real-time processible 3DUS (Aloka, Tokyo, Japan). Upper left image shows the fetal face. Upper right image shows the face and arms. Lower left image shows thorax with scapula (arrow), arm, and hand. Lower right image shows thorax and spine. (With permission from Aloka.)

of 3DUS imaging will play a useful role in prenatal diagnosis and therapy due to the rapid image development and the high image quality.

Fetal Face

3DUS has proven valuable in evaluating development of the fetal face (see Table 6-2) using both multiplanar and volume-rendered images (Figure 6-6; also see Figure 6-1). The best results occur when the mother is placed in an optimal position so that enough amniotic fluid is positioned adjacent to the face to optimize rendering of the face. Rendered and planar images are essential to rapid identification of the fetal face by physicians and nonphysicians (6,11,27–29).

Advantages of 3DUS Face Evaluation

Several advantages of viewing the fetal face with 3DUS have been identified. First, the face can be rotated to a standard anatomic orientation that allows the physician to view

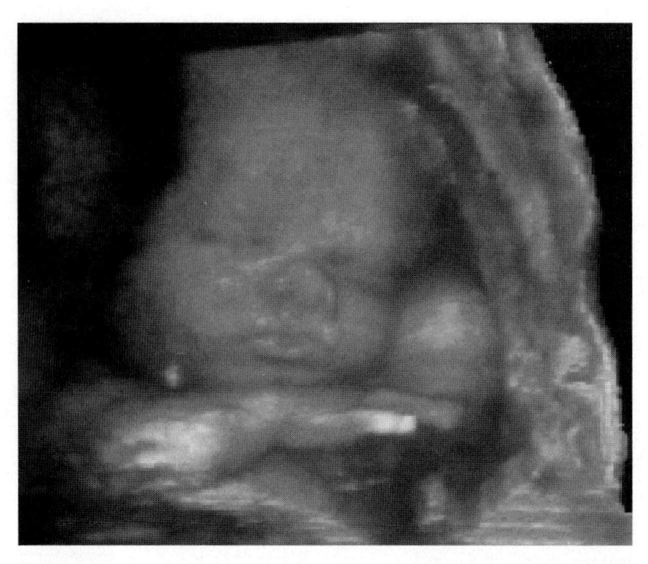

Figure 6-6. Normal fetal face and arm. (Courtesy of Vingmed.)

the upright face in coronal (frontal face), sagittal (profile), and axial views (see Figure 6-1A). Second, rotation of the fetal face along a vertical axis provides depth cues to understand the anatomy and makes it more comprehensible (Figure 6-1B). Third, the volume can be evaluated millimeter by millimeter in symmetrical, anatomic planes. The upper lip can be assessed in both coronal and axial views. The palate can be evaluated easily in the axial plane at the appropriate level. Fourth, the perpendicular planes can be evaluated in relation to each other and to the rendered images. For example, the level of the primary palate can be rapidly identified by referencing to the rendered image or to the planar profile image (see Figure 6-1). Fifth, the facial profile can be identified in all fetuses in which fluid is located in front of the face. Not until 3DUS was available did we realize how often the profile was not appropriately positioned (15). Sixth, two or more anomalies can be viewed in the same rendered image, which requires at least two 2D images to display. For example, a fetus with hypotelorism and medial cleft lip was shown together on the surface-rendered image (Figure 6-7). Seventh, families and clinical colleagues are able to recognize the face without assistance by the sonographer. This is advantageous in both normal and abnormal fetuses.

Frequency of Facial Evaluation

3DUS imaging of the fetal face can be performed rapidly and successfully in most patients. Initial studies by Pretorius and Nelson showed that surface-rendered images were obtained in 24 out of 27 fetuses studied (11). In a follow-up study of 71 fetuses, 3DUS confirmed the presence of a normal lip in 92%, compared with 76% with 2DUS. 3DUS was particularly helpful in confirming the presence of a normal lip in fetuses less than 24 weeks gestational age: 93% with 3DUS, compared with 68% for 2DUS (8). Multiplanar, orthogonal imaging was more helpful than surface rendering in assessing the lips, particularly in the younger fetuses. Merz et al. added significant information to our understanding and expectation of 3D imaging of the face in their report of 618 pregnant women between 9 and 37 weeks gestation. Multiplanar orthogonal images could be obtained on all fetuses, whereas a surface-rendering display could be obtained successfully in only 72% of patients. Unacceptable surface rendering occurred in 28% of the cases due to (a) oligohydramnios, (b) unfavorable fetal position (with adjacent or superimposed structures), or (c) excessive fetal movement. Merz reported that optimal surface display was possible from 20 to 35 weeks.

Using volume rendering without multiplanar capabilities, Hata studied 94 healthy fetuses (15–40 weeks) and reported visualization of facial structures in 65% at less than 24 weeks

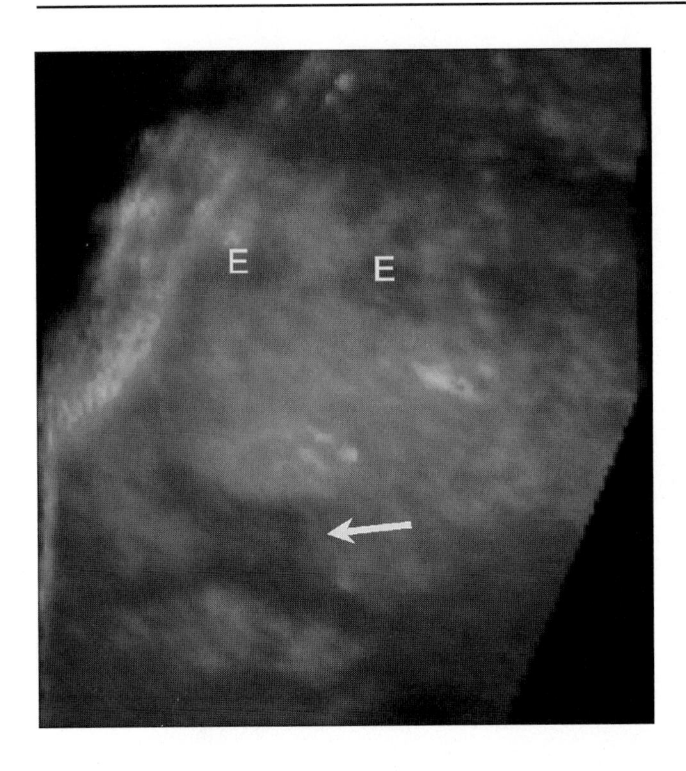

Figure 6-7. Median cleft lip and palate and hypotelorism in a fetus at 35 weeks gestational age. The median cleft lip (*arrow*) and palate are well visualized in this volume-rendered image. In addition, hypotelorism (E, eyes) is shown in this patient with holoprosencephalic facies and anencephaly. 3D imaging allows for the hypotelorism and cleft lip/palate to be identified in a single image.

gestational age and in 84% after 24 weeks; they found no significant difference between 2DUS and 3DUS (25). The fetal ears and their relationship to the face also may be evaluated with 3DUS. Specifically, it may be possible to identify low-set ears (27). The shape and structure may also be evaluated (6).

Facial Profile

The facial profile (Figure 6-8) can be obtained using 3DUS on any fetus in which it is possible to obtain surface-rendered pictures of the fetus and in many fetuses in which the surface-rendered pictures are not optimal due to young gestation age or where there is not sufficient amniotic fluid adjacent to the face. Merz et al. studied 125 fetuses to examine the effect of 3D imaging on the axis of the facial profile found without multiplanar imaging (15). They found that in 30.4% of the results that the profile was off by 3–20 degrees compared with a true profile view. As a result, only 69.6% of the cases had a true profile. The importance of this finding should not be understated. When a true mid-plane is not identified, anomalies may be missed or overcalled. In Figure 6-9 a fetus with Pierre Robin sequence was found to have micrognathia when the profile was in the true midsagittal plane, whereas it appeared normal in a parasagittal plane. It is important to note that the profile can be obtained with both planar and rendered imaging.

Cleft Lip and Palate

Cleft lip and/or palate are often difficult to diagnose with 2DUS by an inexperienced sonographer or physician because of lack of landmarks in the appropriate image plane. In the RADIUS trial, where 7685 low-risk fetuses were scanned, only three out of nine cleft lips were identified prenatally (1). 3DUS allows the examiner to evaluate the anterior alveolar ridge or the primary palate in the appropriate axial plane by using the rendered image or the sagittal planar image as a reference image while scrolling through parallel

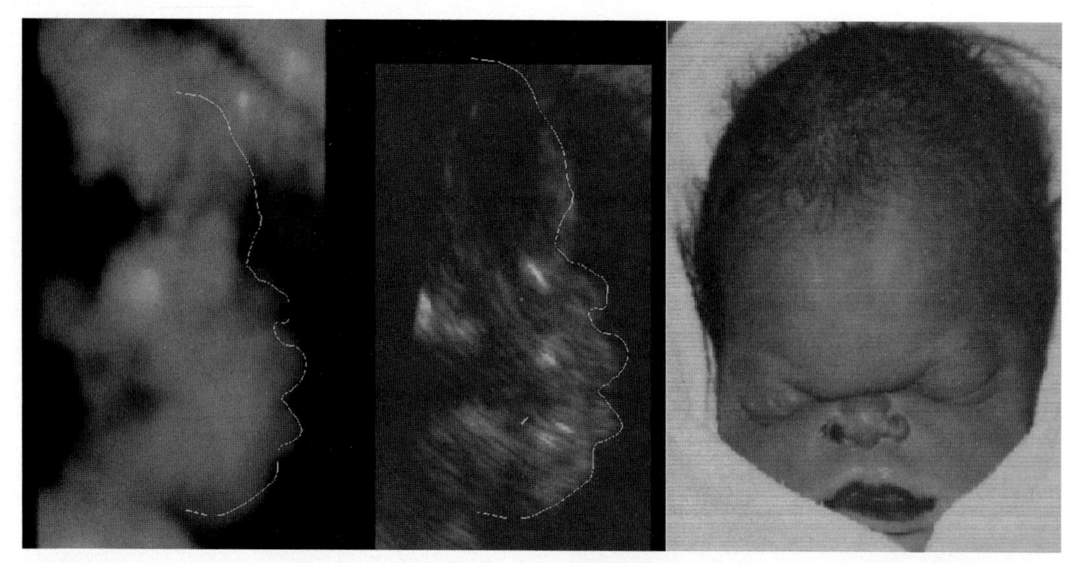

Figure 6-8. Midface hypoplasia in a 24-week-gestational-age fetus. A concavity in the midface is seen in this fetus with thanatophoric dysplasia and is consistent with midface hypoplasia. The image on the left is a volume-rendered picture of the profile; the middle image is a 3D planar image of the profile. The solid line is drawn along the profile for emphasis. A photograph of the fetus is shown on the right.

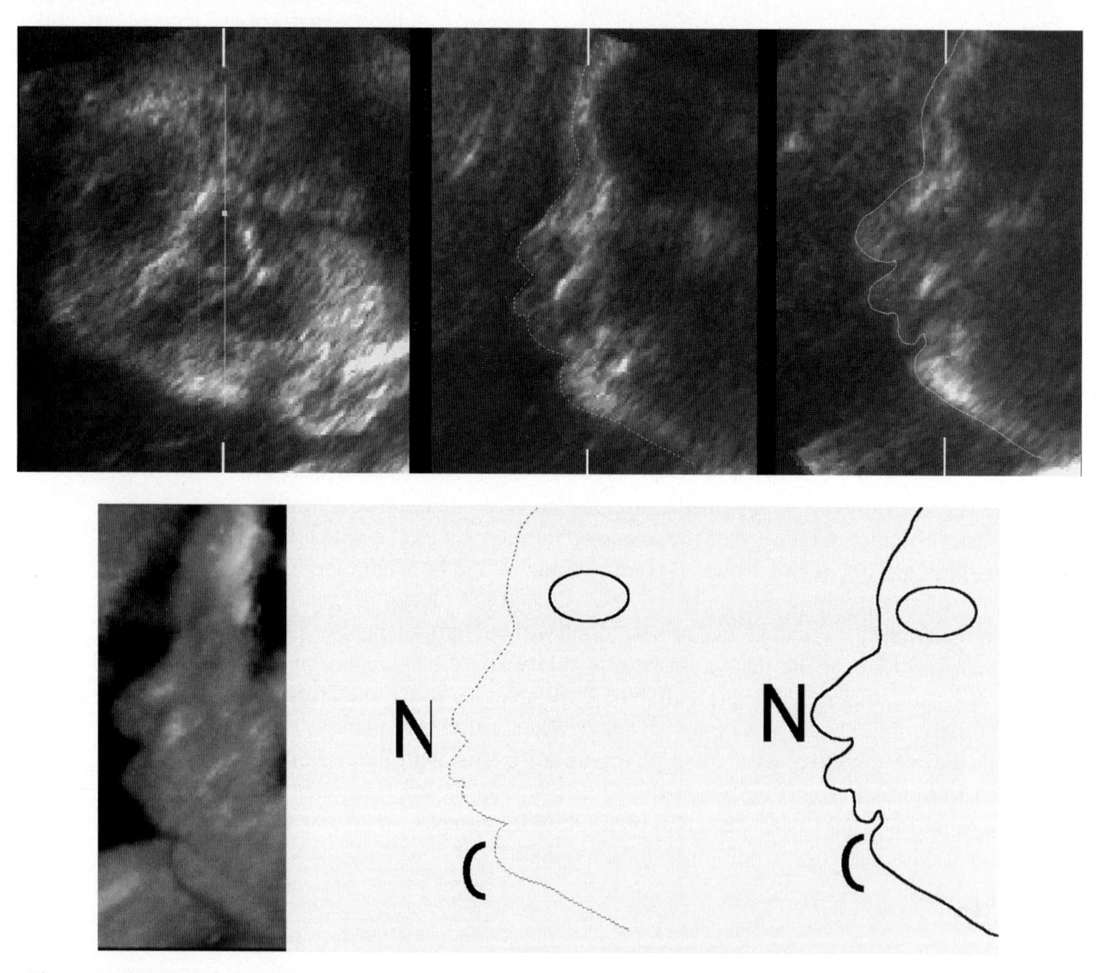

Figure 6-9. Facial profile of a 28-week gestational age fetus with Pierre Robin sequence and micrognathia. The correct diagnosis was made on the 3DUS exam in retrospect; it was not made on the 2DUS exam.The upper left image shows the frontal view of the face with vertical lines showing the location of the sagittal images displayed to the right. The *dashed line* is approximately 3 mm to the left of midline; the sagittal image from this location is displayed in the upper middle image. A line drawing is also displayed below it. The facial profile appears normal. The *solid line* is located perfectly midline and represents the true profile, which is shown in the upper right image. Micrognathia is suspected from the small chin. The line drawing is also displayed below it. The volume-rendered image of the profile is shown on the lower left and is consistent with micrognathia. N, nose; C, chin.

axial planes (Figure 6-10). The ability to scroll through parallel planes in both axial and frontal planes allows for careful evaluation of the face that is not possible with 2DUS imaging. It also allows for demonstration of the anomaly to the family, professional colleagues, and physicians in training.

Rendering of facial clefts is valuable in conveying a sense of the fetal appearance to the physician and family that can affect patient management decisions (see Figure 6-10) (6,8,11,15,17,28,29). Interactive rotation of the fetal face enhances anomaly recognition. In some cases, viewing 3DUS-rendered images of the facial anomaly has directly affected management decisions about carrying or terminating the pregnancy (17).

It is important that rendered data be reviewed with planar data from the same volume to evaluate the direction of the sonographic beam that is more apparent in planar images than in rendered images. An artifact has been reported in which a normal fetus appeared to have a cleft lip on the rendered image that was shown to be due to a shadow from the umbilical cord, which was lying adjacent to the upper lip on the multiplanar images (18).

Central Nervous System

Diagnosis of central nervous system (CNS) abnormalities has a significant impact on obstetric management and parental decisions about termination of pregnancy. Although most spinal abnormalities can be detected with 2DUS by experienced operators, some

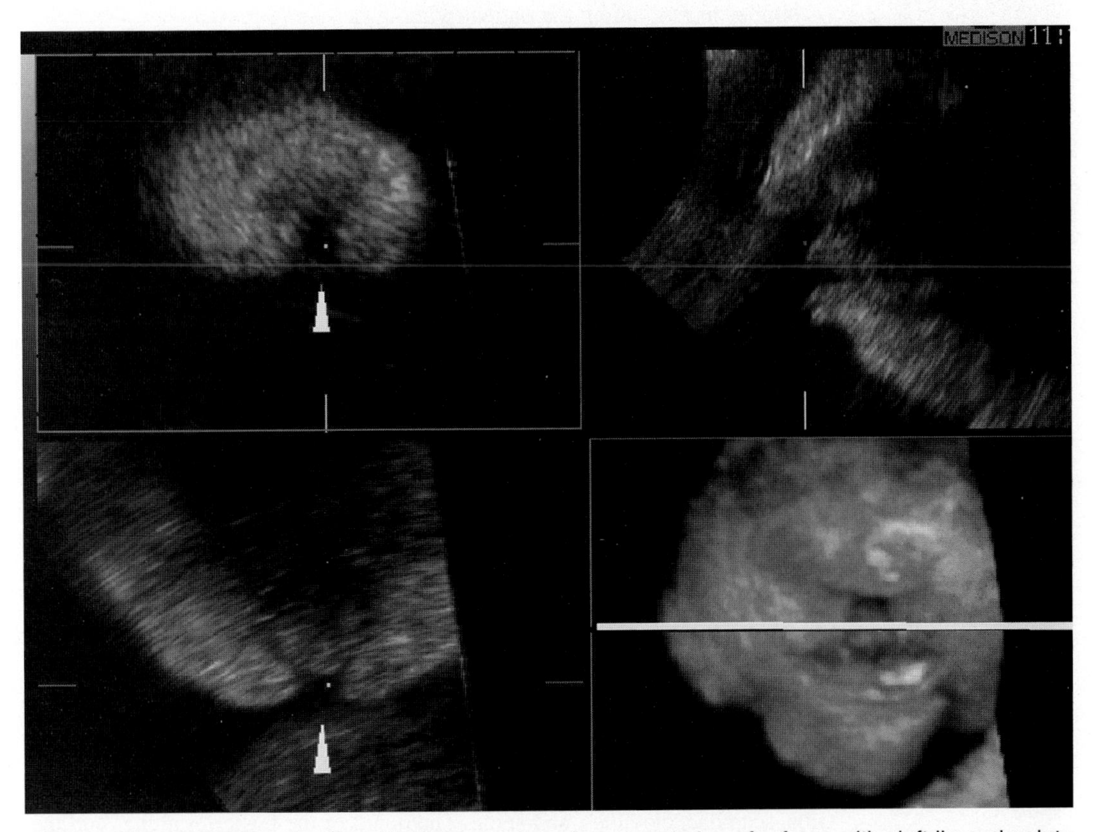

Figure 6-10. Cleft lip and palate. Multiplanar and volume rendering of a fetus with cleft lip and palate at 35 weeks of gestational age. The volume-rendered image in the lower-right-hand image is used as a reference to identify the cleft lip and palate (*arrow*) in the axial plane in the lower-left-hand image. The *solid white line* shows the location of the axial scan on the rendered image. (Image courtesy of 31.)

limitations and difficulties exist in which 3DUS may be of benefit (see Table 6-3): (a) identification of the level of neurotube involvement, (b) subtle neurotube defects with minimal or no bony changes, (c) demonstration of scoliosis, and (d) identification of neurotube defects by less experienced operators. Several investigators (6,16,29,30) have suggested that 3DUS is helpful in demonstrating the extent of neurotube defects. The level of the defect can be determined by using the rendered image, the sagittal and/or coronal planar images as a reference for the assessment of the transverse images through the vertebral bodies (Figure 6-11). Simultaneous display of orthogonal images through the same vertebral body allows the examiner to more accurately identify the specific level of that vertebral body. Johnson et al. reported that 3DUS studies displayed the level of the defect more accurately than 3DUS in three out of five cases (16). Accurate identification of the level of spinal involvement of a neurotube defect is very important in counseling families regarding prognosis, and this information can impact the decision about whether to continue the pregnancy. Johnson et al. also reported a case where 3DUS allowed for identification of a neural tube defect with endovaginal scanning that could not be categorized with 2DUS due to maternal body habitus and fetal position. In that case, arbitrary plane slicing of a volume acquired with an endovaginal transducer could be optimized into a transverse projection to show the exact level of neural tube defect involvement (Figure 6-12) (16).

Review of volume data allows for standardized orientation of the spine for review of the data in desired anatomic planes, millimeter by millimeter. Flight path technology (Medison, Korea) permits the examiner to direct the planar evaluation in a transverse orientation through the spine even though it is curved. This technique allows for a controlled, accurate assessment of the spine without fetal movement. Subtle changes in the spine can be reviewed repeatedly by the examiner and/or with colleagues in a fashion that is similar to rescanning the patient without her being present.

The anatomic relationship of two spines in a case of conjoined twins was displayed clearly with 3DUS; the two spines could not be seen in one plane using 2DUS. The volume-rendered image was displayed using a maximum-intensity setting to optimize the bony skeleton (32).

Several limitations of 3DUS for evaluation of the fetal spine should be considered, such as an insufficient number of original 2DUS image slices, which can reduce 3DUS image quality; fetal motion during the 3DUS acquisition, which can distort 3DUS images; incorrect selection of rendering parameters, which can result in a suboptimal 3DUS display; rendering that obscures some structures because of overlap of adjacent tissue and 2DUS artifacts, shadowing, and poor lateral resolution; and focusing that compromises 3DUS image quality (Figure 6-13) (30,31). Often artifacts are more easily recognized on planar images showing the original 2DUS beam direction. Proper training in scanner operation can improve the quality of the 2DUS and resulting 3DUS scans.

Encephaloceles may be displayed with the multiplanar format, allowing for the location of the mass to be located specifically. Richards et al. found that 3DUS added information compared with 2DUS in two of three encephaloceles (18,23).

Fetal Skeleton

3DUS provides a clear view of skeletal development because it can display curved structures such as the spine, ribs, skull, and extremities in a single rendered image. Developmental anomalies of the skeleton can be identified with more confidence using 3DUS than 2DUS (see Table 6-4). The thorax is a combination of structures, including the curved ribs, scapulae, clavicles, and spine. The depth of the posterior arches and the relationship of the three ossification centers of each vertebral body to each other and to the ribs and spine also can be demonstrated (Figure 6-14) (9). Stereo viewing of the thorax further enhances depth information and improves the separation of overlapping structures, enhancing comprehension of the overall anatomic relationships (9). Magnification of portions of

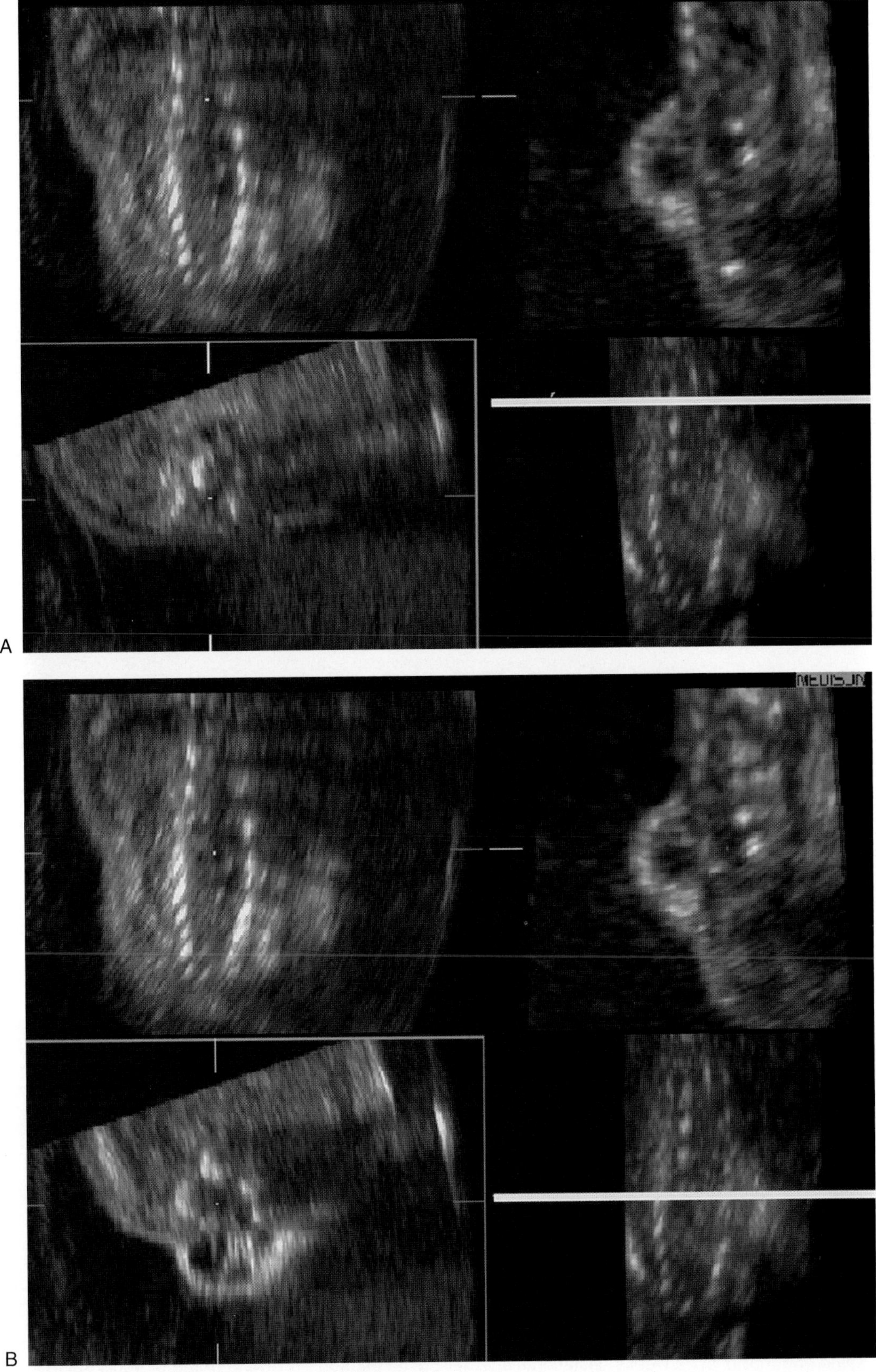

Figure 6-11. Myelocystocele in fetus at 18.5 weeks gestational age. **A:** The myelocystocele is displayed in a multiplanar and volume-rendered format. The axial view through the spine (lower-left-hand image) is identified specifically to be at the second lumbar vertebral body level by referencing the axial plane to the volume-rendered image in the lower-right-hand image, where a *white line* shows the position on the rendered image. **B:** The axial plane is now evaluated at the L5 level, which is determined by referencing to the volume-rendered image. (With permission from Ref. 45.)

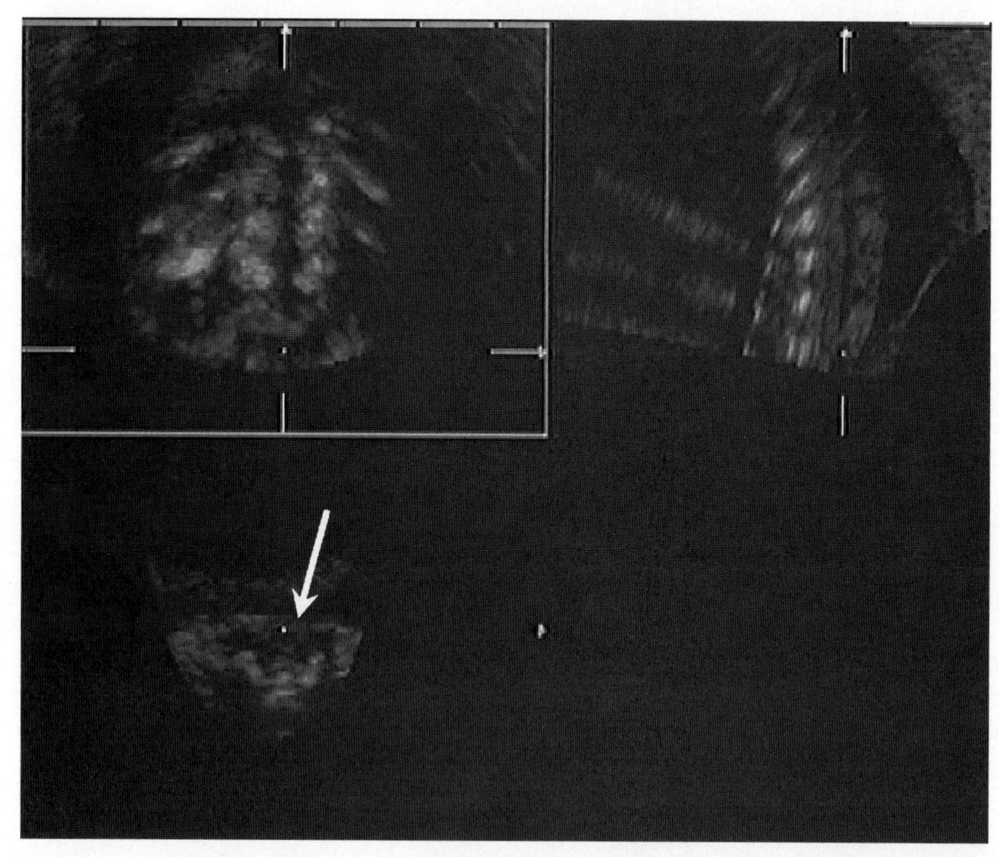

Figure 6-12. Neurotube defect on endovaginal imaging. Multiplanar imaging of this patient with very large maternal body habitus shows widening (*arrow*) at the second lumbar level on the axial plane (lower-left-hand image). Axial images through the spine could not be obtained with conventional 2DUS imaging.

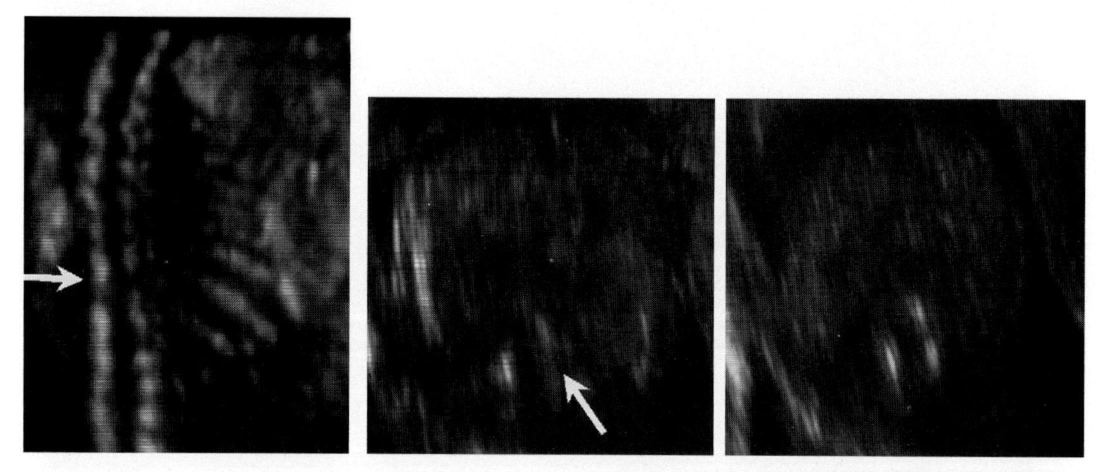

Figure 6-13. Pseudo-narrowing of the fetal spine in a 22-week-gestational-age fetus. 3DUS of a normal spine and thorax at 22 weeks gestational age. **A:** Spinal pseudo-narrowing of the thoracic spine in the rendered image at the level (*arrow*) that reflects the position of the axial plane. **B:** Review of consecutive coronal plane by scrolling through the volume proves normal anatomic structures without narrowing of the thoracic spine. The artifact is created by reduced echogenicity of the ossification center in the lateral arch (*arrow*) and higher echo intensity of the ossification of the central vertebral body. **C:** All three ossification centers of the vertebral body are seen in the lower thoracic spine.

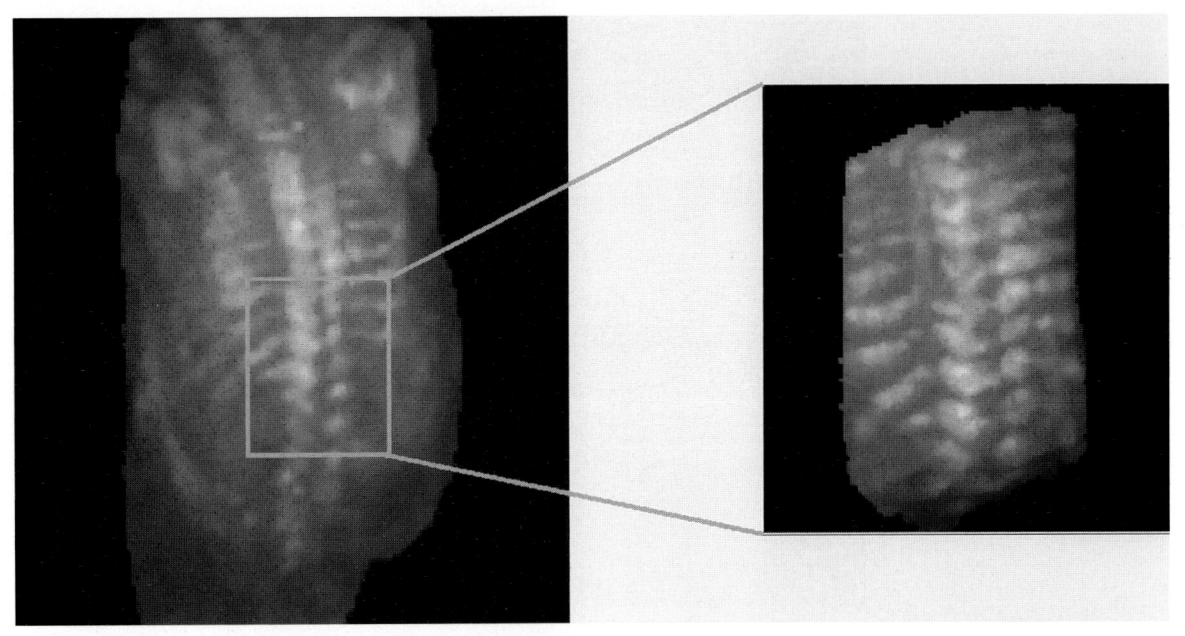

Figure 6-14. Normal fetal spine and thorax of 22-week-gestational-age fetus. Volume-rendered image of the thorax with a magnified view (right) of the lower thoracic region shows the curvature of the ribs and the continuity of the ribs to the vertebrae. (With permission from Ref. 45.)

a volume data set allow for more detailed assessment of special spinal regions (see Figure 6-14).

Fetal Skull

Cranial sutures and fontanelles are spaces between the fetal skull plates that allow for progressive growth of the brain and skull bones in fetal development. The sutures and fontanelles are curved structures that are seen clearly on 3DUS imaging (Figure 6-15) (27,33) but that appear only as slits in the skull on 2DUS imaging. Sutures commonly identified included coronal, lambdoidal, and squamosal, whereas fontanelles included anterior, posterior, mastoid, and sphenoid. 3DUS offers clearer visualization of the sutures and fontanelles in that the skull is positioned in a standard orientation and is viewed as it rotates along a vertical axis from the top of the head to the neck. Maximum-intensity depth-cued volume rendering offered optimal images of the sutures and fontanelles. Stereo viewing further enhanced depth information, particularly when evaluating the area of the temporomandibular joint and nasal cartilage (33).

Abnormalities of the sutures and fontanelles may be related to delayed closure or widening and to craniosynostosis, such as kleeblattschädel. Delayed ossification of the cranium was demonstrated with 3DUS in a fetus with achondroplasia by showing widening of the frontal suture compared with a fetus with similar age (15). Identification of these types of abnormalities may assist in the diagnosis of specific congenital or chromosomal abnormalities and affect management of fetuses at delivery.

Ribs

Ribs can be viewed using 3DUS; the curvature and relationship of the rib ends to the vertebral bodies and the anterior chest wall can be demonstrated as well as the entire length. Shortened ribs have been seen in fetuses with skeletal dysplasias (Figure 6-16); viewing the rotating rendered image of the thorax optimizes this diagnosis. Shortened ribs have been very difficult to appreciate using 2DUS because of the curvature of the ribs and the planar approach to rib evaluation.

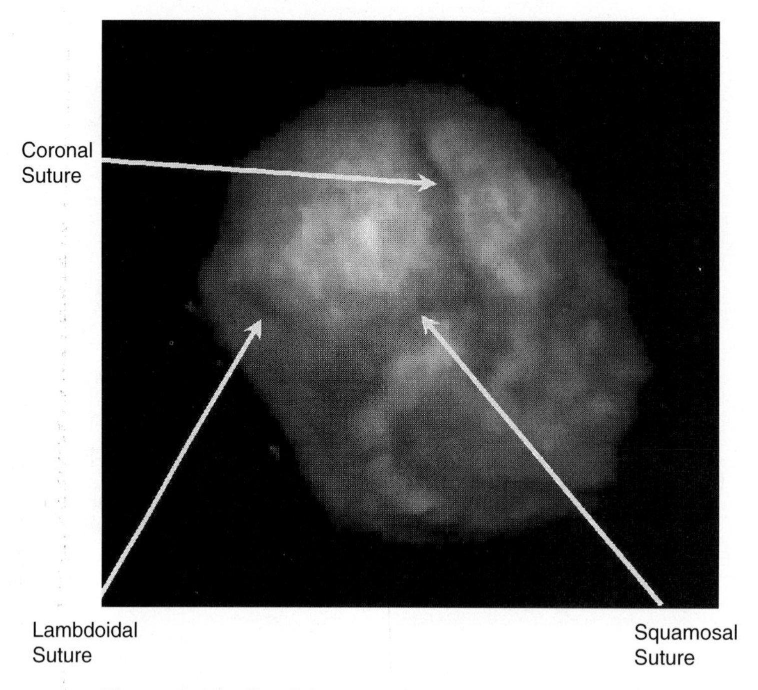

Figure 6-15. Cranial sutures in an 18-week-gestational-age fetus. 3D volume-rendered image shows the cranial sutures with continuity. (With permission from Ref. 45.)

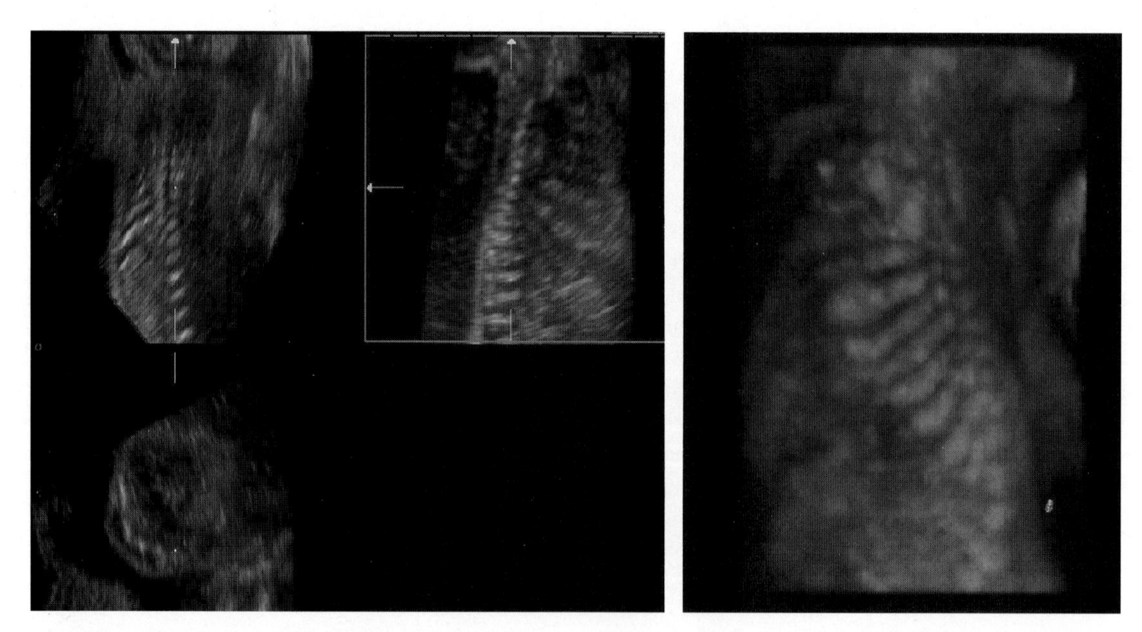

Figure 6-16. Thanatophoric dysplasia in a 24-week-gestational-age fetus. The multiplanar image on the left shows the thorax and spine in coronal (upper left), sagittal (upper right), and axial (lower left) planes. The shortened ribs cannot be appreciated. Short ribs are seen on the volume-rendered image on the right using maximum-intensity weighting.

Scoliosis

Scoliosis of the spine, particularly related to neurotube defects and hemivertebra, may be difficult to demonstrate with conventional planar imaging. Rotation of rendered volume data is very helpful in depicting visually abnormal spinal curvature. Controlled review of volume data in the standardized orientation is also helpful to show hemivertebra.

Upper Extremities

3DUS may allow for improved evaluation of hands and feet, which are often elusive for 2D imaging (Figure 6-17). Ploeckinger-Ulm et al. studied 72 fetuses from low-risk pregnancies and showed that 3DUS allowed a complete evaluation of all digits of the hand in significantly more fetuses than did 2DUS (74.3% versus 52.9%, $p < 0.05$) (34). Fetal digits were optimally visualized at 20–23 weeks gestation, with at least one fetal hand being visualized in 93% of cases using with 2DUS and in 100% of cases using 3DUS. Advantages of 3DUS included (a) the ability to rotate the echoes into any arbitrary plane for planar evaluation of hand digits; (b) rotation of volume-rendered images, allowing for more comprehensive view of fetal fingers; and (c) storage of data that allowed for review of digits (number and anatomy) after the patients had left the scanning area. Limitations of 3DUS in hand evaluation included gestational age of less than 19 weeks, fetal movement, unfavorable fetal position, and maternal obesity.

Budorick et al. studied 44 hands in 40 high-risk fetuses: 32 had normal hands and 12 had abnormal hands (35). Hands were correctly identified on both 2DUS and 3DUS as normal or abnormal. 3DUS provided additional information, including the following: (a) three orthogonal planes could be used to evaluate the hand simultaneously; (b) rotation of the volume allowed for planes to be evaluated, which could not be imaged with 2DUS; (c) the hands with loosely curled fingers could be assessed as normal; (d) the thumb and fingers could be routinely evaluated simultaneously; and (e) abnormal hands had improved assessment using both multiplanar and rendered images. In the study population, all metacarpals and digits were accurately counted in normal hands, even though most (28/32) were loosely curled in position. 3DUS assisted in the assessment of abnormal hands by

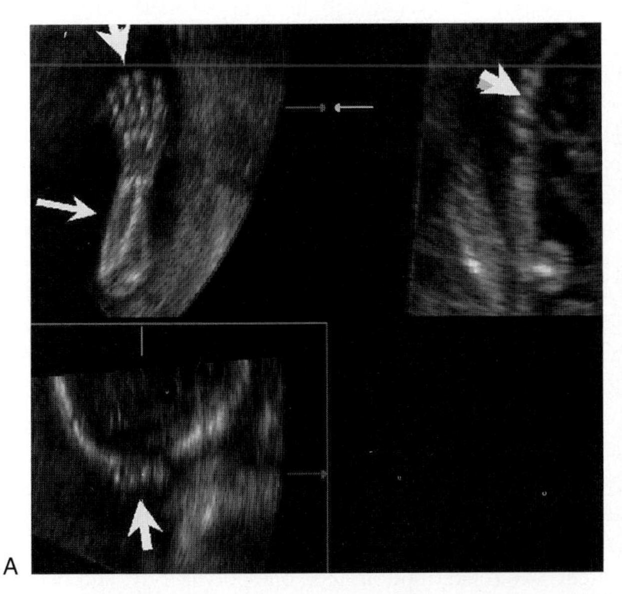

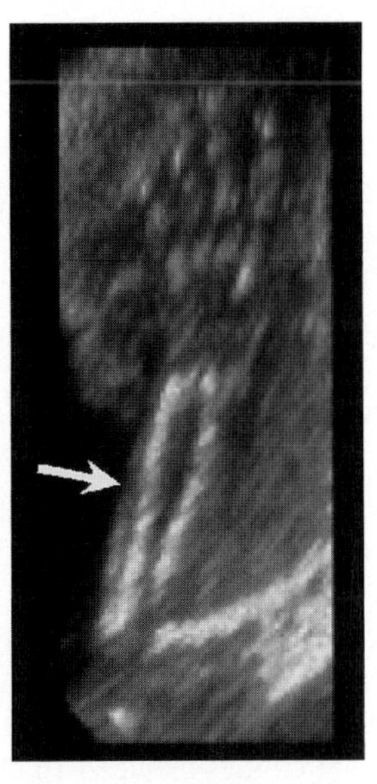

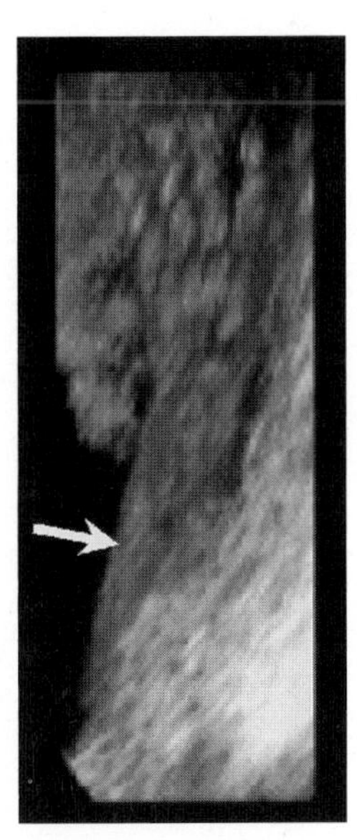

Figure 6-17. Normal hand and forearm in a 21-week-gestational-age fetus. **A:** The forearm (*arrow*) and hand (*arrowheads*) are seen in a multiplanar format. The digits can be counted by scrolling through the volume and examining parallel slices of the data. Upper left, coronal; upper right, sagittal; and lower left, axial. **B:** Volume-rendered image using maximum-intensity display showing the radius and ulna (*arrow*) as well as the digits. **C:** Volume-rendered image using surface display showing the soft tissues of the forearm (*arrow*). (With permission from Ref. 45.)

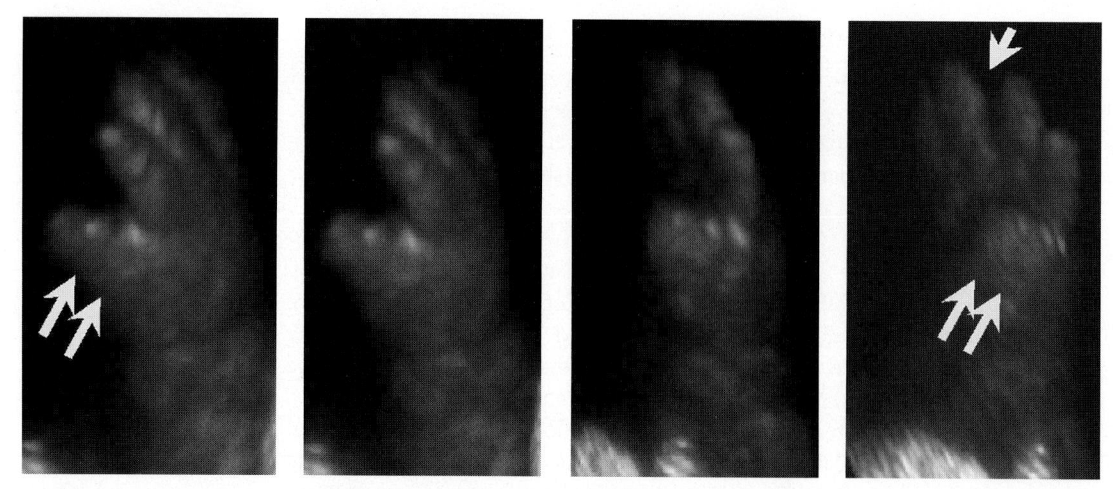

Figure 6-18. Abnormal hand in fetus with thanatophoric dysplasia. Several images from the rotating hand are displayed showing the brachydactyly and abnormal widening between the third and fourth digits (*arrow*). The thumb (*double arrow*) is shown.

facilitating improved understanding of the precise relationships of the wrist, hand, fingers, and thumb (Figure 6-18). The extent of finger overlap (10/12 hands) and the distinction of the thumb from severely overlapped fingers (2/12 hands) were more easily assessed with 3DUS.

Although both Budorick et al. (35) and Ploeckinger-Ulm et al. (34) reported that 3DUS assisted in counting digits of fetal hands accurately, it is important to note that there may be some patients in whom it will not be possible (e.g., fetal body shadowing hand, rapid movement of fetal hands, obese patients).

Lower Extremities

Sonographic evaluation of the distal extremity is frequently difficult because of the necessity of evaluating the tibia and fibula in relation to the hindfoot and forefoot. Optimal 2DUS evaluation of the distal extremity requires obtaining the sagittal plane through the lower leg and foot. Although coronal and axial imaging are helpful, the sagittal view is the critical image needed. 3DUS provides all three orthogonal planes to be evaluated with one volume acquisition. The distal extremity routinely can be evaluated in a standard orientation: coronal, sagittal, and axial (Figure 6-19). Scrolling through the volume (parallel images) in the coronal plane allows the two bones of the lower extremity to be visualized adjacent to each other, with the toes of the foot projecting outward toward the examiner as the volume is scrolled forward. The sagittal view of the distal extremity shows the perpendicular relationship of the tibia and fibula to the ankle. In volume data, the two limb bones can be seen in the same image slice in young fetuses, or they can be seen immediately adjacent to each other when scrolling the volume in older fetuses. Scrolling through the volume (parallel images) in the axial plane allows the two bones of the lower extremity to be visualized in cross-section, followed by a complete image of the sole of the foot that is similar to a ''hang ten'' view (see Figure 6-19).

Budorick et al. studied 46 distal lower extremities in 36 high-risk patients: 22 fetuses with bilaterally normal distal lower extremities, 13 fetuses with bilaterally abnormal distal lower extremities, and one fetus with unilateral polydactyly (36). The abnormal distal extremities included 16 clubfeet, two rocker bottom feet, one polydactyly, and two shortened distal extremities with only a single bone. (One fetus had both a shortened distal extremity and a clubfoot.) Multiplanar evaluation of the distal extremity allowed the position of the leg bones to be assessed in relationship to the foot. In fact, normal features of the normal distal extremity could be demonstrated more often with 3DUS (85%) than with 2DUS (52%). Rotation of the rendered volume of the distal extremities provided

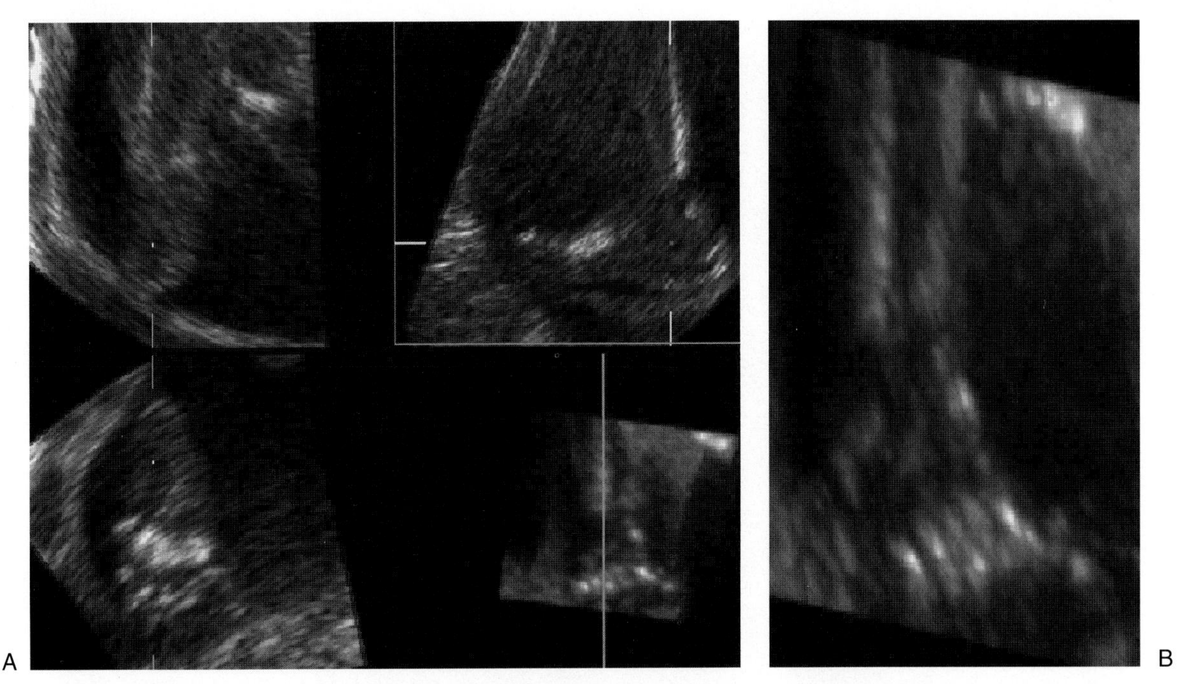

Figure 6-19. Normal fetal ankle at 20 weeks gestational age. **A:** Multiplanar and volume rendering show the coronal view of the ankle in the upper left image, a sagittal view in the upper right image, and an axial "hang ten" view in the lower left image. The volume-rendered image is shown in the lower right image. **B:** A volume-rendered image of the ankle is shown. Rotation of the volume data allows for a more recognizable representation of the ankle.

assistance in assessing foot position (Figure 6-20). Although 3DUS did not offer significant improvement in the ability to diagnose normal from abnormal distal lower extremities in their study population, Budorick et al. suggest that significant differences may be seen in centers less skilled in evaluating fetal anomalies. Experience using 3DUS in evaluating the distal extremity has allowed for more confidence in scanning patients with 2DUS because the relationships between the bones are more readily understood.

Skeletal Dysplasias

3DUS allows for a more comprehensive view of skeletal dysplasia than 2DUS. Steiner et al. have suggested that 3DUS assists in showing the true facial profile in often abnormal faces, showing planes not seen on 2DUS, improving measurement accuracy of abnormally shaped bones, displaying bowing of long bones, demonstrating the anomaly with surface rendering and showing the skeleton with a transparency mode, and emphasizing platyspondyly and rib anomalies (10). Pretorius et al. (37) have also shown that 3DUS is advantageous in demonstrating hypoplastic scapulae (Figure 6-21), shortened ribs (see Figure 6-16), abnormal hands (see Figure 6-18), and abnormal head shape (kleeblattschädel).

Diagnosis of phocomelia using 3DUS in a fetus with probable thrombocytopenia with absent radii syndrome was reported by Lee et al. (38). They reported that 3DUS enabled the extent of the fetal malformation to be in a more photographic style than conventional ultrasound. Volume rendering of the data allowed visualization of both the surface features and internal skeletal structures (using maximum mode).

Abnormalities of the Fetal Thorax, Heart, and Abdomen

3DUS has identified some developmental abnormalities in the fetal thorax, heart, and abdomen (see Table 6-5). Fetal lung volume estimations have been determined with 3DUS techniques and found to be promising. Lee et al. calculated fetal lung volumes by stepping

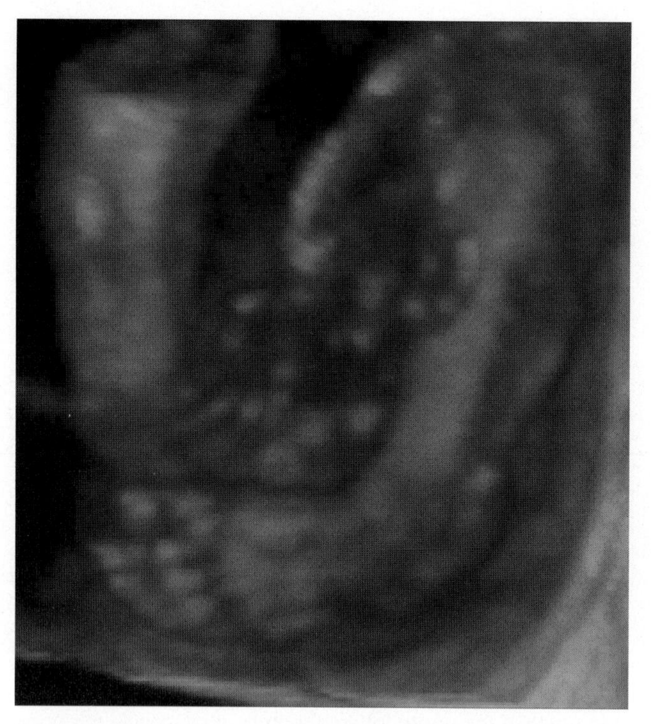

Figure 6-20. Clubfoot. The foot is shown to be abnormally positioned in a conventional club orientation. (Courtesy of Dr. Benoit.)

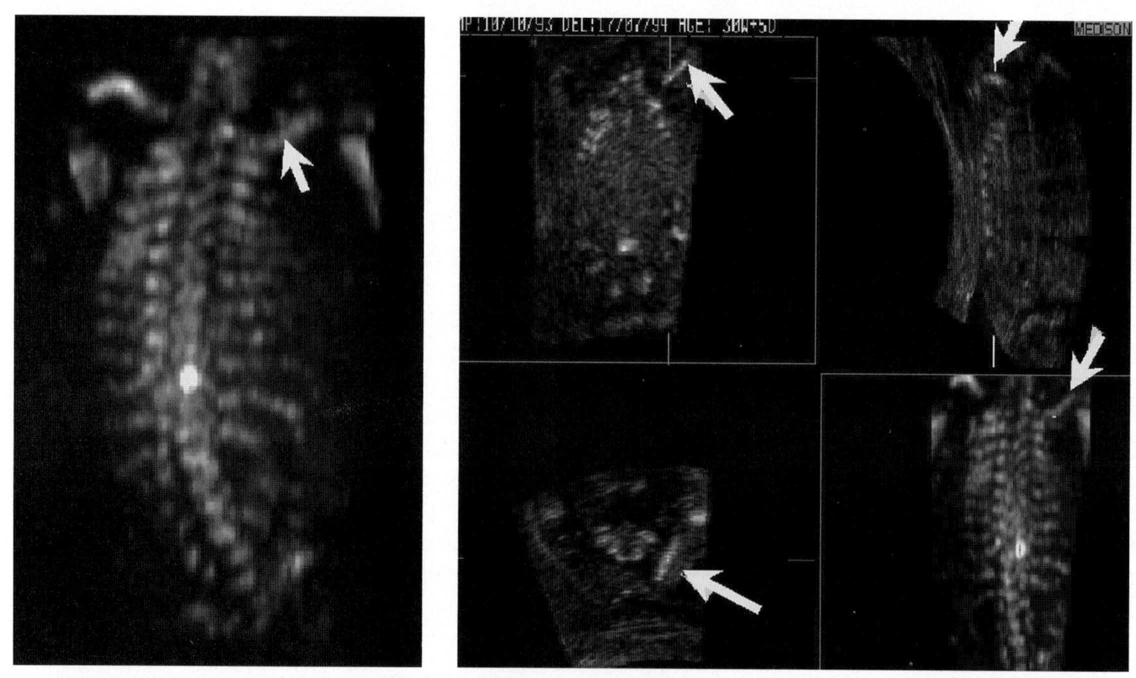

Figure 6-21. Hypoplastic scapula in a 21-week-gestational-age fetus. **A:** Volume-rendered image of thorax and spine showing that the upper portion of the scapula (*arrow*) is identified but the lower portion is hypoplastic and not ossified at this time. **B:** Multiplanar and volume-rendered views of same fetus. *Arrow* points to hypoplastic scapula. The presence of hypoplastic scapula clarified the diagnosis of campomelic dysplasia in this moderately short-limbed fetus.

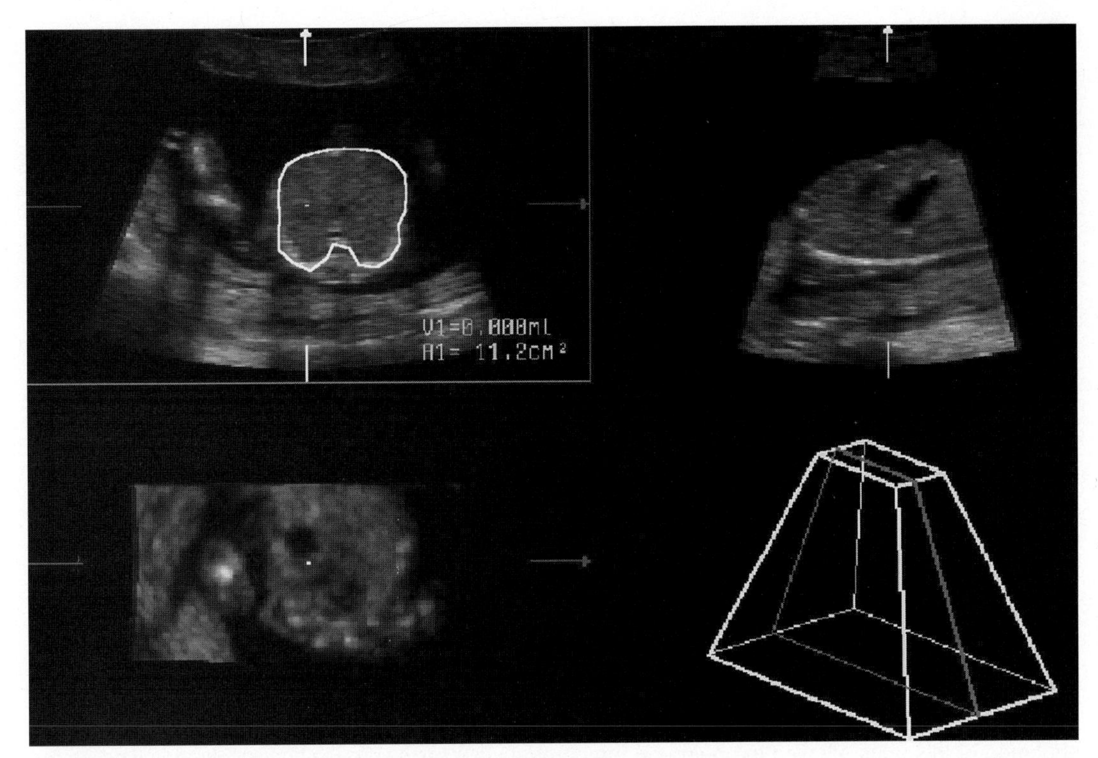

Figure 6-22. Fetal lung volume. Fetal lung volumes may be calculated more accurately with 3D imaging by tracing the lungs on multiple parallel planes. A *dotted line* is drawn around the fetal lung in the upper left image.

through a volume acquired through the thorax and subtracting the fetal heart volume from the fetal thorax volume, which included lungs, vessels, and heart but excluded vertebrae and ribs (39). Volumes were calculated by tracing the areas in cross-section planes of the thorax from the clavicle to the diaphragm (Figure 6-22). It is hoped that this technology will assist in diagnosing pulmonary hypoplasia more accurately than 2DUS.

3DUS assessment of fetal cardiac development is discussed in Chapter 11. Cardiac ectopia and omphalocele have been identified at 10 weeks gestational age using both 3DUS and 2DUS. Reconstruction of volume data from transvaginal data was helpful in identifying the ventral wall defect and in providing a complete view of the anomaly prenatally (40).

3DUS offers the potential to estimate the volume of omphaloceles and gastroschisis. The wall defects should be more apparent on 3DUS than 2DUS for less experienced observers. Applications of 3DUS power Doppler imaging to evaluation of thoracic masses are discussed later.

Abnormalities of the Genitourinary Tract and Genitalia

3DUS has shown that it can identify many anomalies with a modest improvement over 2DUS (see Table 6-6). In a series of congenital abnormalities published by Merz et al., 51 were genitourinary. The 3DUS images were advantageous in 43% and provided the same information as 2DUS in 57% (6).

Multiplanar evaluation of a fetus with bladder exstrophy was assisted by 3DUS evaluation in that what was initially thought to be the scrotum on coronal imaging proved to be an anterior abdominal wall defect on transverse images from the same volume. The multiplanar display assisted in making the diagnosis of bladder exstrophy (Figure 6-23) (18).

Steiner et al. reported a case in which 3DUS was helpful in differentiating a fetal abdominal retroperitoneal cystic structure from mildly dilated ureters (12). This was done

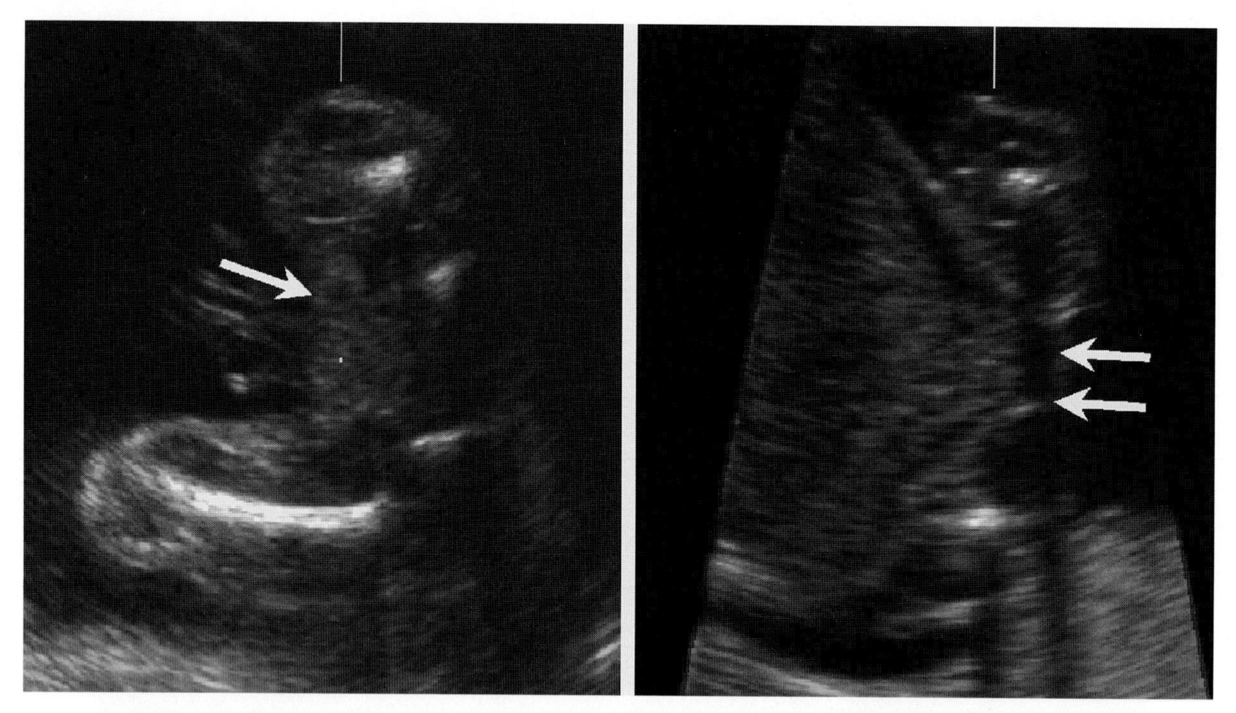

Figure 6-23. Bladder exstrophy in a 22-week gestational-age fetus. Multiplanar imaging through the abdomen of this fetus allowed the bulge on the coronal view (*arrow*) to be accurately identified as an abdominal wall defect on the transverse view (*double arrow*). Initially, the image on the left was thought to be a scrotum rather than an abdominal wall defect. 3D imaging allowed for the diagnosis of bladder exstrophy. (With permission from Ref. 23.)

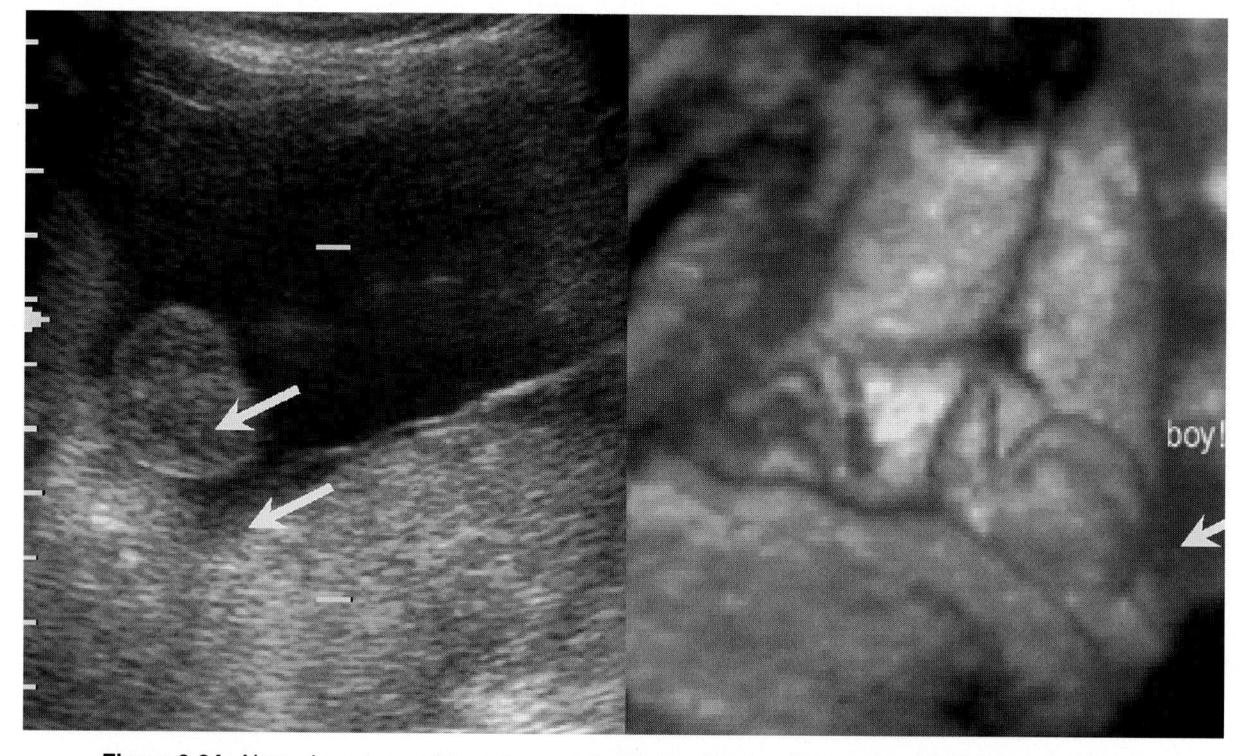

Figure 6-24. Normal scrotum at 34 weeks gestational age. Scrotum shown using 2DUS on left (*arrows*) and 3DUS on right. Surface rendering of the normal scrotum displays the scrotum and the penis. (With permission from Aloka.)

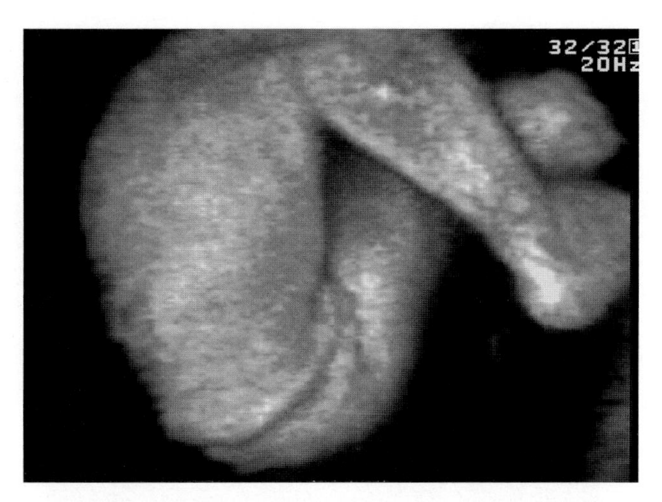

Figure 6-25. Female genitalia. (With permission from Aloka.)

by using the ''flight path'' technique (Medison, Korea) to follow the dilated ureters from the renal pelvis down to the urinary bladder.

Normal genitalia can be displayed with volume rendering that shows the anatomic parts in an easily recognized image (Figures 6-24 and 6-25). Ambiguous genitalia, hypospadias, and bipartite scrotum are often difficult to diagnose with 2DUS. Lee et al. have reported a case in which the diagnosis of bipartite scrotum was made with 3DUS from a surface-rendered image of the genitalia in a 27-week-old fetus (40). The fetus originally had a karyotype that showed male chromosomes yet the 2DUS showed two echogenic areas in the region of the genitalia that looked enlarged. The penis was identified but it could not be seen in the same plane as the scrotum.

Miscellaneous Anomalies Identified with 3DUS

3DUS also provides improved visualization of developmental anomalies for a variety of fetal structures (see Table 6-7).

FETAL GROWTH ASSESSMENT

Volume measurements of irregular objects are improved with 3DUS. Lee et al. studied a series of term fetuses to evaluate the usefulness and accuracy in predicting fetal weight with 3DUS techniques (41). They used abdominal and thigh volumes obtained with 3DUS equipment to predict birth weight. Although the mean systematic error and precision for birth weight predictions by 3DUS ($-0.03 \pm 6.1\%$) were not significantly different from those by 2DUS ($-0.60 \pm 8.8\%$), the conventional prediction methods yielded three birth weights with greater than 15% error, whereas all predictions based on 3DUS were within 11% of true values.

OBSTETRIC IMAGING WITH 3DUS COLOR/POWER DOPPLER IMAGING

Color and power Doppler techniques have been used to display the fetus, placenta, and cord. Ritchie et al. used a commercial scanner and image registration hardware to display volume-rendered images (42). They studied vascularity both in phantoms and in the placenta of patients and demonstrated that spatial relationships between vascular structures could be visualized. Pretorius et al. studied placental vasculature with 3DUS using color and power Doppler techniques in 14 normal patients and one patient with intrauterine growth retardation (43). 3DUS images showed that vessels correlated with known anatomy

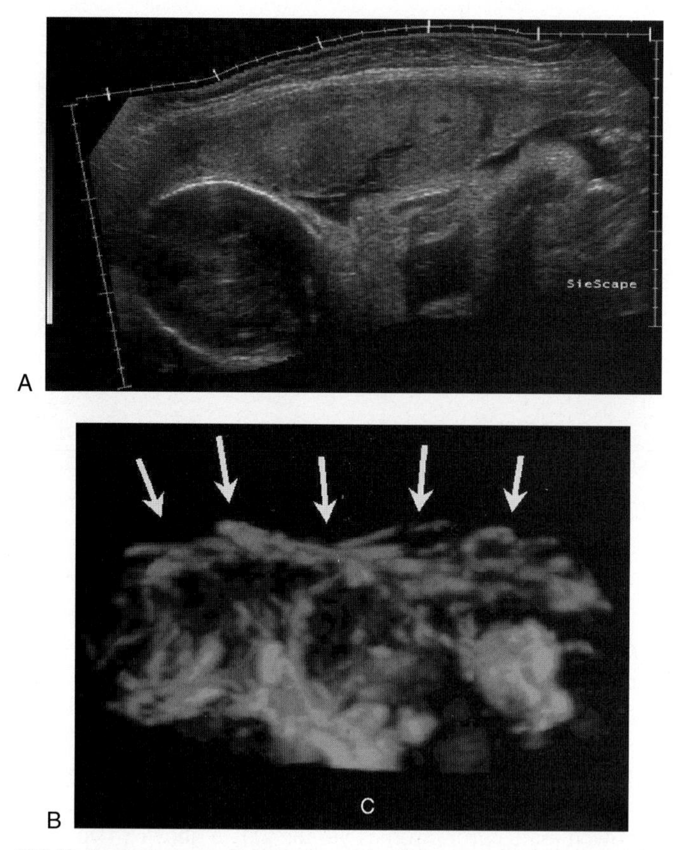

Figure 6-26. 2DUS planar image of placenta and volume-rendered image of placental blood vessels. **A:** Extended gray scale planar view of placenta using Siescape technology. **B:** Volume-rendered image of placental blood supply acquired using power Doppler. The cord insertion (**C**) is seen entering the placenta. Vessels are seen throughout the placenta and along the maternal surface (*arrows*). (With permission from Siemens Ultrasound, Issaquah, WA.) See color version of figure.

(Figure 6-26); visualization of vessels was assessed with regard to (a) the number of vessels seen within the placenta, (b) the branching pattern of the vessels within the placenta, (c) the number of vessels seen along the surface of the placenta, and (d) the number of vessels seen in the maternal circulation. A progressive increase in the number of intraplacental vessels and the number of vascular branches observed was seen with increasing gestational age. Volume data review using three orthogonal planar images had two distinct advantages over 2DUS: they could be obtained from orientations not possible using 2DUS alone, and they could be viewed in conjunction with volume-rendered images to allow for referencing and identification of specific vessels. Volume-rendered images assisted the observer in following the continuity of vessels as they wrapped around and twisted through 3D space. Stereo viewing was helpful in distinguishing overlapping vessels and identifying a velamentous could insertion. Volume imaging combined with color power Doppler imaging methods allowed for individual vessels in the placenta to be identified both in the fetal and the maternal circulations.

Vascular anatomy of the fetus has also been studied with 3DUS, including the umbilical cord, circle of Willis (Figure 6-27), aorta, renal arteries, and lung vessels. Coleman et al. reported on the study of 17 patients with fetal chest masses in which 3D power Doppler studies were performed and correlated the findings with magnetic resonance images in 12 cases (44). Pulmonary vessels were identified in all nine cases of cystic adenomatoid malformation, regardless of the degree of mediastinal shift. The location of the liver within the thorax or abdomen was identified appropriately in all eight cases of diaphragmatic hernia. 3DUS also diagnosed ipsilateral bronchopulmonary sequestration in one case of diaphragmatic hernia. Coleman et al. concluded that power Doppler 3DUS provided an

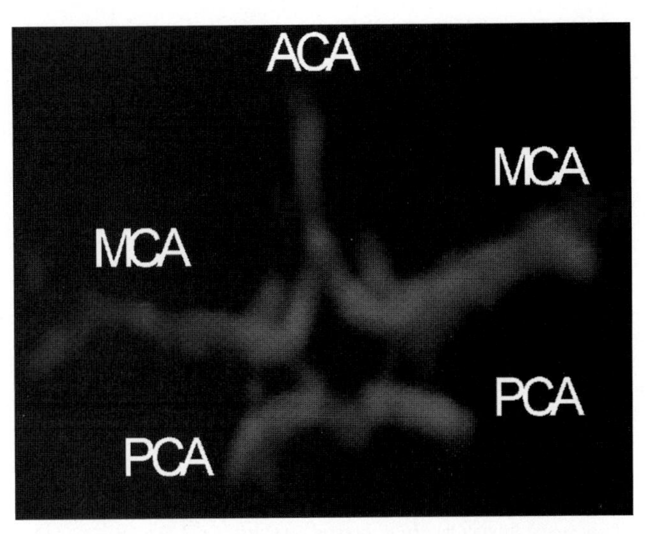

Figure 6-27. Volume-rendered image of the circle of Willis in a fetus at 31 weeks gestational age. The vessels of the circle of Willis are shown, including the anterior cerebral arteries (ACA), the middle cerebral arteries (MCA), and the posterior cerebral arteries (PCA). (With permission from ATL.) See color version of figure.

excellent means of demonstrating the vascular anatomy of fetal chest masses and of planning for surgery.

HOW TO DO IT
Practical Clinical Tips for Obstetric Evaluation

- Prescan fetus with 2DUS prior to selecting the region of interest for the volume acquisition.
- Set scanning parameters to optimize size, depth, and speed of the volume acquisition:
  - Optimize contrast when evaluating surface features.
- Scan area of interest in the desired plane of acquisition then acquire the volume.
- Delete volume if fetus moves during an acquisition.
- Acquire multiple volumes for anomalies.
- Review volume data after acquisition to verify quality:
  - Identify whether motion occurred.
  - Verify that desired region of interest was scanned.
- Reorient organ of interest to a standard orientation prior to storing the volume.
- Evaluate all volumes using the multiplanar display.
- Evaluate selected volumes using volume-rendered display.
- Review orthogonal images with volume-rendered images:
  - To permit cross-referencing locations and between images.
  - To assist in understanding data displayed in unfamiliar plane orientations.
  - To assist in identifying specific anatomic landmarks:
    - Palate, midline facial profile, etc.
- Measure desired structures in volume data:
  - Length, area, volume.
- Archive selected volumes for patient record.

REFERENCES

1. Crane JP, LeFevre ML, Winborn RC, et al. A randomized trial of prenatal ultrasonographic screening: impact on the detection, management, and outcome of anomalous fetuses. *Am J Obstet Gynecol* 1994;171: 392–399.

2. Baba K, Satoh K, Sakamoto S, Okai T, Ishii S. Development of an ultrasonic system for three-dimensional reconstruction of the fetus. *J Perinat Med* 1989;17:19–24.
3. Kuo HC, Chang FM, Wu CH, Yao BL, Liu CH. The primary application of three-dimensional ultrasonography in obstetrics. *Am J Obstet Gynecol* 1992;166:880–886.
4. Nelson TR, Pretorius DH. 3-Dimensional ultrasound of fetal surface features. *Ultrasound Obstet Gynecol* 1992;2:166–174.
5. Hamper UM, Trapanotto V, Sheth S, DeJong MR, Caskey CI. Three-dimensional US: preliminary clinical experience. *Radiology* 1994;191:397–401.
6. Merz E, Bahlmann F, Weber G. Volume scanning in the evaluation of fetal malformations: a new dimension in prenatal diagnosis. *Ultrasound Obstet Gynecol* 1995;5:222–227.
7. Merz E, Bahlmann F, Weber G, Macchiella D. Three-dimensional ultrasonography in prenatal diagnosis. *J Perinat Med* 1995;23:213–222.
8. Pretorius DH, House M, Nelson TR, et al. Evaluation of normal and abnormal lips in fetuses: comparison between three- and two-dimensional sonography. *AJR* 1995;165:1233–1237.
9. Nelson TR, Pretorius DH, et al. Visualization of the fetal thoracic skeleton with three-dimensional sonography: a preliminary report. *AJR* 1995;164:1485–1488.
10. Steiner H, Spitzer D, Weiss-Wichert PH, et al. Three-dimensional ultrasound in prenatal diagnosis of skeletal dysplasia. *Prenat Diagn* 1995;15(4):373–377.
11. Pretorius DH, Nelson TR, et al. Fetal face visualization using three-dimensional ultrasonography. *J Ultrasound Med* 1995;14:349–356.
12. Steiner H, Staudach A, Spitzer D, Schaffer H. Three-dimensional ultrasound in obstetrics and gynaecology: technique, possibilities and limitations. *Hum Reprod* 1994;9:1773–1778.
13. Maier B, Steiner H, Wienerroither H, Staudach A. The psychological impact of three-dimensional fetal imaging on the fetomaternal relationship. In: Baba K, Jurkovic D, eds. *Three-dimensional ultrasound in obstetrics and gynecology*. New York: Parthenon, 1997:67–74.
14. Pretorius DH. Maternal smoking habit modification via fetal visualization. University of California Tobacco Related Disease Research Program. Annual Report to the California State Legislature, 1996:76.
15. Merz E, Weber G, Bahlmann F, Miric-Tesanic D. Application of transvaginal and abdominal three-dimensional ultrasound for the detection or exclusion of malformations of the fetal face. *Ultrasound Obstet Gynecol* 1997;9:237–243.
16. Johnson DD, Pretorius DH, Riccabona M, Budorick NE, Nelson TR. Three-dimensional ultrasound of the fetal spine. *Obstet Gynecol* 1997;89:434–438.
17. Pretorius DH, Johnson DD, Budorick NE, Jones MC, Lou KV, Nelson TR. Three-dimensional ultrasound of the fetal lip and palate. *Radiology* 1997;205(P)(suppl):245.
18. Pretorius DH, Richards RD, Budorick NE, et al. Three-dimensional ultrasound in the evaluation of fetal anomalies. *Radiology* 1997;205(P)(suppl):245.
19. Steiner H, Gregg AR, Bogner G, Graf AH, Weiner CP. First trimester three-dimensional ultrasound volumetry of the gestational sac. *Arch Gynecol Obstet* 1994;255:165–170.
20. Feichtinger W. Editorial: transvaginal three-dimensional imaging. *Ultrasound Obstet Gynecol* 1993;3:375–378.
21. Blaas HG, Eik-Nes SH, Kiserud T, Berg S, Angelsen B, Olstad B. Three-dimensional imaging of the brain cavities in human embryos. *Ultrasound Obstet Gynecol* 1995;5(4):228–232.
22. Blaas HG, Eik-Nes SH, Berg S, Torp H. *In vivo* three-dimensional ultrasound reconstructions of embryos and early fetuses. *Lancet* 1998;352:1182–1186.
23. Richards RD, Pretorius DH, Budorick NE, et al. Three-dimensional ultrasound in the evaluation of fetal anomalies. *Ultrasound Obstet Gynecol* 1998 (*submitted*).
24. Baba K, Okai T, Kozuma S, Taketani Y, Mochizuki T, Akahane M. Real-time processable three-dimensional US in obstetrics. *Radiology* 1997;203:571–574.
25. Hata T, Yonehara T, Aoki S, Manabe A, Hata K, Miyazaki K. Three-dimensional sonographic visualization of the fetal face. *AJR* 1998;170:481–483.
26. Hata T, Aoki S, Hata K, Miyazaki K, Akaahane M, Mochizuki T. Three-dimensional ultrasonographic assessments of fetal development. *Obstet Gynecol* 1998;91:218–223.
27. Devonald K, Ellwood DA, Griffiths KA, et al. Volume imaging: three-dimensional appreciation of the fetal head and face. *J Ultrasound Med* 1995;14:919–925.
28. Lee A, Deutinger J, Bernaschek G. Three dimensional ultrasound: abnormalities of the fetal face in surface and volume rendering mode. *Br J Obstet Gynaecol* 1995;102(4):302–306.
29. Mueller GM, Weiner CP, Yankowitz J. Three-dimensional ultrasound in the evaluation of fetal head and spine anomalies. *Obstet Gynecol* 1996;88(3):372–378.
30. Riccabona M, Johnson D, Pretorius DH, et al. Three-dimensional ultrasound: display modalities in the fetal spine and thorax. *Eur J Radiol* 1996;22:141–145.
31. Riccabona M, Pretorius DH, Nelson TR, et al. Three-dimensional ultrasound: display modalities in obstetrics. *J Clin Ultrasound* 1997;25(4):157–167.
32. Johnson DD, Pretorius DH, Budorick NE. Three-dimensional ultrasound of conjoined twins. *Obstet Gynecol* 1997;90:701–702.
33. Pretorius DH, Nelson TR, et al. Prenatal visualization of cranial sutures and fontanelles with three-dimensional ultrasonography. *J Ultrasound Med* 1994;13:871–876.
34. Ploeckinger-Ulm B, Ulm MR, Lee A, Kratochwil A, Bernaschek G. Antenatal depiction of fetal digits with three-dimensional ultrasonography. *Am J Obstet Gynecol* 1996;175:571–574.
35. Budorick NE, Pretorius DH, Tartar MK, Johnson DD, Nelson TR, Lou KV. Three-dimensional US of the fetal hands: normal and abnormal findings. *Radiology* 1996;201(P):160.
36. Budorick NE, Pretorius DH, Tartar MK, Nelson TR. 3-Dimensional US of the fetal ankle. *Radiology* 1995;197(P):197.
37. Garjian KV, Pretorius DH, Budorick NE, Cantrell CJ, Johnson DD, Nelson TR. Three-dimensional ultrasound of fetal skeletal dysplasia. *Radiology* 1999 (in press).

38. Lee A, Kratochwil A, Deutinger J, Bernaschek G. Three-dimensional ultrasound in diagnosing phocomelia. *Ultrasound Obstet Gynecol* 1995;5:238–240.
39. Lee A, Kratochwil A, Stumpflen I, Deutinger J, Bernaschek G. Fetal lung volume determination by three-dimensional ultrasonography. *Am J Obstet Gynecol* 1996;175(3, Pt 1):588–592.
40. Liang RI, Huang SE, Chang FM. Prenatal diagnosis of ectopia cordis at 10 weeks of gestation using two-dimensional and three-dimensional ultrasonography. *Ultrasound Obstet Gynecol* 1997;10:137–139.
41. Lee A, Deutinger J, Bernaschek G. Volvision: three-dimensional ultrasonography of fetal malformations. *Am J Obstet Gynecol* 1994;170:1312–1314.
42. Lee W, Comstock CH, Kirk JS, et al. Birthweight predictions by three-dimensional ultrasonographic volumes of the fetal thighs and abdomen. *JUM* 1997;161(12):799–805.
43. Ritchie CJ, Edwards WS, Mack LA, Cyr DR, Kim Y. Three-dimensional ultrasonic angiography using power-mode Doppler. *Ultrasound Med Biol* 1996;22:277–286.
44. Pretorius DH, Nelson TR, Baergen RN, Pai E, Cantrell C. Imaging of placental vasculature using three-dimensional ultrasound and color power Doppler: a preliminary study. *Ultrasound Obstet Gynecol* 1998; 12:45–49.
45. Coleman B, Hubbard AM, Jackson GM, Howell LJ, Crombleholm TC, Adzick NS. Three-dimensional power Doppler of fetal chest masses. *J Ultrasound Med* 1998.
46. Pretorius DH, Nelson TR. 3-Dimensional ultrasound in gynecology and obstetrics: a review. *Ultrasound Q* 1998;14:218–233.

7

Gynecology

OVERVIEW

This chapter will review applications of three-dimensional ultrasound (3DUS) in gynecologic imaging, which shows considerable promise in evaluating the pelvic organs such as the uterus or the adnexa. 3DUS offers improved visualization by permitting optimization of viewing plans in volume data acquired using both transabdominal and transvaginal probes. 3DUS also offers more accurate volume estimation, retrospective review of stored data, more complete viewing of pathology using rendered images leading to more accurate identification of location of abnormalities needing surgical intervention, and assessment of tumor invasion.

KEY CONCEPTS

- Potential applications in the uterus:
  - Congenital abnormalities.
  - Endometrial cancer:
    - Screening.
    - Invasion.
  - Endometrial hyperplasia.
  - Location and size of fibroids.
  - Sonohysterography:
    - Endometrial polyps.
    - Adhesions.
  - IUD localization.
- Potential applications in the adnexa:
  - Adnexal masses.
  - Infertility:
    - Follicular volume estimations.
    - Fallopian tube patency.
  - Ectopic pregnancy.

3DUS is usually used as an adjunct to 2DUS in the pelvis; specifically, it is used to evaluate a particular area of interest in the pelvis such as the uterus or the adnexa. 3DUS imaging offers considerable promise in evaluating the pelvic organs because of the possibility of arbitrary plane viewing, more accurate volume estimation, viewing of contrast movement within the fallopian tubes, retrospective review of stored data, more accurate viewing of pathology, accurate identification of location of abnormalities needing surgical intervention, and assessment of tumor invasion. Examination of the anatomy in the pelvis with 2DUS has been limited predominantly by the orientation of the ultrasound beam from the abdominal–pelvic wall, the vagina, and the perineum. This has meant that the pelvic structures could only be evaluated in two primary planes (sagittal and coronal), with minor tilting of the transducer resulting in oblique plane viewing. 3DUS offers new viewing windows by allowing for arbitrary plane evaluation through a volume data set acquired from the pelvis using both transabdominal and transvaginal probes.

3DUS OF THE UTERUS

Normal Uterus

Three-dimensional ultrasound imaging of the uterus permits arbitrary planes of the uterus to be evaluated that cannot be seen with 2DUS due to limitations of movement of the transducer. Viewing of the transverse plane of the uterus allows the physician to view both horns of the endometrium and the cervix at the same time (Figure 7-1). The normal uterus is generally thought to have a convex shape of the endometrium and myometrium in the fundus, and an arcuate uterus has a rounded, concave upper line of the endometrium

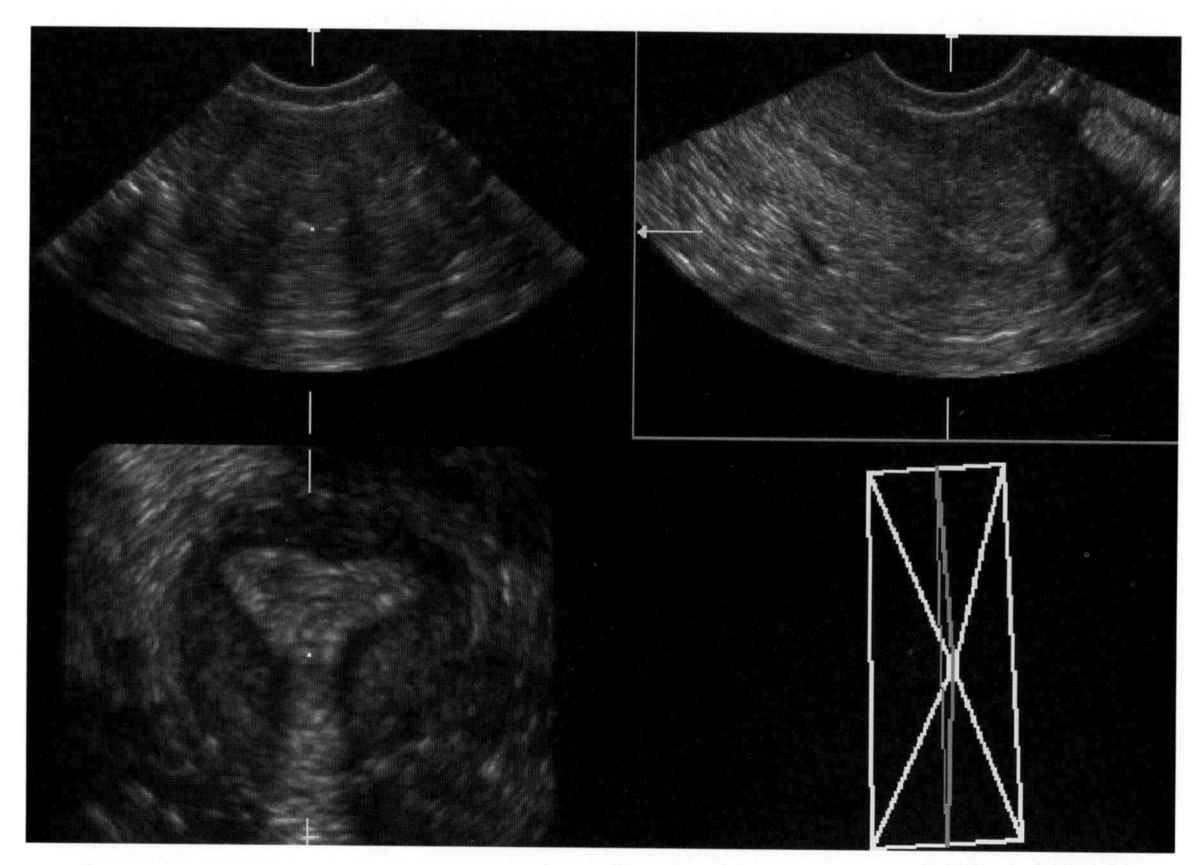

Figure 7-1. 3DUS study of a normal uterus. Multiplanar display of uterus with the upper-left-hand box showing the coronal plane, upper-right-hand box showing the sagittal plane, lower-left-hand box showing the transverse plane, and lower-right-hand box showing an orientation diagram.

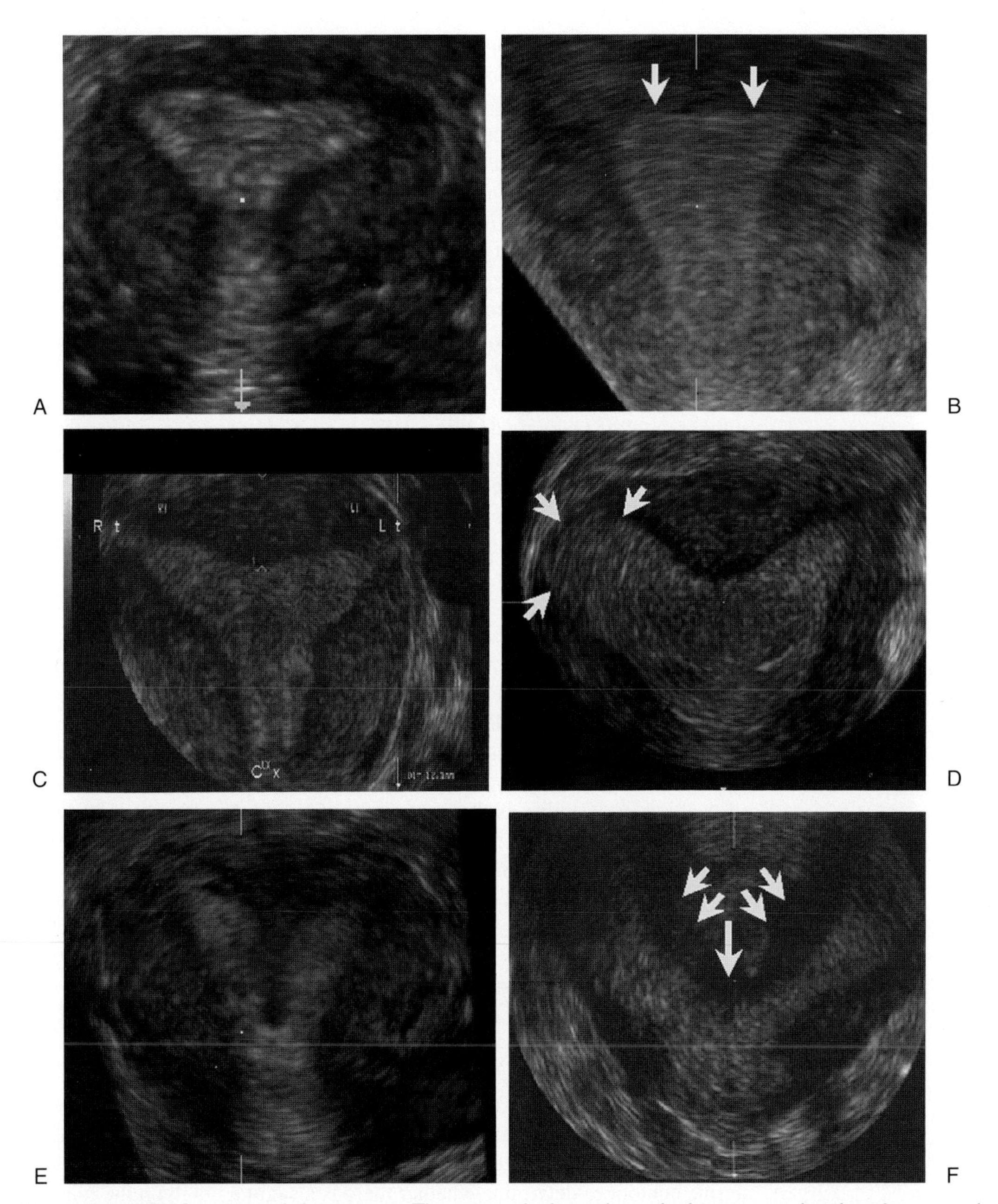

Figure 7-2. 3DUS scans of the uterus. The coronal plane through the uterus showing the normal endometrium and myometrium is shown for several different morphometries. The coronal plane cannot be obtained using conventional 2DUS. **A:** Normal uterus. **B:** Normal uterus showing slight variation in the upper border of the endometrium (*arrows*). **C:** Nearly symmetrical uterine horns and cervix, occasionally called arcuate. **D:** Septate uterus showing septation within the uterus and asymmetric horns of the endometrium, with the right horn (*arrows*) being larger and closer to the wall of the uterus. **E:** Septate uterus showing the septate appearance of the endometrium and the normal myometrium. **F:** Bicornuate uterus showing the two horns of the endometrium and the heart-shaped appearance of the myometrium (*arrows*). Rt, right horn; Lt, left horn; Cx, cervix.

with a normal outer shape of the uterus (1). In clinical practice, many normal variations on these two shapes are seen (Figure 7-2).

Congenital Uterine Anomalies

The endometrium presents with a spectrum of different shapes from normal to abnormal, which can have an important impact on patient management (see Figure 7-2). 3DUS permits the shape of the uterus to be more clearly demonstrated and thus allows more appropriate management for the patient based on better knowledge of the uterine shape. Identification of uterine anomalies is important in the work-up of infertility patients, but such anomalies are also important during pregnancy, because of such obstetric complications as second-trimester abortion, premature birth, intrauterine growth retardation, fetal malpresentation, and retained placenta (2). Congenital anomalies of the uterus are displayed optimally with 3DUS because of the ability of the examiner to manipulate the 3D image to view the transverse plane through the uterus (see Figure 7-2).

2DUS is helpful in identifying the two horns of the uterus, but the distinction between septate uterus and bicornuate uterus generally cannot be made because the contour of the myometrium cannot be imaged in the transverse plane. Hysterosalpingography (HSG) displays the uterine cavity very well, but the myometrium and external contour of the uterus cannot be evaluated, and thus the differentiation between septate and bicornuate uterus cannot be made. In addition, HSG is an invasive test, requiring the use of iodinated contrast agents and exposure to radiation. Jurkovic et al. point out that lateral fusion disorders of the uterus often require laparoscopy or laparotomy to make a definitive diagnosis (1). Magnetic resonance imaging (MRI) can be performed to assess congenital anomalies of the uterus as well. Although MRI offers many of the advantages of 3DUS, including multiplanar evaluation, it is rarely used in many regions of the world because of its relative expense.

Jurkovic et al. investigated 61 patients with a history of recurrent miscarriage or infertility who had a HSG and who underwent 2DUS and 3DUS (1). They reported that 2DUS showed five false-positive diagnoses of arcuate uterus and three of major uterine anomalies, whereas 3DUS agreed with HSG in all nine cases of arcuate uterus and three cases of major congenital anomalies. They commented that 3DUS was able to easily differentiate septate (see Figures 7-2D and 7-2E) and bicornuate (see Figure 7-2F) uterus as a result of its ability to visualize both the uterine cavity and the myometrium. Jurkovic et al. also reported that good-quality 3DUS volumes were obtained in 95% of the patients; suboptimal exams occurred in women with fibroids. In their series, Jurkovic et al. (1) reported that there were no false-positive or false-negative diagnoses of congenital uterine anomalies with 3DUS.

Raga et al. have also reported that 3DUS was superior to 2DUS in evaluating congenital uterine anomalies in a study group of 12 anomalies and 30 normal uteri (3). Transvaginal 2DUS detected Mullerian anomalies in 75% of cases (missed two out of five septate uteri and one arcuate uterus), whereas 3DUS detected 11 of the 12 anomalies (missing one due to a ''strategically'' located uterine myoma).

Multiplanar display of the uterus is emerging as the optimal method for identifying congenital uterine anomalies (1,3–5), with additional rendered images not generally necessary. However, Raga suggests that surface reconstruction may allow for study of the outer uterine contour to determine whether there is a sagittal notch on the fundus or to measure the depth of the myometrial spur in the wall that separates the two hemicavities of a didelphic uterus. They also report that the endometrial cavity can be displayed using a transparent maximum–minimum mode to differentiate it from the myometrium (3).

Endometrial Cancer

Endometrial cancer is the most common malignancy involving the gynecologic organs. Of women presenting with postmenopausal bleeding, approximately 10% will have endometrial cancer. Invasive tests such as endometrial biopsy and dilatation and curettage are often required to exclude the diagnosis of cancer in these patients. Conventional 2DUS has been used as a screening test to evaluate patients for the possibility of endometrial cancer, particularly in postmenopausal patients who are bleeding. Traditionally, the thickness of the endometrium has been measured on a sagittal plane through the uterus. Unfortunately, there have been difficulties with this approach, as the endometrium may have varying thickness, and other pathology, such as endometrial polyps, may be displayed as

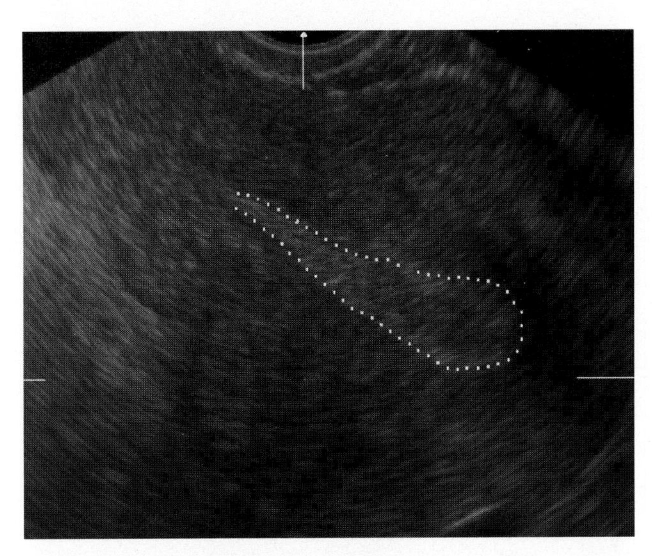

Figure 7-3. Endometrial volume. The volume of the endometrium is measured by delineating the uterine cavity (*dotted line*) on parallel longitudinal sections 1–2 mm apart. The sections are added together using a built-in computer program to calculate the volume.

a focal area of thickening. In addition, there is significant overlap in the endometrial thickness in patients with cancer and endometrial hyperplasia.

Volume measurements of the endometrium may be much more helpful in distinguishing endometrial cancer from benign pathology. Gruboeck et al. have reported on a series of 103 patients presenting with postmenopausal bleeding in which they performed both conventional 2DUS and 3DUS imaging of the uterus (6). Measurements of endometrial volume were estimated from outlining the endometrium (Figure 7-3) on serial parallel slices from the 3DUS studies and were compared with the greatest thickness (Figure 7-4) of the endometrium obtained on the longitudinal plane from the 2DUS study. Both 2DUS and 3DUS studies were performed with transvaginal transducers, and adequate studies were obtained on both in 97 cases. Their results are shown in Figures 7-5 and 7-6. Receiver operating curve data showed that endometrial volume is superior to endometrial thickness for the diagnosis of endometrial cancer. The sensitivity and positive predictive

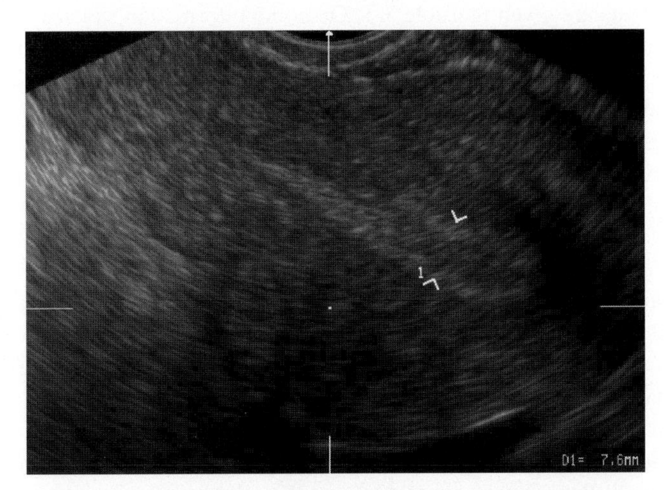

Figure 7-4. Endometrial thickness. The thickness of the endometrium is 7.6 mm and is measured on a longitudinal scan through the uterus. Calipers mark the thickness.

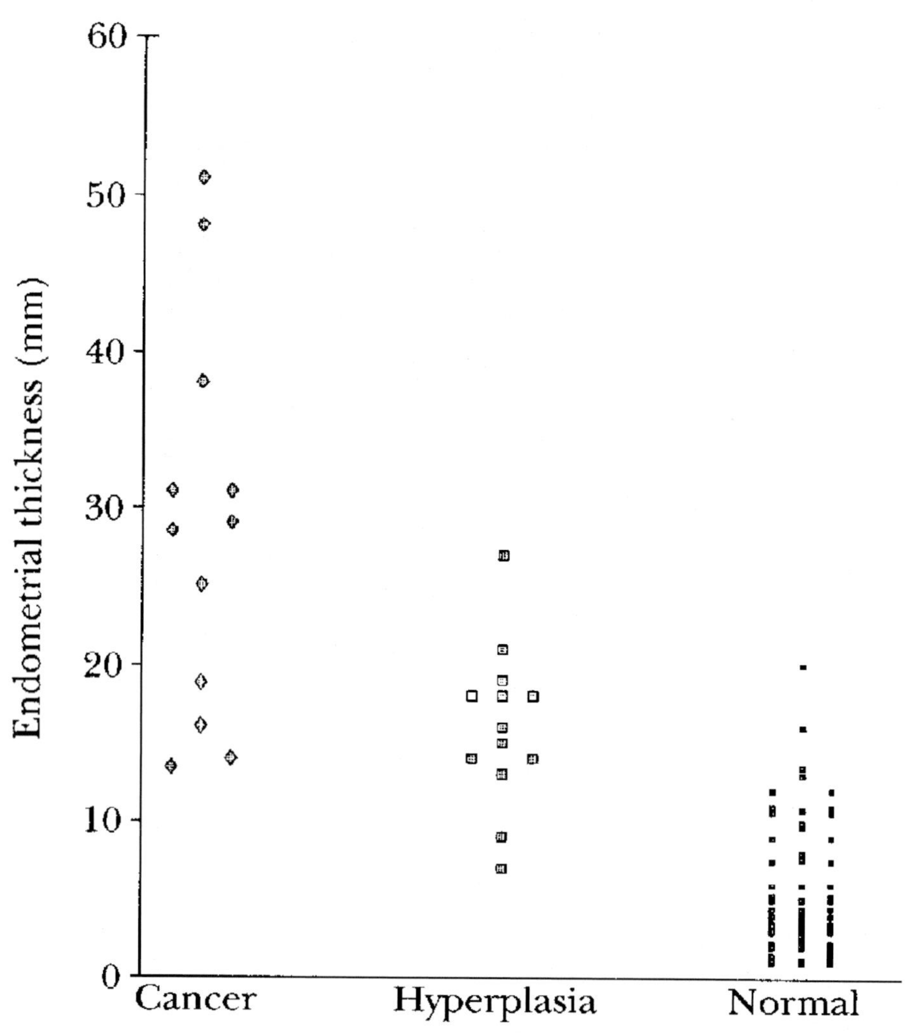

Figure 7-5. Endometrial thickness in 97 patients with cancer (*diamonds*), hyperplasia or polyps (*large squares*), and normal or atrophic endometrium (*small squares*). (With permission from Ref. 6.)

value of these tests are given in Table 7-1. Outcome information showed that 11 patients had endometrial cancer, eight patients had endometrial hyperplasia, seven patients had endometrial polyps (Figure 7-7), and 71 patients had an atrophic or normal endometrium. Olstad et al. also have reported on volume measurements of the endometrium using an automated edge-detection method (7).

The severity or grade of the endometrial cancer was also related to the size of the endometrium. Tumors that were moderately or poorly differentiated generally had larger volumes and thicknesses as compared with those that were well differentiated (Table 7-2). Increasing volume size was associated with progressive myometrial invasion, but this was not statistically significant in their small series. However, the data are encouraging, and additional work needs to be performed in this area.

Invasion of endometrial carcinoma into the myometrium may also be examined with 3DUS. Bonilla et al. suggest that 3D hysterosonography allowed for better visualization of myometrial invasion in three patients they studied and that this technique may play a significant role of staging malignant tumors in the future (8). They were able to see endophytic growth toward the endometrial cavity and the myometrium.

Because of the simultaneous display of the transverse plane with 3DUS, it should be possible to visualize infiltration of cervical or endometrial carcinoma into the bladder or

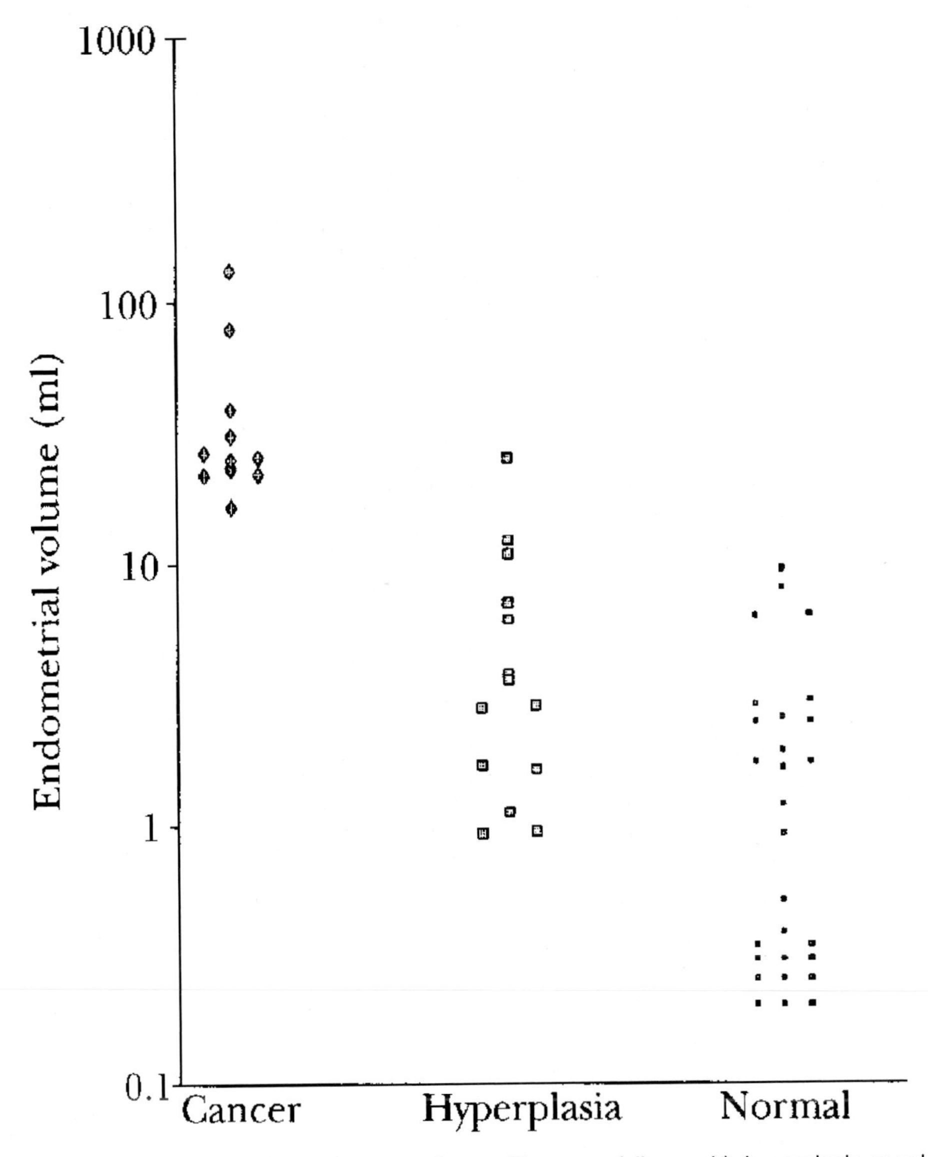

Figure 7-6. Endometrial volume in 97 patients with cancer (*diamonds*), hyperplasia or polyps (*large squares*), and normal or atrophic endometrium (*small squares*). (With permission from Ref. 6.)

rectum (9). Future work is needed in this area, as 3DUS may be able to replace some CT scans currently performed for staging tumors.

Fibroids

The exact location of fibroids can be demonstrated within the uterus by using simultaneous display of three perpendicular planes. The relationship of the fibroids to the endometrium can be assessed more accurately with 3DUS than with 2DUS. Use of sonohysterography may be of value in demonstrating submucosal fibroids (Figure 7-8) (5,8,10,11).

Table 7-1. *Optimal cut-off value, sensitivity, and positive predictive value (PPV) of endometrial thickness and volume for diagnosis of endometrial cancer (6)*

	Cut-off	Sensitivity	PPV
Thickness	15 mm	83%	55%
Volume	13 ml	100%	92%

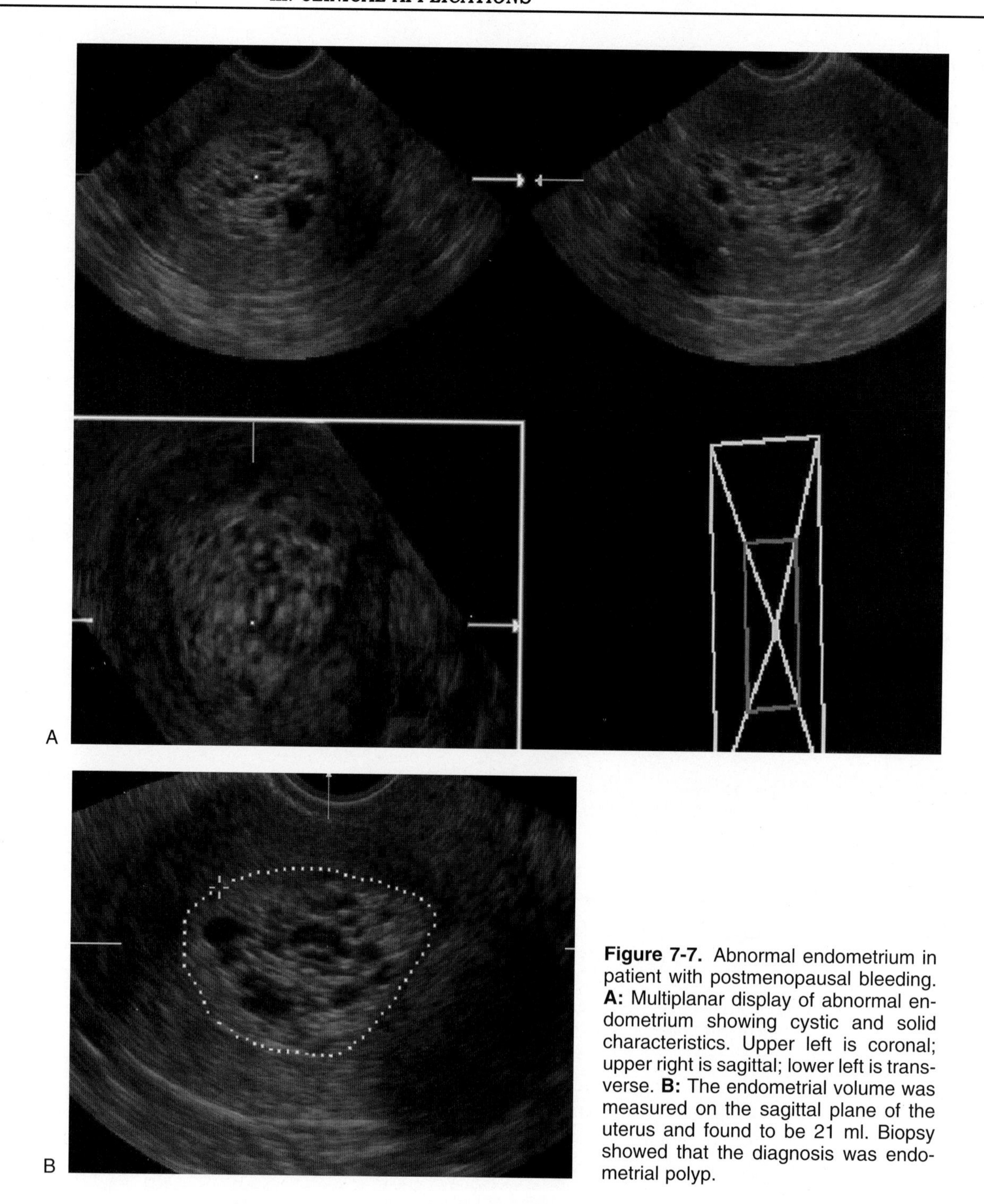

Figure 7-7. Abnormal endometrium in patient with postmenopausal bleeding. **A:** Multiplanar display of abnormal endometrium showing cystic and solid characteristics. Upper left is coronal; upper right is sagittal; lower left is transverse. **B:** The endometrial volume was measured on the sagittal plane of the uterus and found to be 21 ml. Biopsy showed that the diagnosis was endometrial polyp.

Table 7-2. *Endometrial thickness and volume in relation to the histological grade of endometrial cancer*

Grade	*n*	Thickness (mm)	Volume (ml)
I	4	21.4 (13.4–31.0)	19.6 (13.5–23.0)
II	3	31.2 (14.0–51.0)	59.8 (24.5–129.0)
III	4	36.5 (29.0–48.0)	41.9 (22.4–77.1)

Values are mean and range (6).

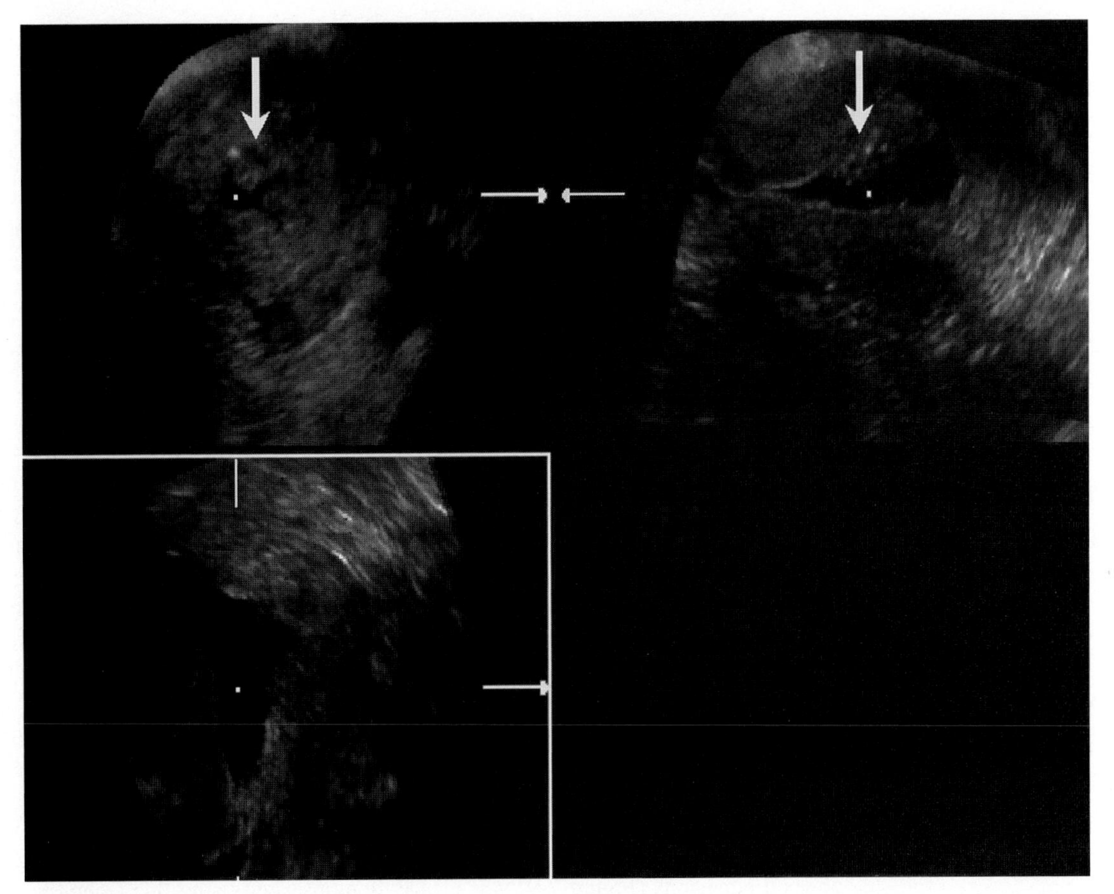

Figure 7-8. Sonohysterography of submucosal fibroids (*arrow*). Upper left upper is coronal; upper right is sagittal; and lower left is transverse. (Courtesy of Dr. Anna Lev-Toaff.)

The size of fibroids should be evaluated more accurately with 3DUS than with 2DUS because the accuracy of volume measurements in other organs, such as the liver (12) and bladder (13), has been established. Patients on medical therapy such as gonadotropin-releasing hormone may be followed with serial 3DUS scans to estimate fibroid size and effectiveness of the therapy. One limitation of scanning a uterus with fibroids occurs when there is significant shadowing from calcification; this is true for both 2DUS and 3DUS.

Sonohysterography

Sonohysterography (SH) is a technique in which contrast material, either saline (negative contrast) or contrast bubbles (positive contrast), is injected into the uterine cavity to distend and evaluate the endometrial cavity. SH is helpful in evaluating uterine fibroids, endometrial polyps, focal hyperplasia, generalized hyperplasia, carcinoma, and adhesions. Typically, sonohysterography is performed with conventional 2DUS, but 3DUS has several advantages. 3DUS provides more accurate information about the location of abnormalities that may be very important for preoperative assessment and distinguishing pathologies. A 3DUS volume can be acquired in only a few seconds and reviewed later, thus allowing the uterus to be distended for a shorter period of time than 2DUS exams. This is a significant advantage for the patient because uterine distention may be uncomfortable.

Bonilla-Musoles et al. suggest that 3DUS may assist in distinguishing polyps, hyperplasia, and carcinoma after studying 36 women with postmenopausal bleeding (8). They found that visualization of the uterine cavity and the endometrial thickness was better with 3D-hysterosonography than with transvaginal sonography, transvaginal sonohysterography, transvaginal color Doppler, or hysteroscopy. 3DUS studies were able to identify the location of focal endometrial hyperplasia and polyps. In patients on hormone replacement therapy or tamoxifen, 3D sonohysterography allowed for differentiation of normal proliferative from hyperplastic endometrium. They also noted that 3D-SH improved ultrasound determination of myometrial and cervical invasion in women with endometrial adenocarcinoma.

Balen et al. studied 10 patients with 3DUS and SH and found that 3DUS was useful in demonstrating the position of submucosal fibroids (see Figure 7-8) (10). Balen studied both saline and a positive ultrasound contrast agent (Echovist) and found the positive contrast superior when looking at the cavity wall. Weinraub et al. found saline to be better for evaluating the actual contents of the uterine cavity (5). They studied 32 patients and found that the negative contrast (saline) allowed for delineation of the outer surface of lesions, whereas positive contrast only created a cast of the cavity.

Weinraub et al. found that visualization of 3DUS data for SH was valuable with both multiplanar imaging and surface rendering (5). Polypoid structures were visualized using the multiplanar views, allowing for the optimal plane to identify their pedicle, whereas surface rendering suppressed unwanted echoes and allowed for the polypoid mass to be seen in continuity with the endometrial lining. Submucous fibroids were well seen on the planar images, and intrauterine synechiae (adhesions) were seen on both multiplanar and rendered imaging. The adhesions could be seen traversing the cavity with the rendered display. Weinraub et al. concluded that surface rendering ''confirmed the presence of pathological findings in equivocal cases, and characterized their appearance, actual size, volume and relationship to the surrounding structures'' (5). They suggested that this was important for surgical planning and in determining whether surgical or conservative management was appropriate. The simultaneous display of three perpendicular planes gave a more comprehensive overview of the uterus and allowed for access to planes not obtainable by conventional 2DUS.

Lev-Toaff et al. had similar results in examining 15 women with 3D SH. The exact location of polyps, residual submucous fibroids after myomectomy, and intrauterine/periadnexal adhesions (Figure 7-9) was clearly demonstrated using this technique (11).

Assessment of Intrauterine Devices

2DUS has been used to identify the location of IUDs within the uterus for many years. Occasionally, it may be difficult to locate an IUD accurately with 2DUS. 3DUS allows for accurate location using the multiplanar display; the transverse plane often allows a large portion of the IUD to be visualized (Figure 7-10). Bonilla-Musoles et al. compared the results of 2D and 3D transvaginal scans performed on 66 asymptomatic women to determine whether 3DUS offered advantages over 2DUS (14). Hysteroscopy was performed in 14 cases in which there was a discrepancy between the information obtained from the two methods. In all cases the IUDs were correctly identified and located with 3DUS. 2DUS was not nearly as accurate. The IUD was misidentified in eight cases, in six the models were not seen, and in two the position of the device was not identified. Abnormal insertions of IUDs were more accurately demonstrated with 3DUS than with 2DUS. In addition, 3DUS was as accurate as hysteroscopy in identifying abnormal insertions. It is interesting that in two cases, 2DUS suggested that the IUDs were laterally displaced, whereas both 3DUS and hysteroscopy showed that they were inverted. Two partially incarcerated IUDs were identified with 3DUS and hysteroscopy and were missed by 2DUS.

Lee et al. investigated the role of 3DUS visualization of IUDs (Tcu380A) soon after insertion in 96 women (15). They found that all parts of the IUD could be imaged in 95% of cases using 3DUS, 64% were visualized completely using the multiplanar display, and an additional 31% were seen only after volume rendering. Two IUDs had incomplete opening of the two arms of the device, and both were displayed with 3DUS. In another case, the IUD was correctly identified on 3DUS as being displaced in the cervical canal.

Bonilla-Musoles et al. suggest that the transparent maximum–minimum mode of rendering was most useful in identifying IUDs, and Lee et al. suggest than the transparent–maximum mode was best (14,15). Rotation of the rendered images of the IUDs allowed for recognition of the small parts of the IUDs (15).

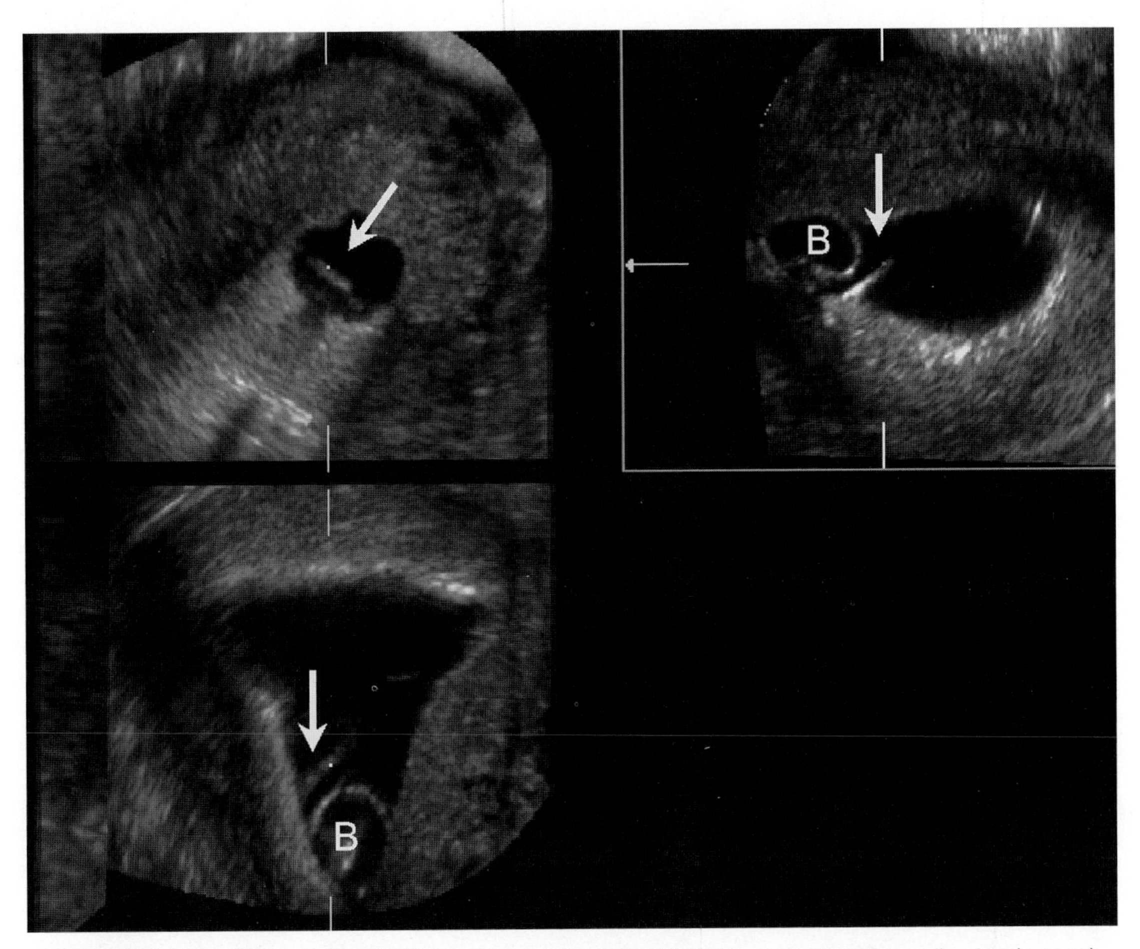

Figure 7-9. Sonohysterography of intrauterine adhesions. Multiplanar images of an adhesion (*arrows*) are shown after injection of saline into the uterine cavity. The balloon (B) from the catheter is also present inferior to the adhesion. (Courtesy of Dr. Anna Lev-Toaff.)

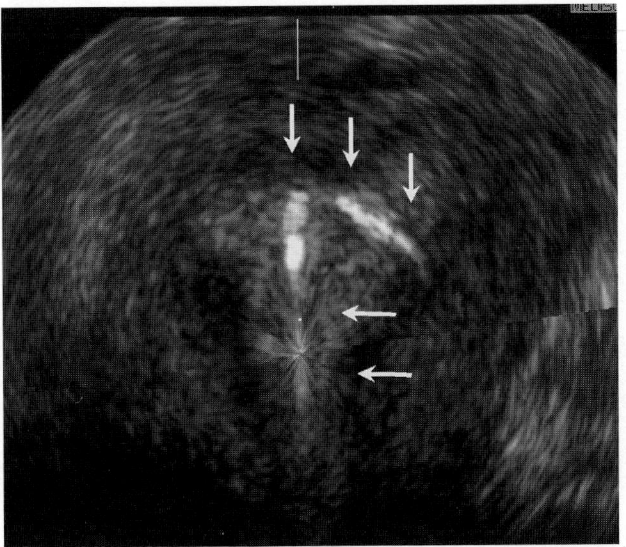

Figure 7-10. Intrauterine device. A ParaGuard intrauterine device is shown in the multiplanar display. The transverse plane shows an image of nearly the entire IUD (*arrows*), which is correctly positioned in the endometrium.

3DUS OF THE ADNEXA

Adnexal Masses

Volume imaging may assist in assessing and displaying tissue characterization within adnexal masses (Figure 7-11). Bonilla-Musoles et al. evaluated 76 women with ovarian masses with both 2DUS and 3DUS and found that 3DUS was superior in evaluating papillary projections, showing characteristics of cystic walls, identifying the extent of capsular infiltration of tumors, and calculating ovarian volume (16). In their study, four of five ovarian malignancies were detected with 2DUS, whereas all five cases were detected with 3DUS. Papillary projections were identified on 3DUS that were missed on 2DUS. Rotation of the volume and planar assessment of the mass in three perpendicular planes allowed for the papillary projection to be identified. We have had a similar experience at UCSD when a mural nodule was seen in an adnexal mass on 3DUS that was not appreciated on 2DUS (Figure 7-12). Bonilla-Musoles et al. also suggest that the meticulous evaluation of masses from any planar orientation may allow for determination of the extent to which a malignant lesion invades the capsule.

In a small series of eight women with adnexal masses, Chan et al. found that 3DUS imaging allowed for better characterization of the mass using a morphologic scoring index developed by Sassone and co-workers (17). The scoring index was judged as a 9 using 2DUS (suggestive of malignancy due to identification of a mural nodule) and as a 13 using 3DUS (more suggestive of malignancy), which was attributed to additional images and views available with volume scanning. Transparent maximum–minimum mode of

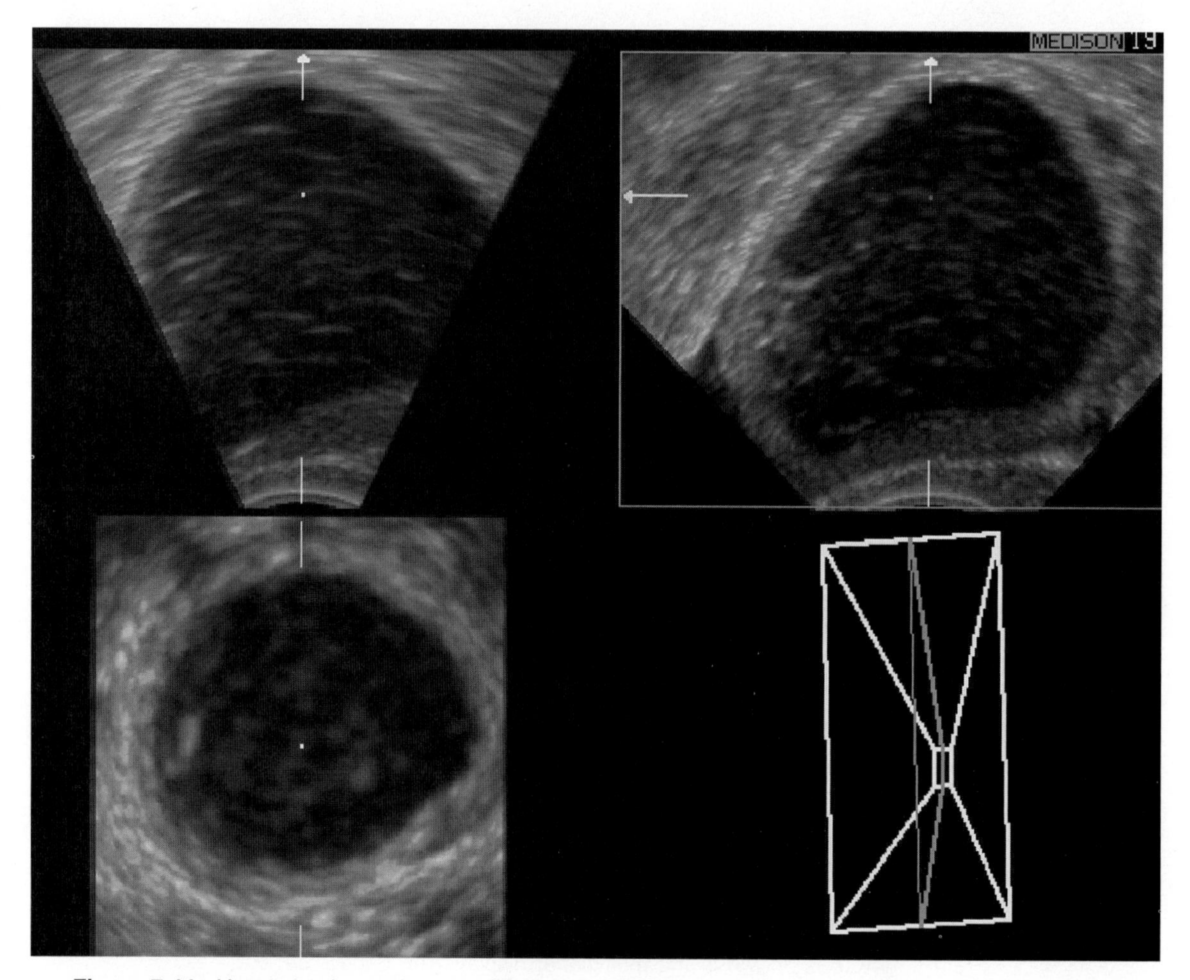

Figure 7-11. Hemorrhagic ovarian cyst. Multiplanar display of hemorrhagic ovarian cyst with echoes inside the cyst.

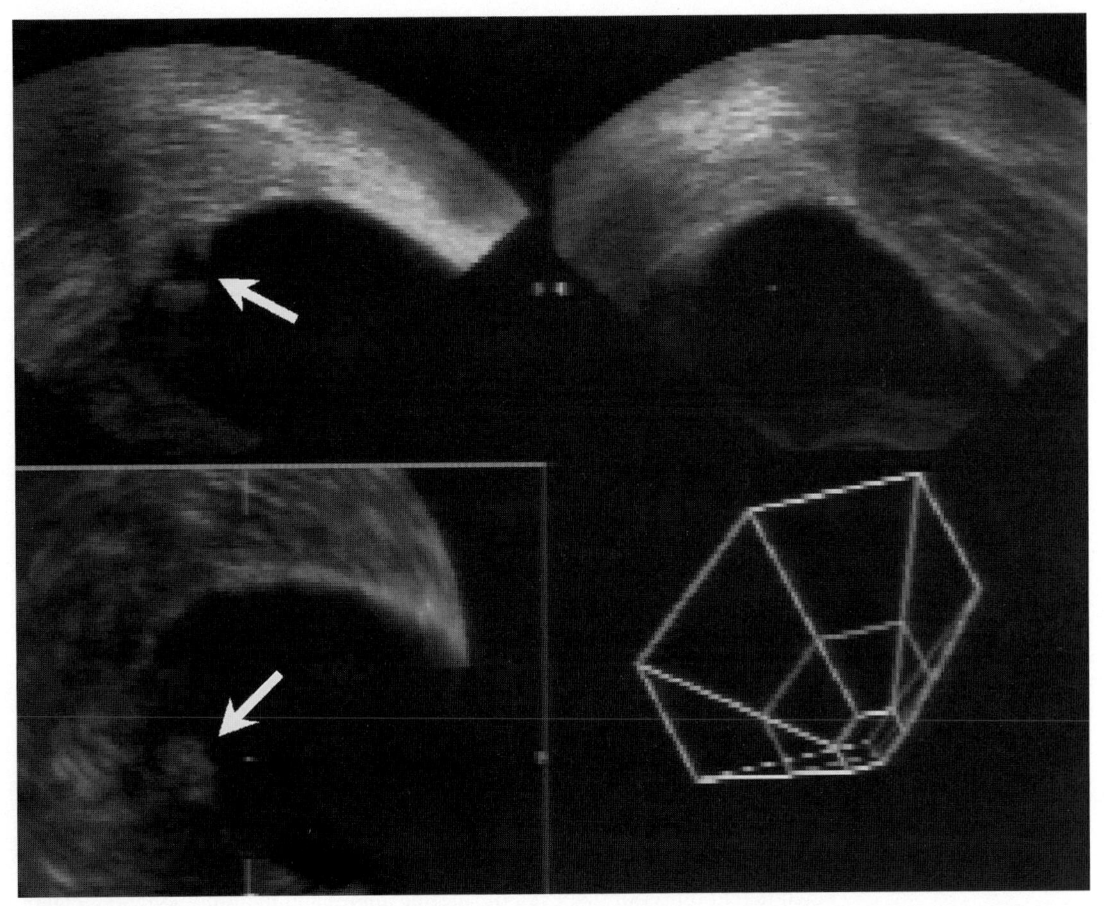

Figure 7-12. Mural nodule in adnexal mass. Multiplanar display of cystic mass with mural nodule. The mural nodule (*arrow*) was seen only as thickening on 2DUS. The mass was a benign serous adenofibroma on pathology.

surface rendering may be useful in evaluating dermoid tumors, because it makes it easier to identify teeth, bones, and calcification (16).

Infertility Assessment

3DUS is also valuable in evaluating ovarian enlargement from multiple follicles and lutein cysts. Bonilla suggests that 3DUS is helpful in predicting ovarian hyperstimulation following hCG administration because of the greater capacity of 3DUS to evaluate the growth of small and intermediate follicles (Figure 7-13) (16). It also has been reported that 3DUS can assist in differentiating follicular cysts from endometriotic cysts (9).

Follicular volume measurements are also important in managing infertility patients safely and effectively. Optimal follicular volumes for oocyte recovery, fertilization, and cleavage are desired and ultrasound can be used for cycle monitoring without hormonal estimations (18). 3DUS allows for accurate measurement of irregularly shaped follicles, which are often distorted by adjacent follicles. Kyei-Mensah et al. measured follicles with both 2DUS and 3DUS and correlated them with the volume of follicular aspirate. They found that the volume was more accurately determined by 3DUS (+0.96 to −0.43 ml) compared with 2DUS (+3.47 to −2.42 ml). They suggested that additional studies are needed to determine the optimal follicular volumes for the best outcome of IVF-ET treatment cycles.

3DUS also is useful in providing guidance for follicular aspiration and transfer of *in vitro* fertilized embryos. The ability of 3DUS to scroll through a volume in multiple planes simultaneously permits the exact location of the needle or air bubbles to be identified.

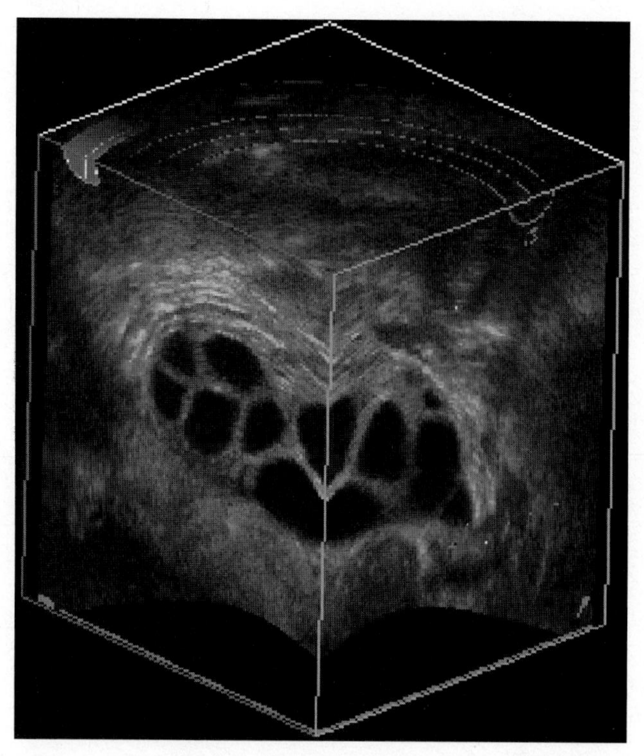

Figure 7-13. Hyperstimulation of ovary with multiple follicular cysts.

Feichtinger points out that even the uterine angles and the fallopian tubes can be evaluated to a greater extent than with 2DUS (9). He also suggests that laser surgery under sonographic 3D control may replace several procedures currently performed by hysteroscopic and laparoscopic surgery (19).

3DUS may be used to evaluate the follicular tubes for patency. The entire tube can be evaluated as the data are collected for the volume, rather than a slice, and reviewed later from any arbitrary plane. 2DUS requires freezing of the image; thus the dynamic process of bubbles passing out a tube may be overlooked (4).

Ectopic Pregnancies

Feichtinger suggests that 3DUS should assist in diagnosing ectopic pregnancies, both in the tubes and in the interstitial regions of the uterus. Evaluation using the three perpendicular planes from the volume allows for more comprehensive evaluation of the adnexa and the uterus. 3DUS can also be used for accurate guidance of the needle for treatment of ectopic pregnancies with methotrexate (9).

HOW TO DO IT
Practical Clinical Tips for Gynecology

- Volume data can be acquired using either transabdominal or transvaginal probes:
  - Transvaginal probes produce higher-quality data for measurements.
- Prescan with 2DUS to identify a region of interest:
  - Identify evidence of peristalsis in bowel adjacent to uterus and adnexa.
    - No movement will be seen in 3DUS study.
- Evaluate premenopausal patients in luteal phase of menstrual cycle if endometrium is important.

HOW TO DO IT, *continued*

- Secretory endometrium is much more prominent during luteal phase (day 14–28).
- Uterine volume data may be acquired through the entire field of view or a subvolume:
  - Optimally acquired from the midline longitudinal plane of the uterus.
- Body of uterus generally can be imaged in one volume:
  - Assessment of cervix may require second volume.
- Uterine anomalies such as uterine didelphis may require two volumes.
- Measure endometrial volumes on longitudinal sections of uterus at 1–2 mm intervals:
  - Requires approximately 3–5 minutes.
  - Follicles with a mean diameter of <10 mm may not be measured accurately.
  - Automatic segmentation using edge detection techniques show promise.
- Display uterus in a standard anatomic orientation:
  - Use coronal, sagittal, and transverse planes.
  - Transverse plane rarely obtainable with 2DUS methods.
  - Very important for assessing uterine congenital anomalies.
- Identification of uterine anatomy can be difficult in some cases:
  - If uterus is arcuate.
  - If uterus appears symmetrical from right to left, and also right to cervix and left to cervix.
- Help diagram may assist anatomic identification.
- Experience with 3DUS equipment facilitates rapid orientation of study in most patients.
- Limited (but important) role for volume rendering in evaluation of the pelvis:
  - 3DUS hysterosonography provides comprehensive view of uterine anatomy surface area:
    - May assist in evaluating an IUD location.
    - Assesses vascularity of pelvic masses.
- Increased transmission behind cystic structures may be hard to see in reconstructed planes:
  - Ovarian cysts may be difficult to see or absent in reconstructed planes.
- Evaluate entire volume to determine whether increased through transmission exists.
  - Evaluate anechoic/hypoechoic masses in original plane of acquisition.

REFERENCES

1. Jurkovic D, Geipel A, Gruboeck K, et al. Three-dimensional ultrasound for the assessment of uterine anatomy and detection of congenital anomalies: a comparison with hysterosalpingography and two-dimensional sonography. *Ultrasound Obstet Gynecol* 1995;5:233–237.
2. Acien P. Reproductive performance of women with uterine malformations. *Hum Reprod* 1993;8:122–126.
3. Raga F, Bonilla-Musoles F, Blanes J, Osborne NG. Congenital Mullerian anomalies: diagnostic accuracy of three-dimensional ultrasound. *Fertil Steril* 1996;65(3):523–528.
4. Steiner H, Staudach A, Spitzer D, Schaffer H. Three-dimensional ultrasound in obstetrics and gynaecology: technique, possibilities and limitations. *Hum Reprod* 1994;9:1773–1778.
5. Weinraub Z, Maymon R, Shulman A, et al. Three-dimensional saline contrast hysterosonography and surface rendering of uterine cavity pathology. *Ultrasound Obstet Gynecol* 1996;8(4):277–282.
6. Gruboeck K, Jurkovic D, Lawton F, Savvas M, Tailor A, Campbell S. The diagnostic value of endometrial thickness and volume measurements by three-dimensional ultrasound in patients with postmenopausal bleeding. *Ultrasound Obstet Gynecol* 1996;8(4):272–276.
7. Olstad B, Berg S, Torp AH, Schipper KP, Eik-Nes SH. 3D transvaginal ultrasound imaging for identification of endometrial abnormality. Proceedings SPIE medical imaging. *Phys Med Imaging* 1995;2432(S):543–553.
8. Bonilla-Musoles F, Raga F, Osborne N, Blanes J, Coelho F. Three-dimensional hysterosonography for the study of endometrial tumors: comparison with conventional transvaginal sonography, hystersalpingography, and hysteroscopy. *Gynecol Oncol* 1997;65:245–252.
9. Feichtinger W. Transvaginal three-dimensional imaging. *Ultrasound Obstet Gynecol* 1993;3:375–378.

10. Balen FG, Allen CM, Gardener JE, Siddle NC, Lees WR. 3-Dimensional reconstruction of ultrasound images of the uterine cavity. *Br J Radiol* 1993;66(787):588–591.
11. Lev-Toaff AS, Rawool NM, Kurtz, AB, Forssberg F, Goldberg BB. Three-dimensional sonography and 3D transvaginal US: a problem-solving tool in complex gynecological cases. *Radiology* 1996;201(P):384.
12. Liess H, Roth C, Umgelter A, et al. Improvements in volumetric quantification of circumscribed hepatic lesions by three-dimensional sonography. *Zeitschrift fur Gastroenterologie* 1994;32(9):488–492.
13. Riccabona M, Nelson TR, Pretorius DH, Davidson TE. *In vivo* three-dimensional sonographic measurement of organ volume: validation in the urinary bladder. *J Ultrasound Med* 1996;15:637–632.
14. Bonilla-Musoles F, Raga F, Osborne NG, Blanes J. Control of intrauterine device insertion with three-dimensional ultrasound: is it the future? *J Clin Ultrasound* 1996;24(5):263–267.
15. Lee A, Eppel W, Sam C, Kratochwil A, Deutinger J, Bernaschek G. Intrauterine device localization by three-dimensional transvaginal sonography. *Ultrasound Obstet Gynecol* 1997;10(4):289–292.
16. Bonilla-Musoles F, Raga F, Osborne NG. Three-dimensional ultrasound evaluation of ovarian masses. *Gynecol Oncol* 1995;59:129–135.
17. Chan L, Lin WM, Uerpairojkit B, Hartman D, Reece EA, Helm W. Evaluation of adnexal masses using three-dimensional ultrasonographic technology: preliminary report. *J Ultrasound Med* 1997;16(5):349–354.
18. Kyei-Mensah A, Zaidi J, Pittrof R, Shaker A, Campbell S, Tan SL. Transvaginal three-dimensional ultrasound: accuracy of follicular volume measurements. *Fertil Steril* 1996;65(2):371–376.
19. Feichtinger W, Strohmer H, Feldner-Busztin M. Laser surgery under sonographic control: preliminary experimental investigations. *Ultrasound Obstet Gynecol* 1993;3:264–267.

8

Genitourinary System

OVERVIEW

This chapter reviews the current clinical utility and limitations of three-dimensional ultrasound (3DUS) imaging in the genitourinary tract, including the kidney, ureters and urethra, bladder, prostate, scrotum, penis, and seminal vesicles but excluding gynecologic imaging, which is discussed in Chapter 7. Potential future benefits of the technology also are discussed.

KEY CONCEPTS

- 3DUS of the genitourinary (GU) tract is still at an early stage of development.
- 3DUS images obtained with two-dimensional ultrasound (2DUS) intraluminal and intracavity scans are usually of higher quality when they are obtained from the skin surface because:
  - Higher-frequency transducers may be used.
  - Structures being evaluated usually are closer to the transducer.
  - Interference from bowel gas, fat, and bones is limited.
  - Structures being evaluated are usually smaller and less technically challenging.
  - Motion usually is less of a problem.
  - Intracavitary transducers are usually mechanically driven.
- 3DUS provides more accurate and more repeatable volumetric assessment, which improves clinical utility in assessing diseases of the prostate, kidneys, and bladder.
- 3DUS provides perspectives unavailable with 2DUS that assist in making diagnoses and guiding interventional procedures.

2DUS has a well-established clinical utility in evaluating diseases of the kidneys, the bladder, and the male and female reproductive organs. It also has a developing role in assessing the male and female urethras and the ureters. Early studies suggest that 3DUS will allow more accurate and repeatable measurements of GU structures, which may improve our understanding of complex pathologies and help in planning and performing interventional procedures. This chapter will discuss the role of 3DUS in the evaluation of each GU organ in turn, rather than considering the GU tract as a whole.

3DUS OF THE KIDNEY

The clinical value of 3DUS in evaluation of the kidney is under investigation. Some studies have shown that 3DUS measurement of kidney volume is more accurate and more reproducible than 2DUS, which may help in the diagnosis and follow-up of several diseases of the kidney that rely on accurate volumetric assessments for their management. It also may be useful as a preoperative planning tool for patients undergoing kidney-sparing surgeries and for assessing rejection in renal transplants.

Real-time 2DUS excels at demonstrating most renal pathologies in a rapid, cost-effective, painless, and noninvasive manner (1–3). It is excellent for the detection and characterization of intrarenal masses, the depiction of diffuse renal disease, and the measurement of renal tissue. In addition, it has a 98% sensitivity in the detection of pelvicaliectasis (2), it accurately guides biopsies and other renal interventions (4,5), and it has an increasingly important role in diagnosing renal vascular disorders (6). It is the imaging modality of choice for most pediatric urinary tract problems (7) and is increasingly assuming that role in many adult centers. The advantages of 3DUS over the standard intravenous urogram (IVU) include little patient preparation, lower costs, no radiation, noninvasive and improved detection of intrarenal masses, and improved bladder assessment (1).

Obtaining high-quality 3DUS Doppler exams of the kidney is also challenging, given the generally lower frame rate that arises from most machines once they are put in Doppler mode, and the poor signal-to-noise of Doppler signal traveling deep within the body. As technology continues to improve (Figure 8-1), vascular evaluation with 3DUS will certainly become an important part of the diagnostic work-up. Typically, transplant kidneys are easier to evaluate than native kidneys.

CLINICAL BENEFITS OF 3DUS

Understanding of Complex Anatomy

Although the official policy of the American Urological Association is to perform a radical nephrectomy on all proven renal cell carcinomas regardless of size, smaller tumors can be successfully cured with less radical surgery (8,9). 3DUS imaging has proved helpful in showing complex anatomy in an easily understandable way to surgeons (Figures 8-2, 8-3) (10,11). The complex relationship of some smaller masses to the remainder of the kidney would probably be better appreciated by surgeons in three dimensions, although this has not yet been proved. This would be of most benefit in imaging tumors in solitary kidneys (12).

Renal Volume Measurement

In vivo and *in vitro* studies have shown that 3DUS measures the kidney volume more accurately and more reproducibly than two-dimensional (2D) assessments (13–16). Gilja et al. showed excellent correlation (r = .998) between 3DUS measurements and true measurements in studies of resected porcine kidneys and stomachs (13). Similar findings have also been obtained with cadaver kidneys (15) and with *in vivo* comparisons with the volumes as assessed by magnetic resonance imaging (MRI) (14).

Accurately assessing the volume of the renal parenchyma helps diagnosis in a variety of clinical circumstances. It determines if a patient presenting with renal failure has acute or chronic disease, a crucial diagnosis in ensuring that the patient gets appropriate care (17). It also helps in the follow-up of people with diffuse renal disease like lymphoma. Pruthi found ultrasound area assessments of renal parenchymal volume in children with

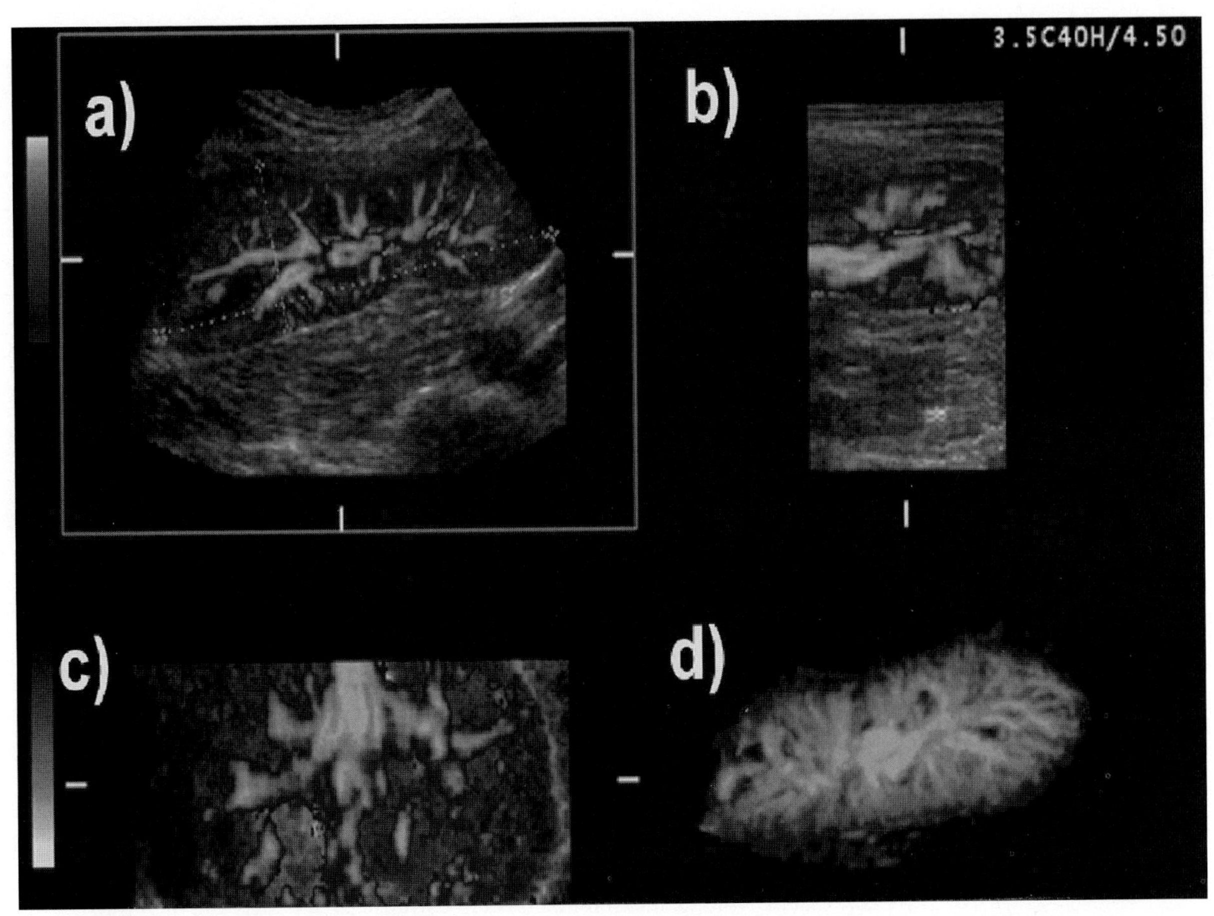

Figure 8-1. 3DUS power Doppler image kidney of a normal volunteer. (**a**) Longitudinal, (**b**) transaxial, (**c**) coronal, (**d**) maximum-intensity projection (MIP) mode power Doppler rendered image. (Image courtesy of Siemens Ultrasound, Issaquah, WA.) See color version of figure.

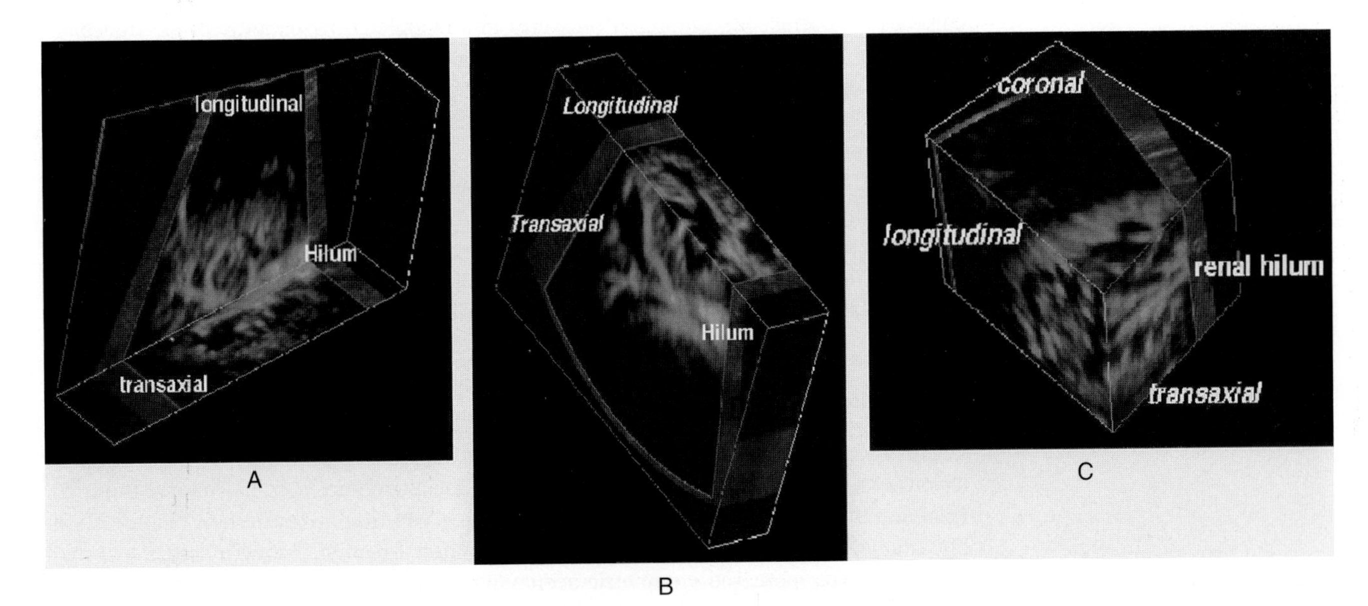

Figure 8-2. Three different perspectives of 3DUS power Doppler imaging of a normal kidney without contrast enhancement.

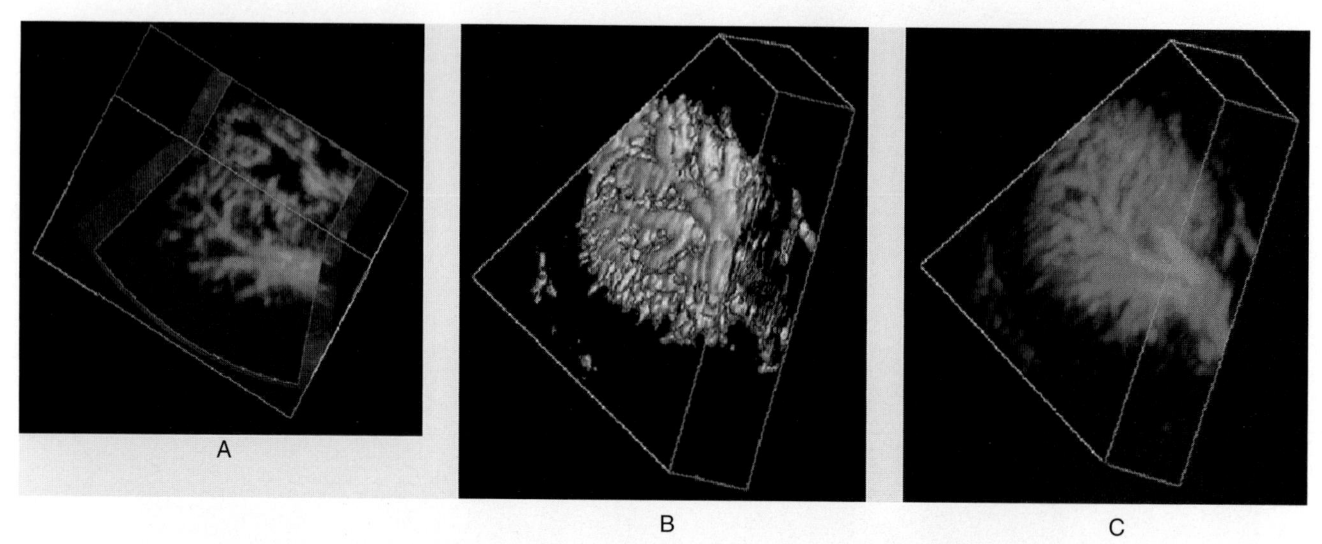

Figure 8-3. A: Linear translation 3DUS power Doppler multiplanar format view of the right kidney of a 26-year-old volunteer. A multiplanar reconstruction showing several orders of arterial branching in the kidney. **B:** Surface rendering. Some shading has been added to aid visual detection of the contour of the vessels within the kidney. **C:** Volume rendering—maximum-intensity projecting following ray casting. The maximum intensity in this projection is displayed. (With permission from Ref. 98.)

grade 4 or 5 primary vesicoureteric reflux (VUR) correlate well with differential renal function assessments, whereas single-plane ultrasound measurements did not (18). They suggest that area assessment may be a sensitive way of following these children. It is likely that 3D volume measurement would correlate even better, and therefore prove more clinically useful.

Much work has been done on early diagnosis of renal transplant rejection (19,20). Changes occur in both the renal anatomic appearance and the pattern of blood flow in the kidney during rejection. Absy showed that the transplant kidney volume correlated positively with renal function (19). His group also showed that renal volume increased with acute rejection and returned to normal after successful treatment (19). Acute rejection episodes also affect the intrarenal pressures and Doppler flow patterns in the transplant kidney (5). This relationship is much more complicated than was originally thought (20), and Doppler evaluation alone rarely obviates the need for a biopsy in a nonobstructed transplant kidney displaying deteriorating function (21). 3DUS offers the potential to combine volumetric data with Doppler perfusion data. This might produce a more clinically useful test, possibly with ultrasound contrast material.

3DUS also improves visibility of the contour of anatomic structures (22). The combination of these two abilities will probably improve radiologic diagnosis, as most popular diagnostic algorithms used in diagnosing renal diseases are based on three criteria: renal volumes, laterality of disease, and appearance of the renal contours (23).

Focal Renal Lesions

Cross-sectional imaging is increasingly diagnosing smaller and smaller noncystic renal lesions (8). Their management is complex, as not all require surgical intervention (8). Knowing the exact size of some of these masses is very important in their classification. For example, renal adenomas are currently considered benign if they measure less than 2 cm (24). The increased volumetric accuracy of 3DUS may help in their diagnosis. As most of these tiny lesions are reimaged at standardized intervals, the improved precision of 3DUS over 2DUS may be valuable. 3DUS power Doppler also may assist in identifying abnormal vessels within or tracking alongside renal masses.

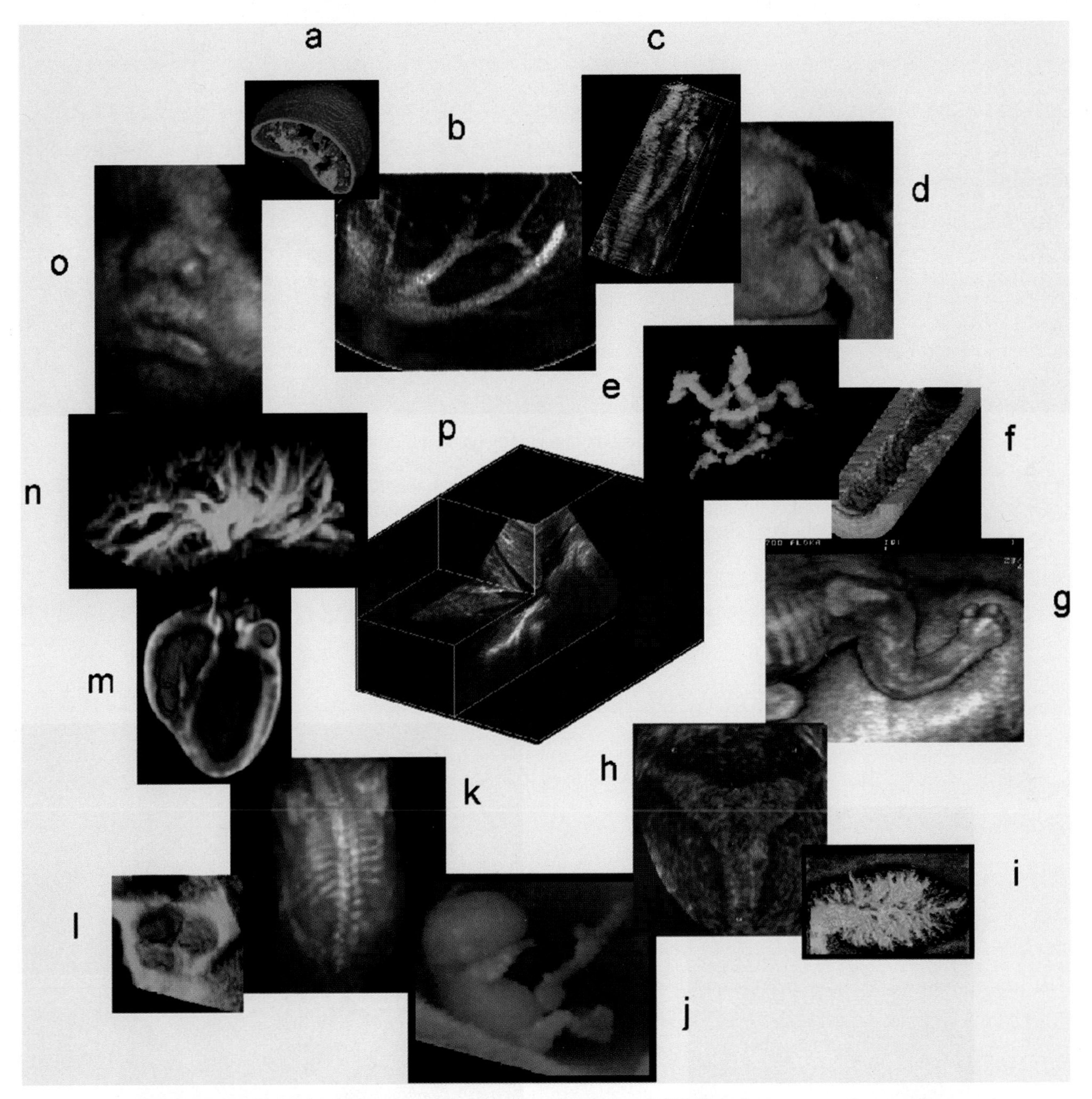

Figure 1–2. Collage demonstrating several applications of 3DUS to image vascular, cardiac, and obstetric patients. a, prostate cancer; b, liver vasculature; c, carotid arteries; d, fetal face; e, adult circle of Willis; f, coronary artery; g, fetal skeleton; h, uterus; i, lymph node; j, 10-week embryo; k, fetal spine; l, aortic valve; m, dog heart; n, kidney vasculature; o, fetal face; p, liver anatomy. See black and white version of figure in text.

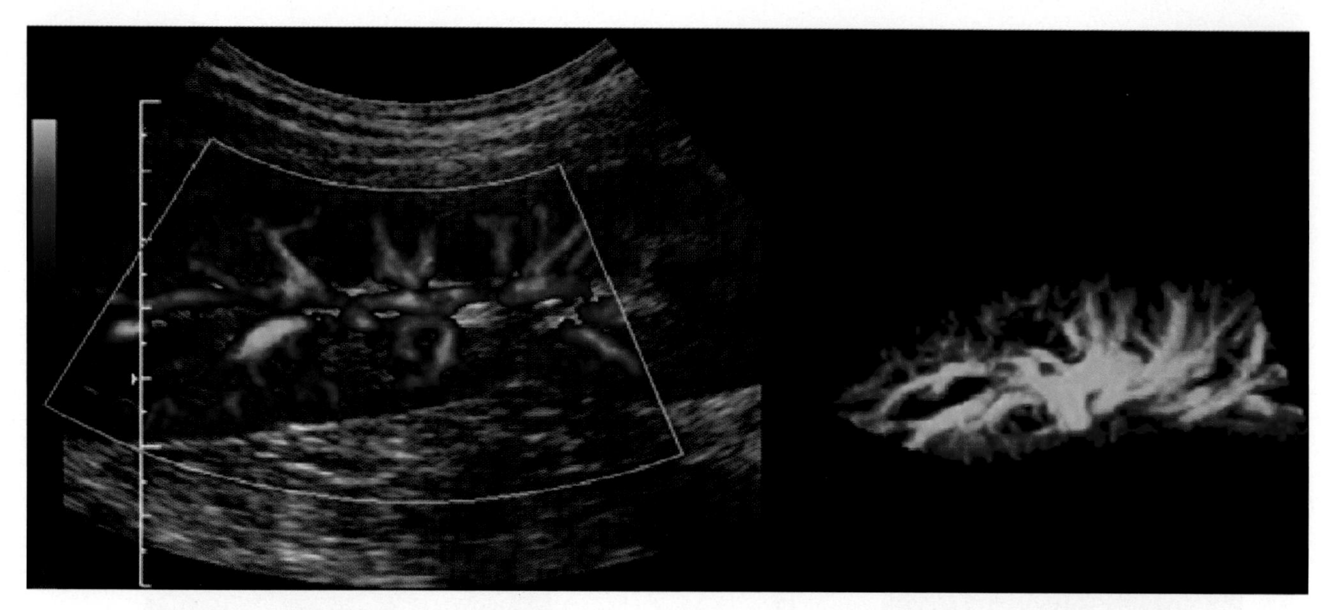

Figure 2–10. 3DUS power Doppler acquisition of a kidney using image-based position sensing to geometrically register 2DUS slices during real-time acquisition. Acquisitions can be performed with a linear, tilt, or rotational technique. The right image is one frame from the 2DUS acquisition. The left frame is from the volume data after segmentation based on the power Doppler signal. (Courtesy of Siemens Ultrasound, Issaquah, WA.) See black and white version of figure in text.

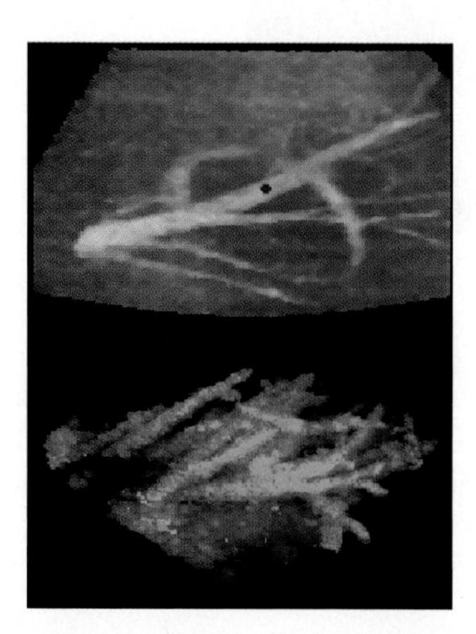

Figure 3–1. Visualizing 3DUS data. Application of volume-rendering methods to reveal internal structure of the liver; the upper image is inverted minimum-intensity projection; the lower image is maximum-intensity projection after segmenting using the power Doppler signal. See black and white version of figure in text.

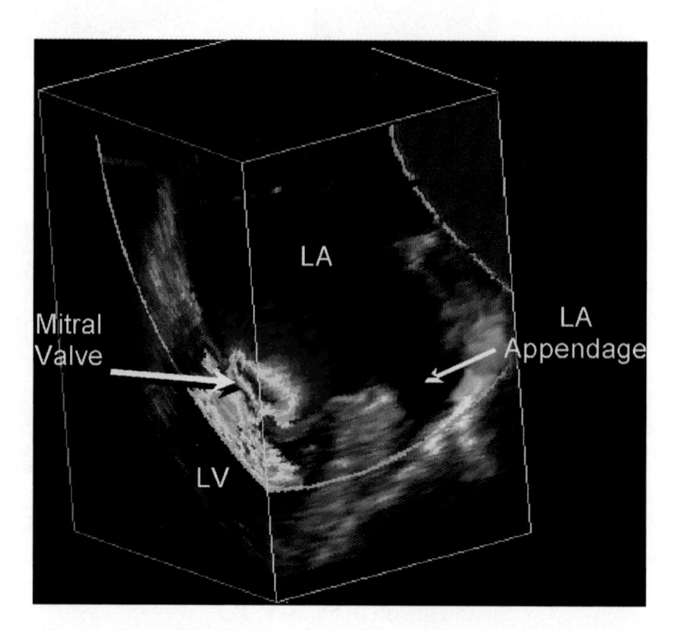

Figure 3–3. 3DUS color Doppler echocardiographic study of a patient with mitral stenosis. Cube view color Doppler data in the image shows some mild mitral regurgitation. Image LA, left atrium; LV, left ventricle. (Image courtesy of Dr. D. Boughner.) See black and white version of figure in text.

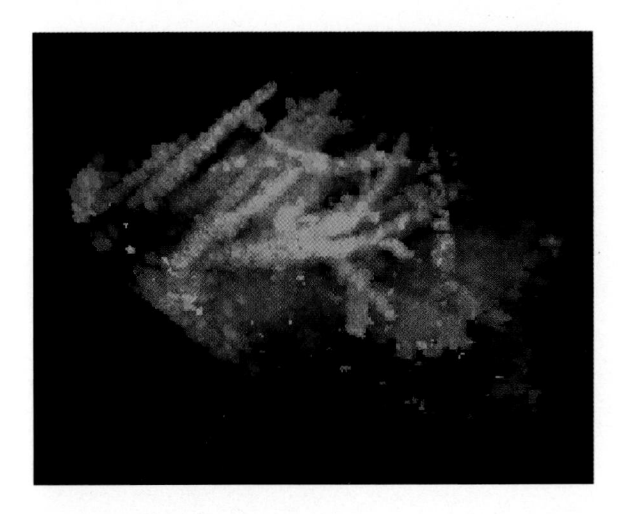

Figure 3–6. Segmentation of vessels from the gray-scale data based on the power Doppler signal. Volume rendering of the flow signal produces a clear picture of the continuity of the vessel flows within the liver. See black and white version of figure in text.

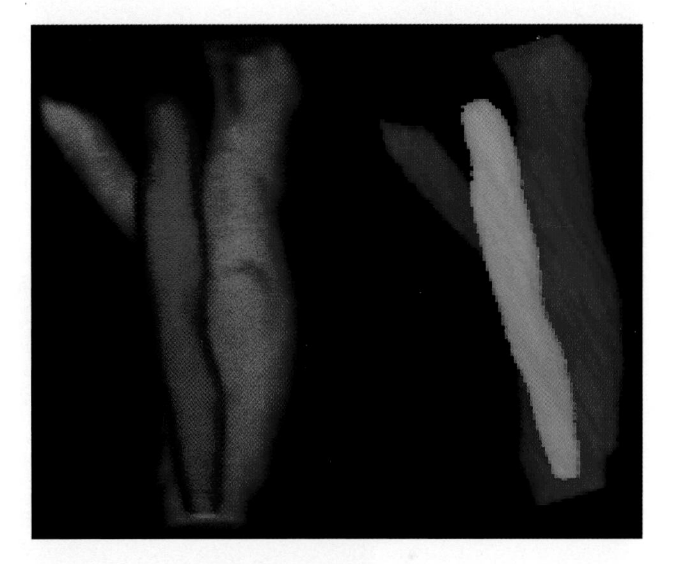

Figure 3–7. Volume data from a velocity Doppler study of the carotid artery. The left image is a surface-rendered image based on a defined threshold for the color velocity data. The right image is a volume-rendered, ray-cast image of the same data. See black and white version of figure in text.

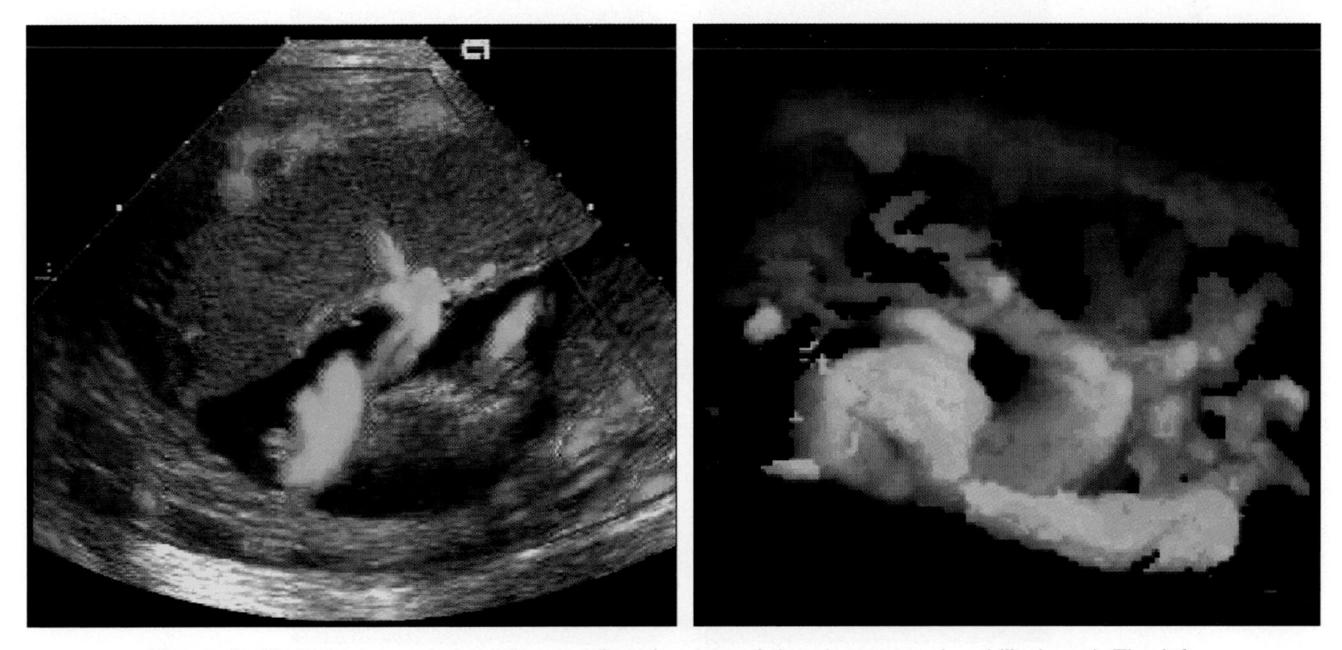

Figure 3–10. Volume rendering of power Doppler data of the placenta and umbilical cord. The left image is a single slice from the 2DUS acquisition. The right image is a volume-rendered image of the vascular anatomy. The gray-scale signal has been removed. The rendering is a modified maximum-intensity method with depth coding. See black and white version of figure in text.

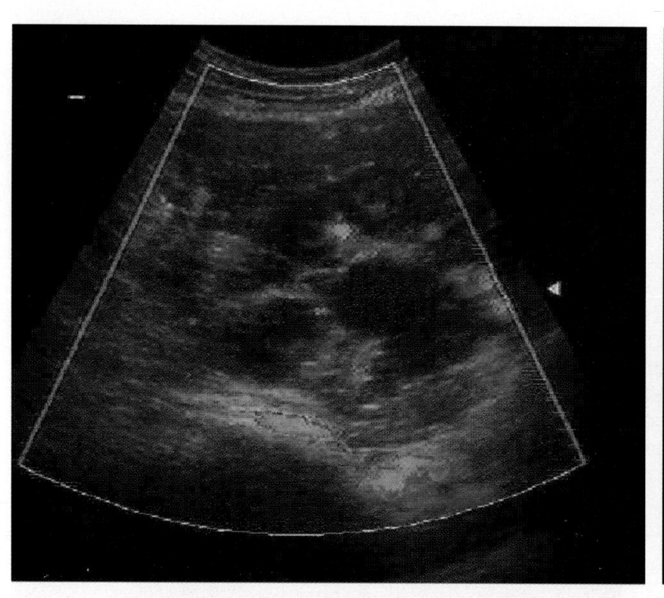

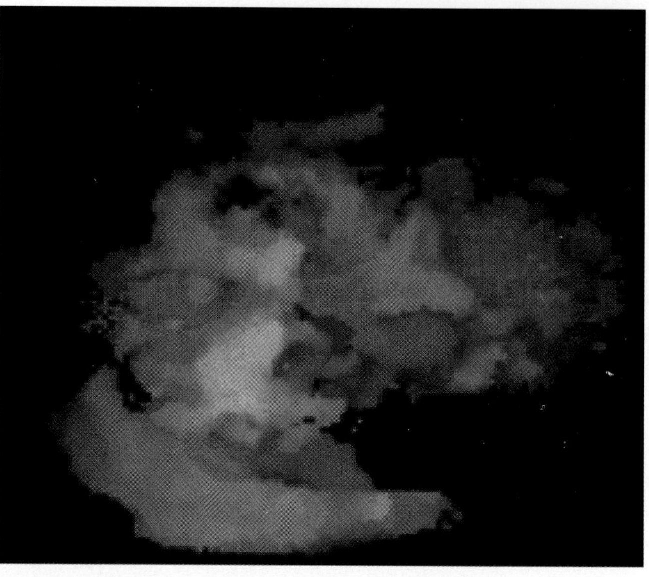

Figure 3–11. Volume rendering of power Doppler data of a polycystic kidney. The left image is a single slice from the 2DUS acquisition. The signal voids represent the cysts in the kidney. The right image is a volume-rendered image segmented using both the power Doppler signal from the vascular anatomy and the signal void from the cysts. Both methods are a modified maximum-intensity method with depth coding. Note that the relative position of the cysts and vessels is clearly shown in the volume-rendered image using two different segmentation methods on the same data. See black and white version of figure in text.

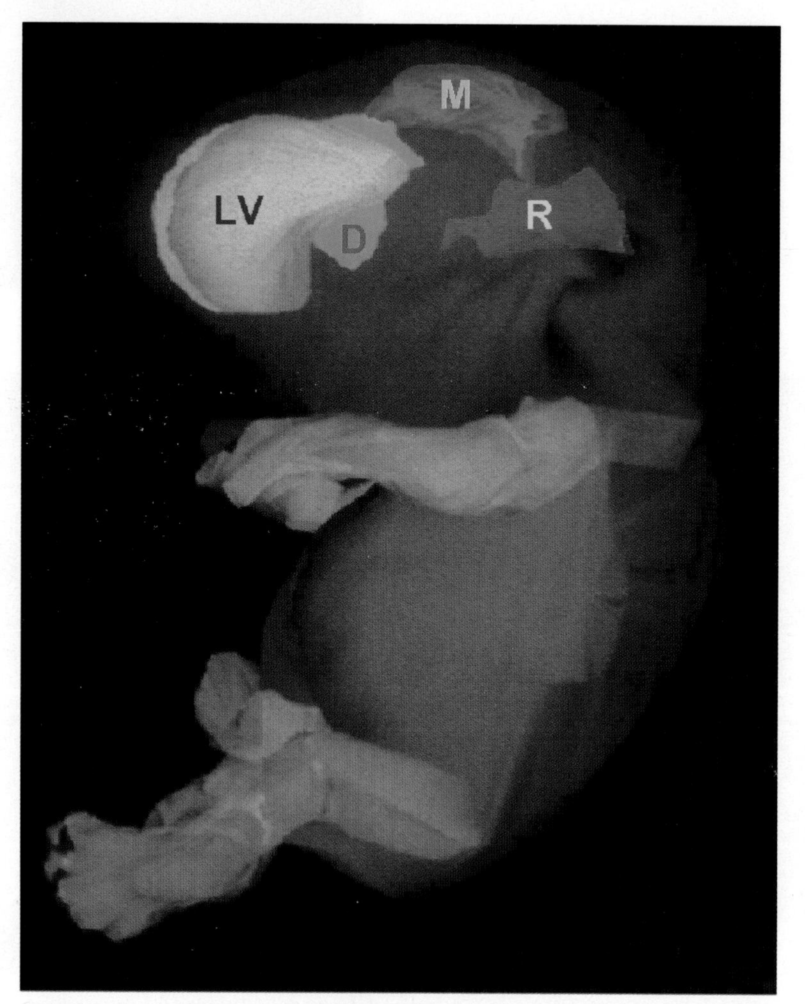

Figure 6–4. Embryo at 9 weeks, 6 days gestational age (29 mm crown-rump length). Surface rendering of the intraventricular cavities is displayed, which may assist in understanding embryologic development. LV, lateral ventricles (yellow); D, diencephalon (green); M, mesencephalon (red); R, rhombencephalon (blue). (With permission from Dr. H. Blaas.) See black and white version of figure in text.

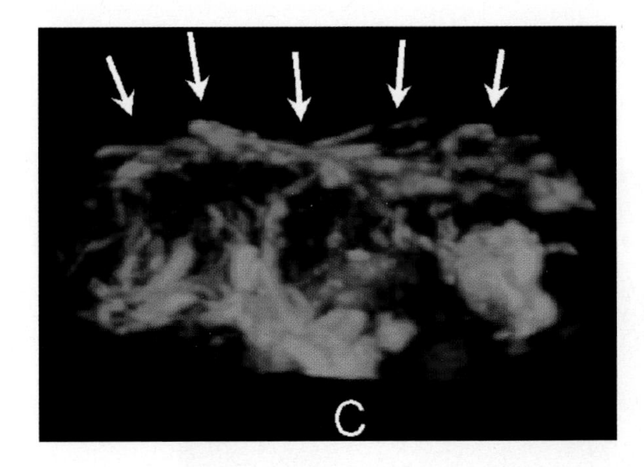

Figure 6–26. Figure 6-26. Volume-rendered image of placental blood supply acquired using power Doppler. The cord insertion (C) is seen entering the placenta. Vessels are seen throughout the placenta and along the maternal surface (arrows). (With permission from Siemens.) See black and white version of figure in text.

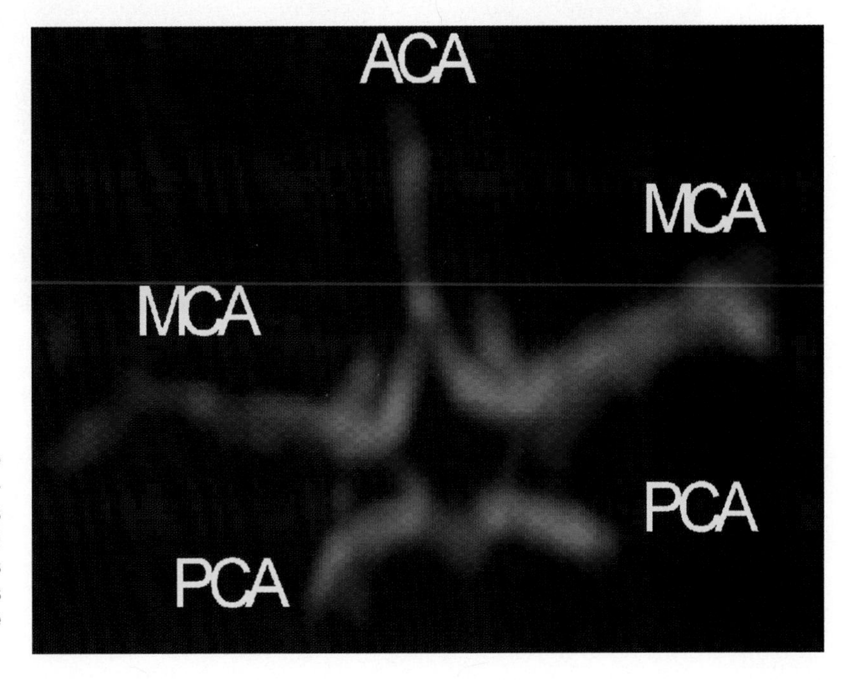

Figure 6–27. Volume-rendered image of the circle of Willis in a fetus at 31 weeks gestational age. The vessels of the circle of Willis are shown, including the anterior cerebral arteries (ACA), the middle cerebral arteries (MCA), and the posterior cerebral arteries (PCA). (With permission from ATL.) See black and white version of figure in text.

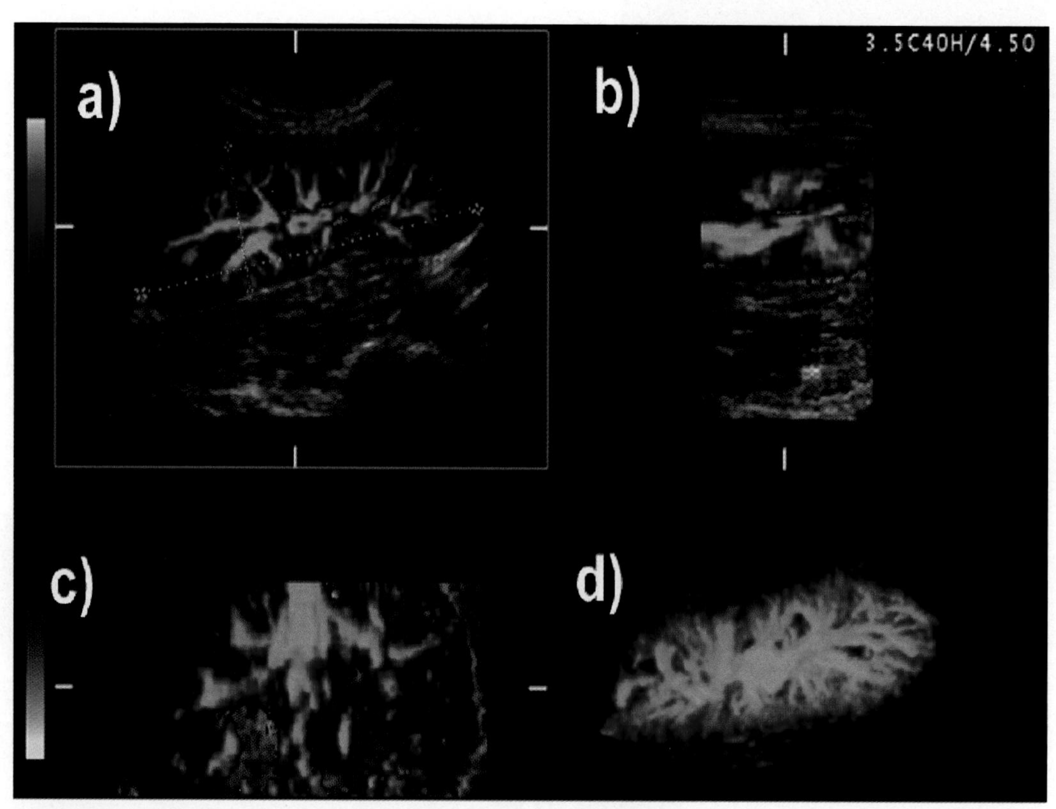

Figure 8–1. 3DUS power Doppler image kidney of a normal volunteer. (a) Longitudinal, (b) transaxial, (c) coronal, (d) maximum-intensity projection (MIP) mode power Doppler rendered image. (Image courtesy of Siemens Ultrasound, Issaquah, WA.) See black and white version of figure in text.

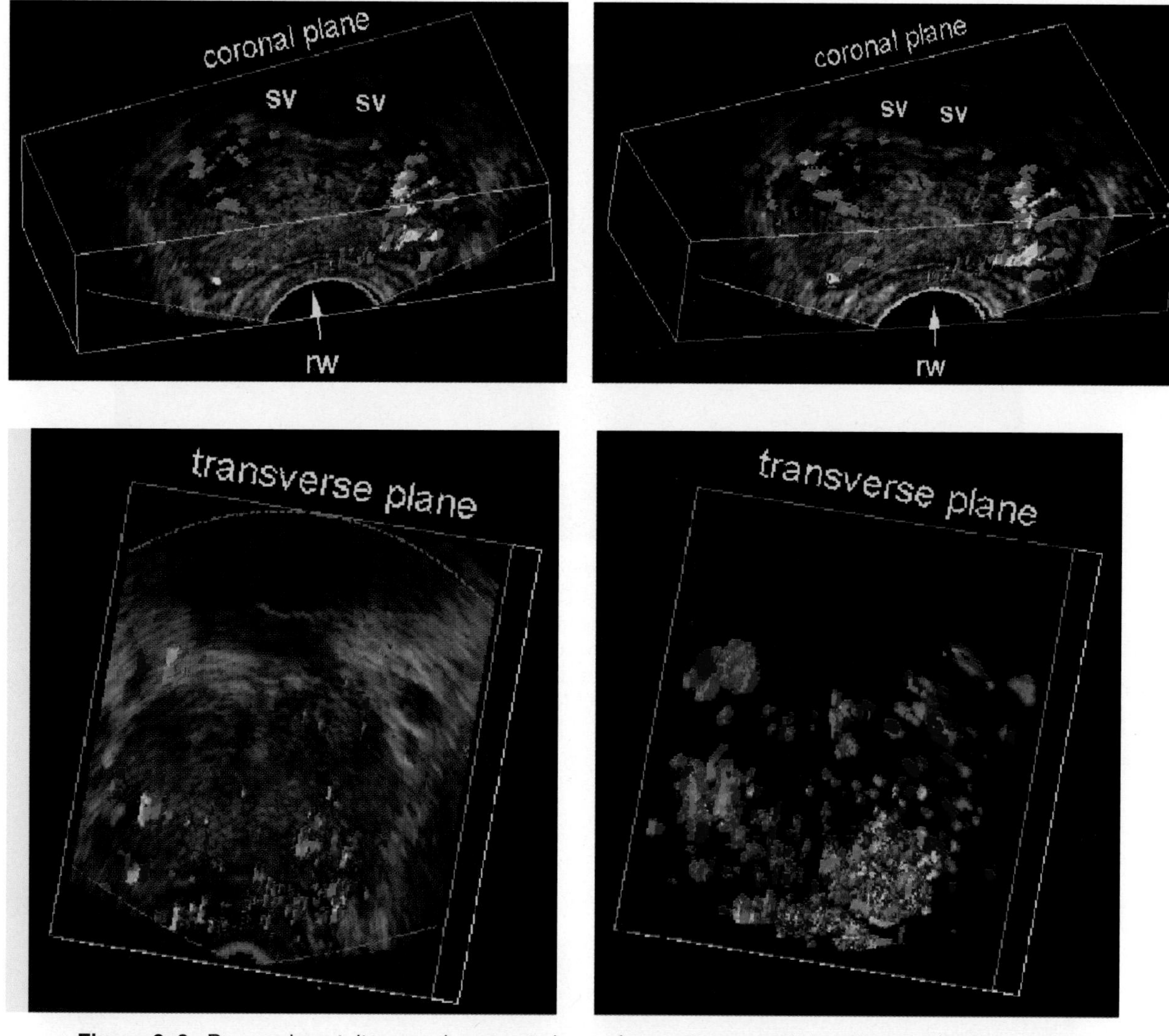

Figure 8–9. Pre- and postultrasound contrast views of a prostate containing Gleason Grade 6 cancer at the right base. Precontrast image shows vascular prominence at the right base where the cancer is located. On an equivalent image obtained several minutes following injection an increased Doppler signal is seen from vessels in both the cancer and the remainder of the prostate. A thin-section transverse plane B mode and color Doppler image of the same prostate obtained slightly more caudally in the prostate. Note the slight decrease in echogenicity and the increased vascularity of the cancer at the right base. Equivalent projected image with the B-mode removed and the color Doppler signal summed. Note the vascularity is markedly increased in the region of the tumor. sv, seminal vesicles; rw, rectal wall. (Echogen Contrast courtesy of Sonus Pharmaceuticals.) See black and white version of figure in text.

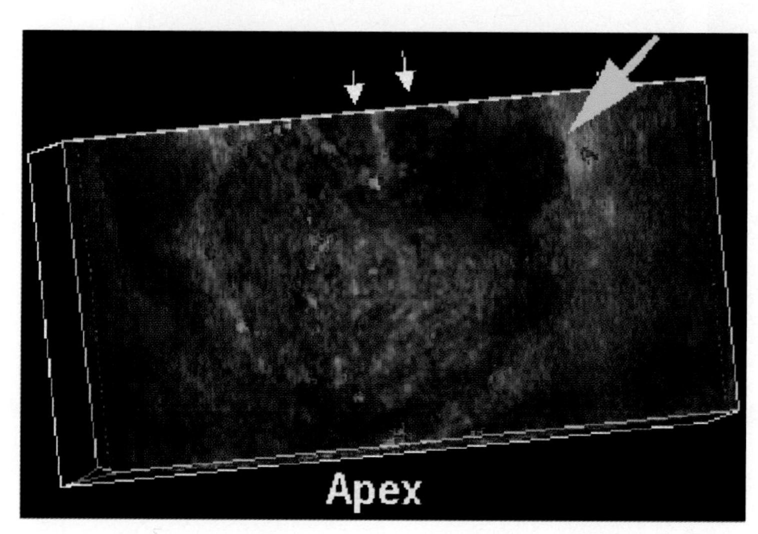

Figure 8–11. Coronal view of left base prostate cancer. An extensive hypoechoic cancer that extends along the left lateral aspect of the prostate toward the apex is clearly shown (large arrows). Ejaculatory ducts are noted (small arrows). The tumor clearly extends outside of the prostate. The tumor is mildly more vascular than the remainder of the prostate. See black and white version of figure in text.

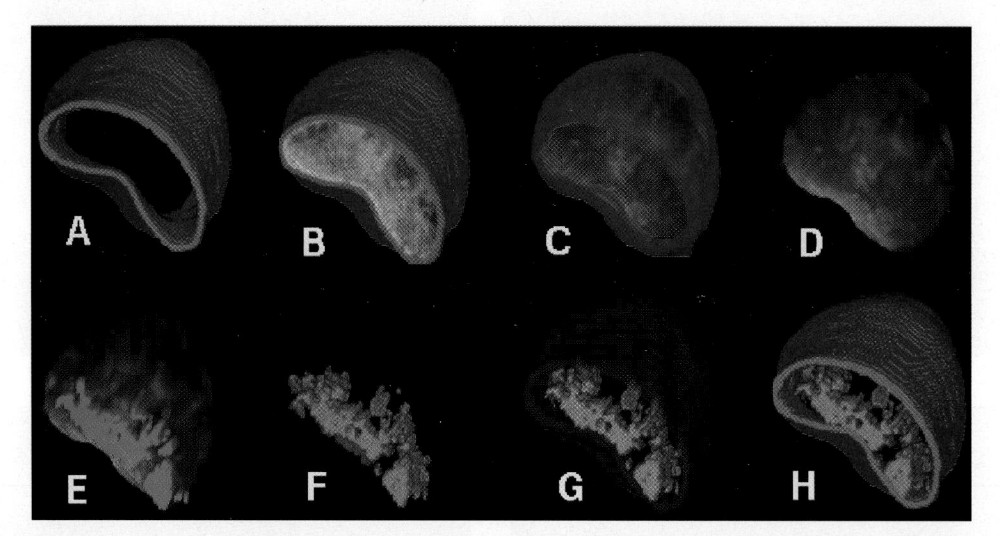

Figure 8–13. Eight 3D renderings of a biopsy proven cancerous prostate viewed from the apex. (A) Opaque capsule; (B) parenchyma within capsule; (C) transparent parenchyma with capsule; (D) transparent parenchyma without capsule; (E) parenchyma with cancerous lesions coded in red; (F) cancerous lesions alone; (G) cancerous lesions with transparent capsule; and (H) cancerous lesions with opaque capsule. The analysis was automatic after identification of capsule boundaries. (With permission from Ref. 94.) See black and white version of figure in text.

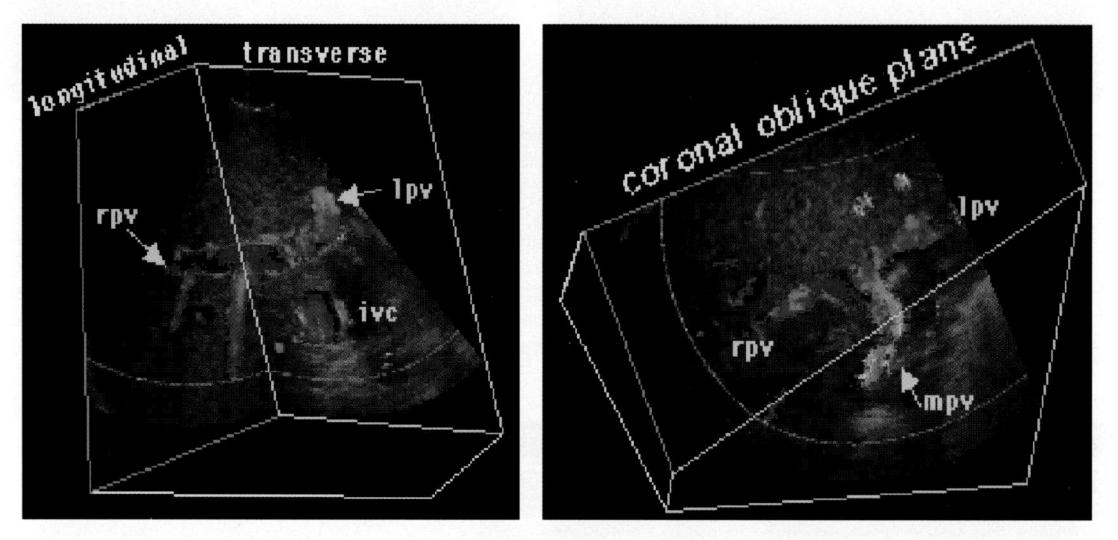

Figure 9–2. Normal portal venous anatomy. The complex anatomy of the portal veins can be viewed from a variety of different perspectives by interactively slicing into the volume cube. (rpv, right portal vein; mpv, main portal vein; lpv, left portal vein; ivc, inferior vena cava.) See black and white version of figure in text.

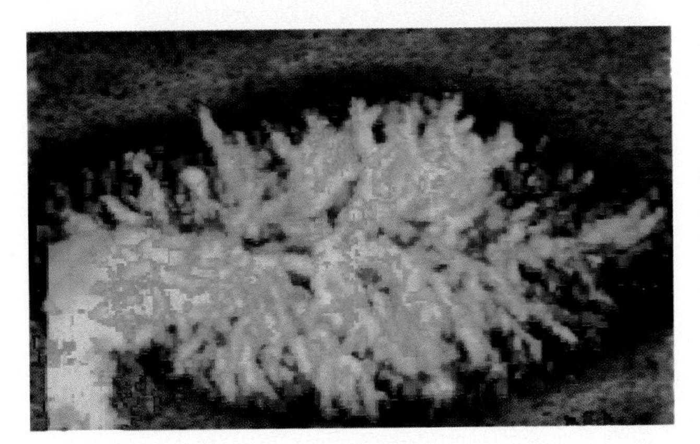

Figure 9–8. 3DUS power Doppler study of an enlarged lymph node. This image helps to appreciate the 3D anatomy of the node. (Image courtesy of GE, Milwaukee, WI.) See black and white version of figure in text.

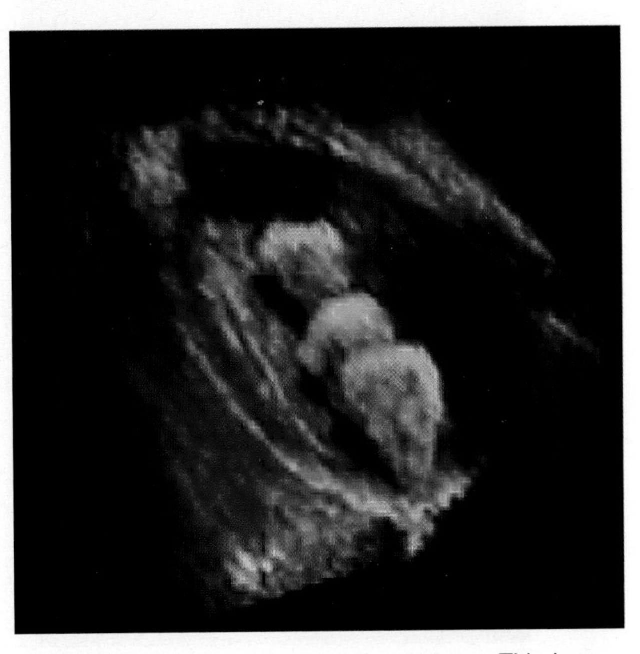

Figure 9–9. Volume rendering of gallstones. This images helps one appreciate the 3D nature of the stones. (Image courtesy of ATL, Bothell, WA.) See black and white version of figure in text.

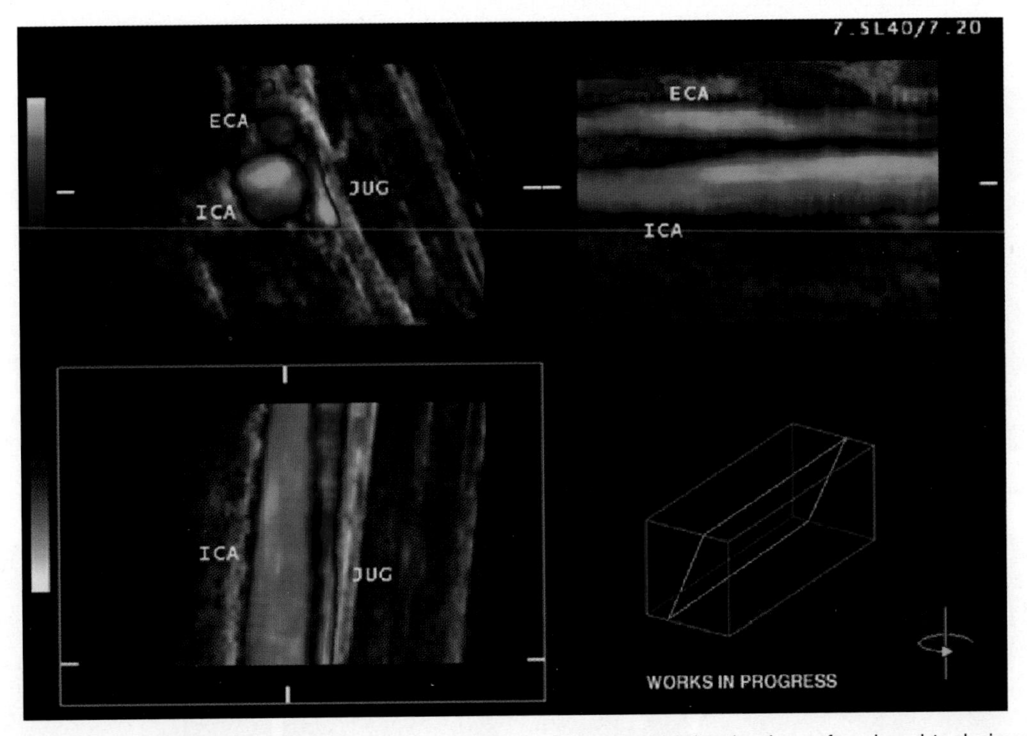

Figure 10–8. 3DUS power Doppler study of the carotid artery obtained using a free-hand technique. Doppler data helps define the edges of the vessels. (Image courtesy of Siemens Ultrasound, Issaquah, WA.) See black and white version of figure in text.

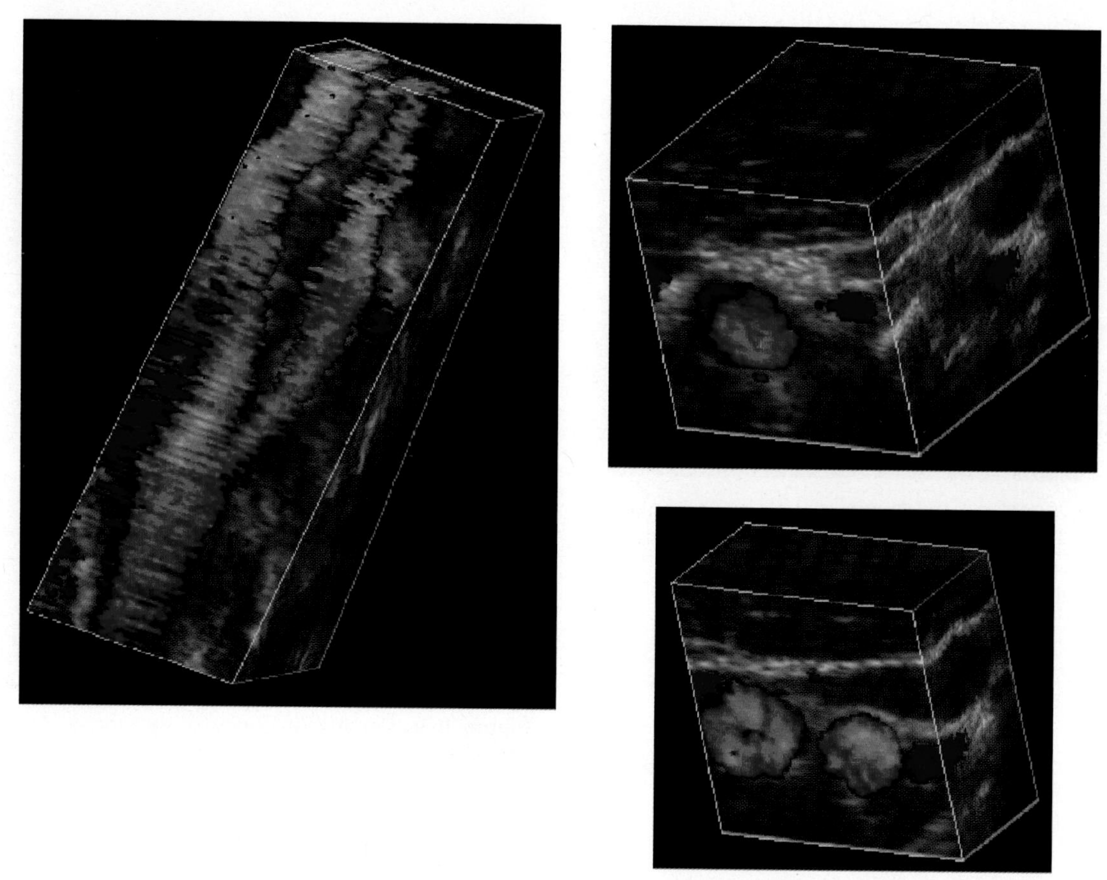

Figure 10–9. 3DUS acquisition of a normal volunteers carotid artery combining color Doppler and B-mode data. The entire course of the carotid artery can be followed. Note the slight irregularity of the boundary of the color caused by artifact. Image was obtained following cardiac gating. (Image courtesy of Life Imaging Systems, Inc.) See black and white version of figure in text.

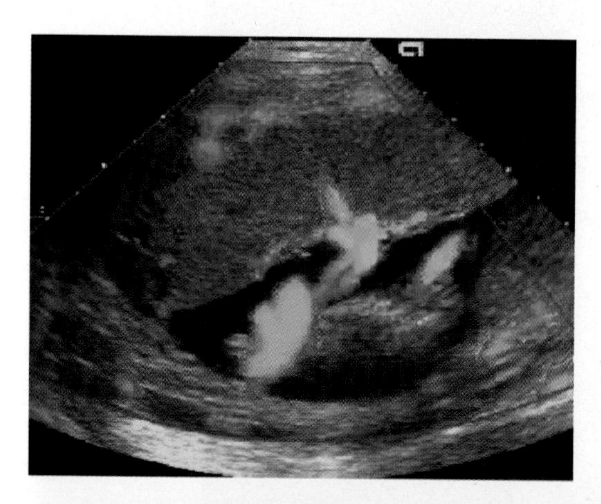

Figure 10–10. 3DUS power Doppler study of the placental circulation using free-hand scanning with position sensing. A: One frame from the 2DUS acquisition showing the point of umbilical cord insertion. B: Three-volume rendered images without the B-mode data showing the vasculature of the placental circulation at progressively higher magnifications. (Arrow, cord insertion.) See black and white version of figure in text.

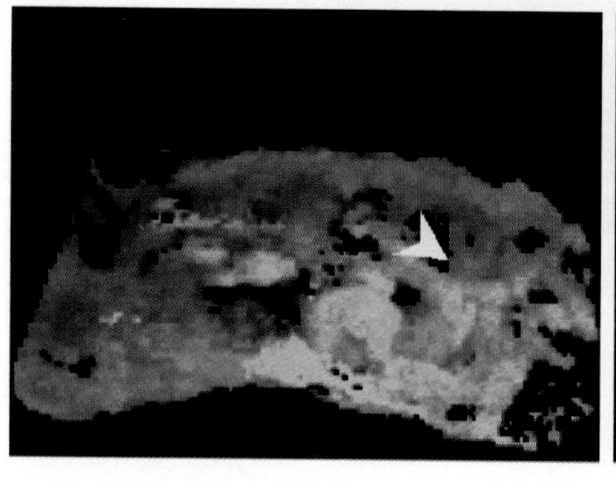

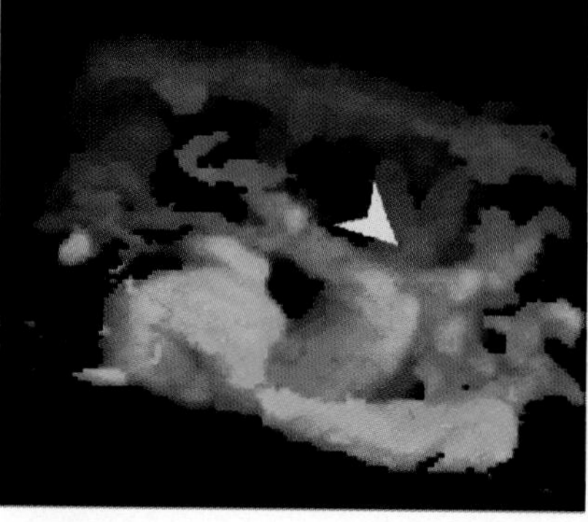

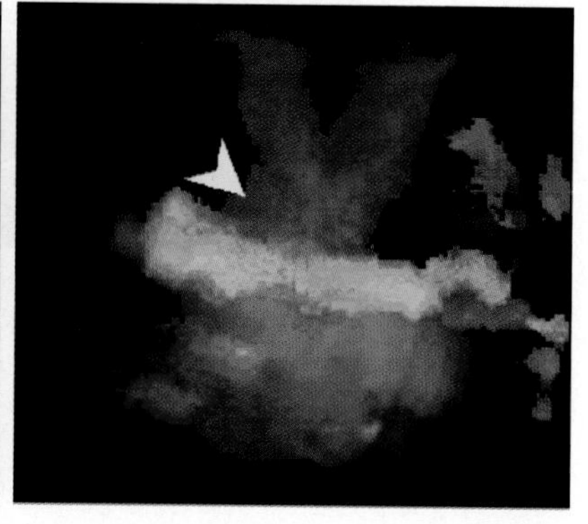

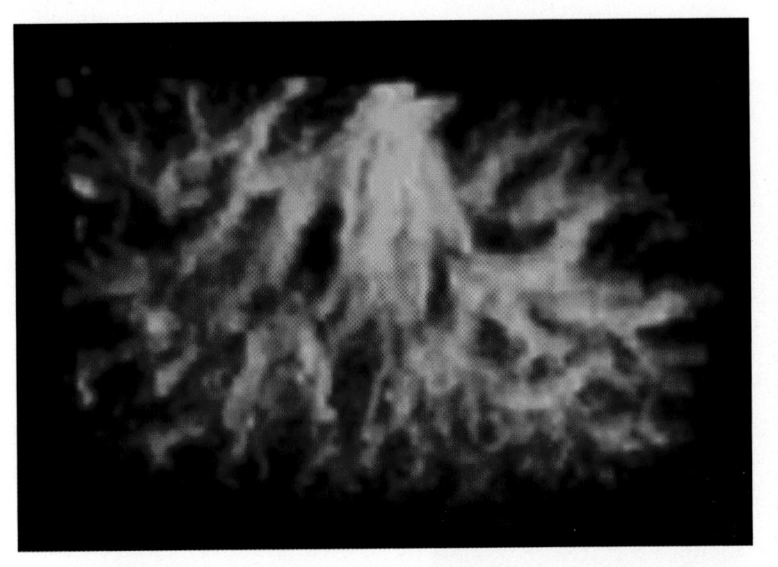

Figure 10–11. 3DUS power Doppler studies of the renal vasculature. Free-hand scan of renal vasculature using image-based position sensing. (Image courtesy of Siemens Ultrasound, Issaquah, WA.) See black and white version of figure in text.

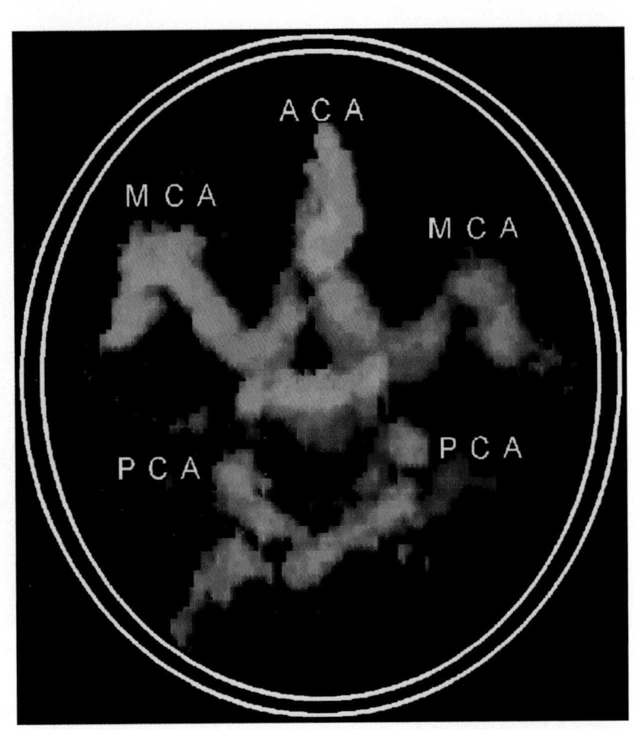

Figure 10–12. Figure 10-12. 3DUS power Doppler study of the cerebral circulation using free-hand scanning with position sensing. 2DUS image data were acquired from both sides of the head through the transtemporal window and combined using the position data. See black and white version of figure in text.

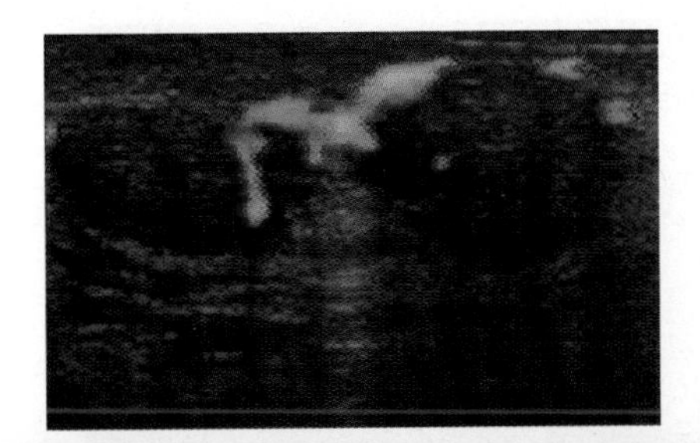

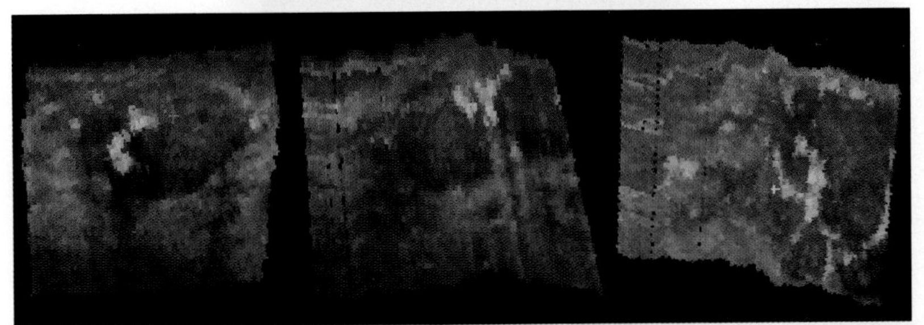

Figure 12–2. Fibroadenoma in breast. An image of the breast mass from a 2DUS examination shows a bilobed mass with vascularity seen anteriorly (arrow) as well as within the mass. The middle row of images shows three orthogonal planar images from the 3DUS scan. The borders are slightly irregular. Volume-rendered images from two orientations are seen in the lower row of images, showing continuity of the vascularity of the lesion. See black and white version of figure in text.

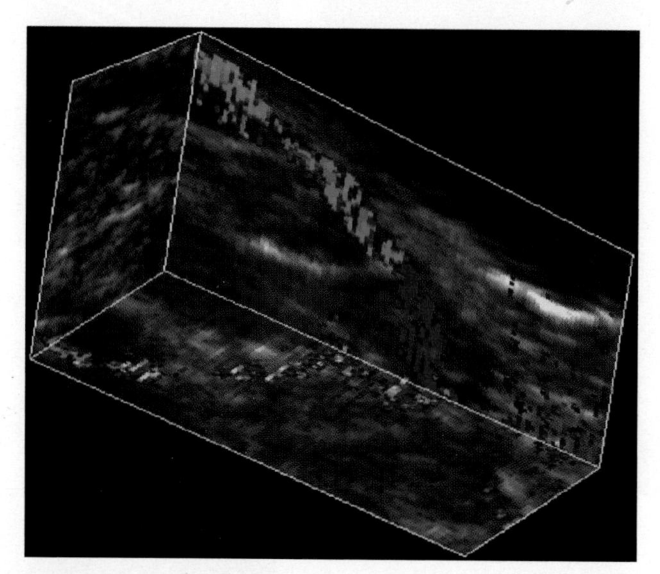

Figure 12–3. Figure 12-3. 3DUS breast combined B-mode and color Doppler images of a normal volunteer. The course of the vessel can be followed by slicing obliquely into the volume. The course of the vessel could not be seen in a single image with conventional 2DUS. (Image courtesy of Life Imaging Systems, Inc.) See black and white version of figure in text.

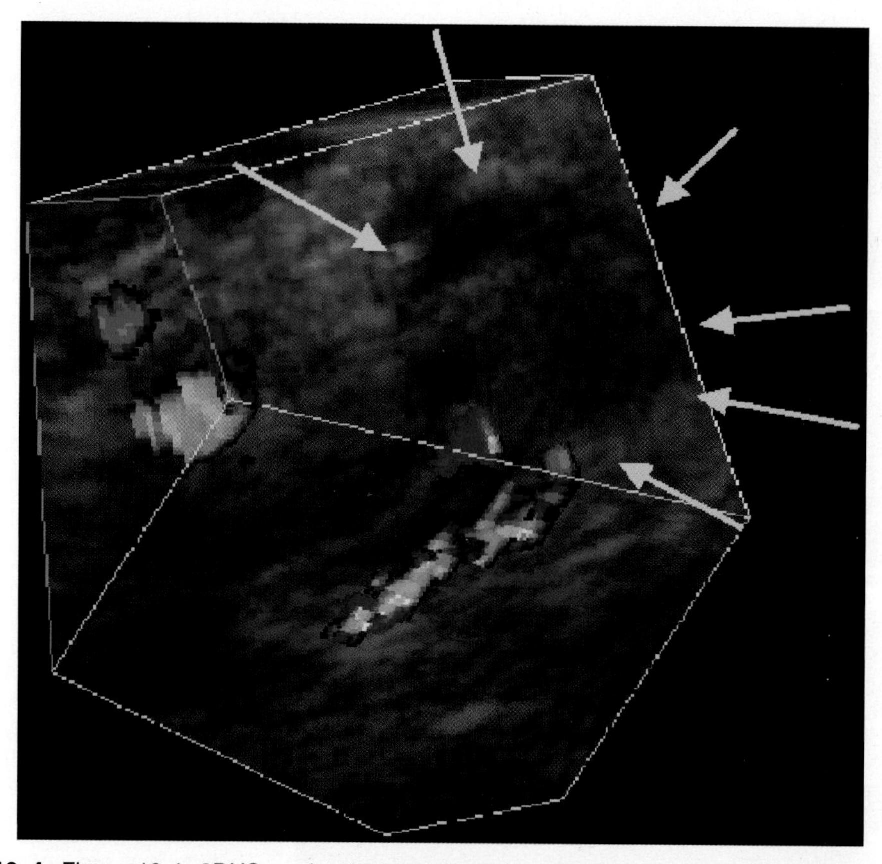

Figure 12–4. Figure 12-4. 3DUS study of a ductal carcinoma using a cube view showing various planes of the ductal carcinoma. Note the vessels approaching the irregular, jagged border of the mass. (Courtesy of Dr. D. B. Downey.) See black and white version of figure in text.

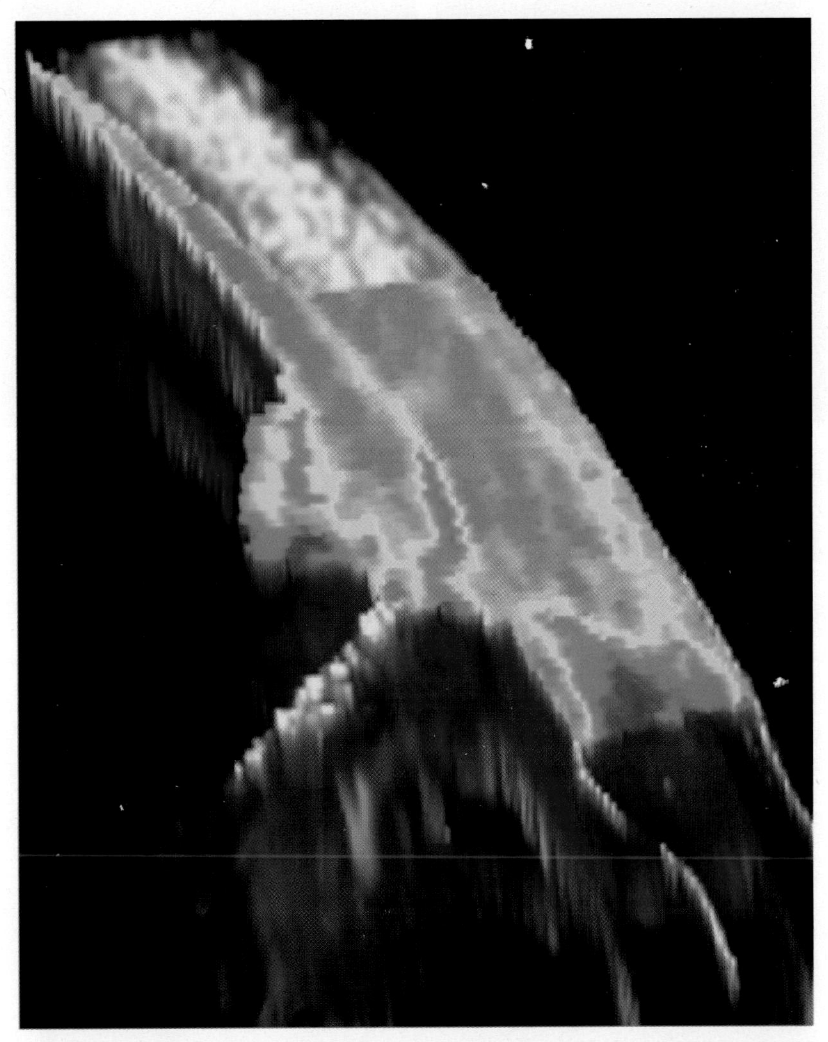

Figure 13–4. 3DUS image of anterior ocular segment obtained from digital 50-MHz ultrasonic data. The image depicts the sclera (upper right, white and red) joining the lower cornea (blue-green). The thin inner ciliary body (red) and the iris (white) lie to the left. Colors represent a derived parameter related to the effective concentration and scattering strength of constituent scatterers. (Image courtesy of Dr. F. Lizzi. Used with permission from Ref. 2.) See black and white version of figure in text.

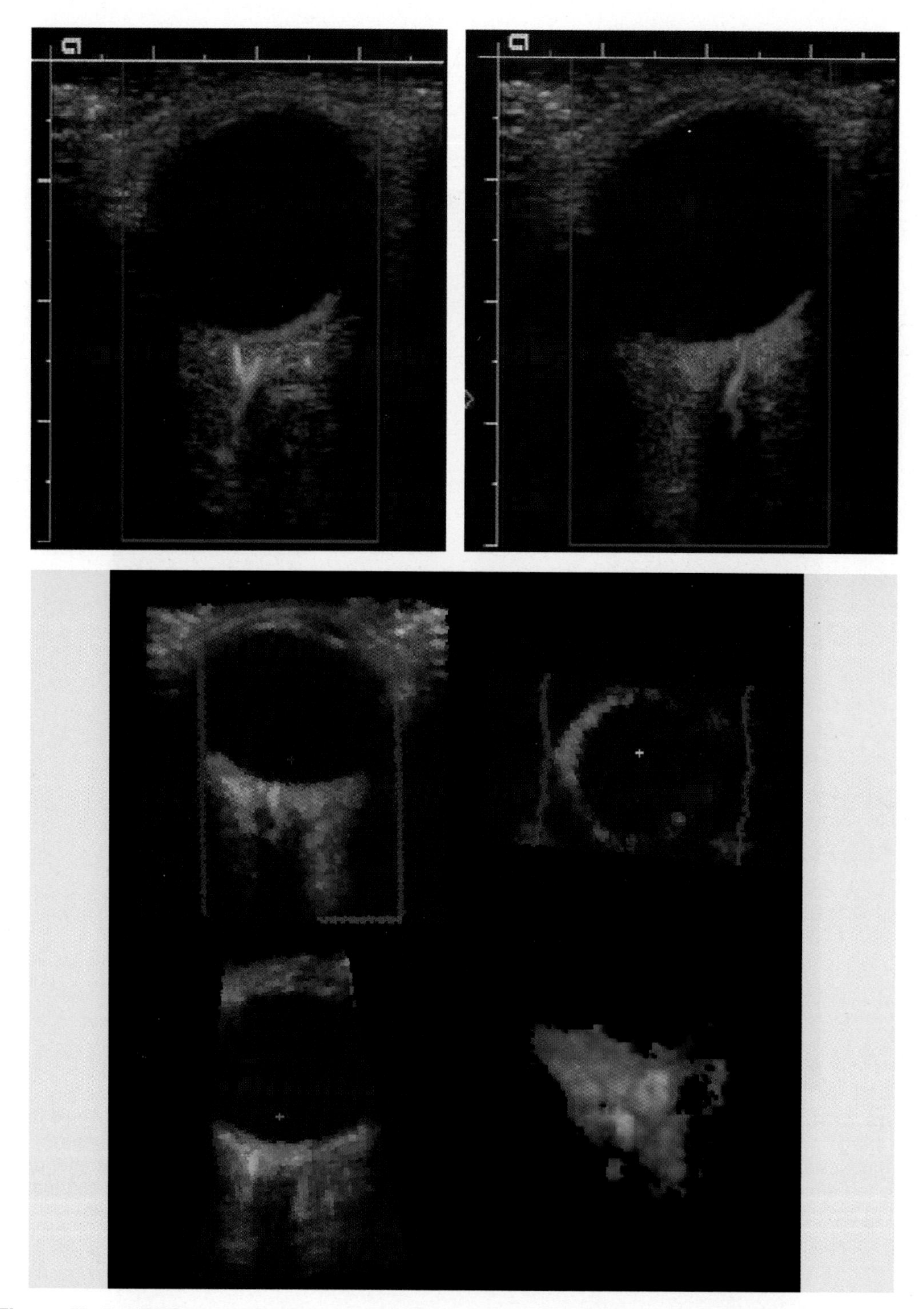

Figure 13–5. 3DUS power Doppler images of the vascular bed of the eye at the retina. The upper panel shows two frames from the 2DUS power Doppler acquisition. The lower panel shows three orthogonal planes through the eye together with a rendered image of the retinal underlying vasculature. The cross marker shows the intersection of the three planes. See black and white version of figure in text.

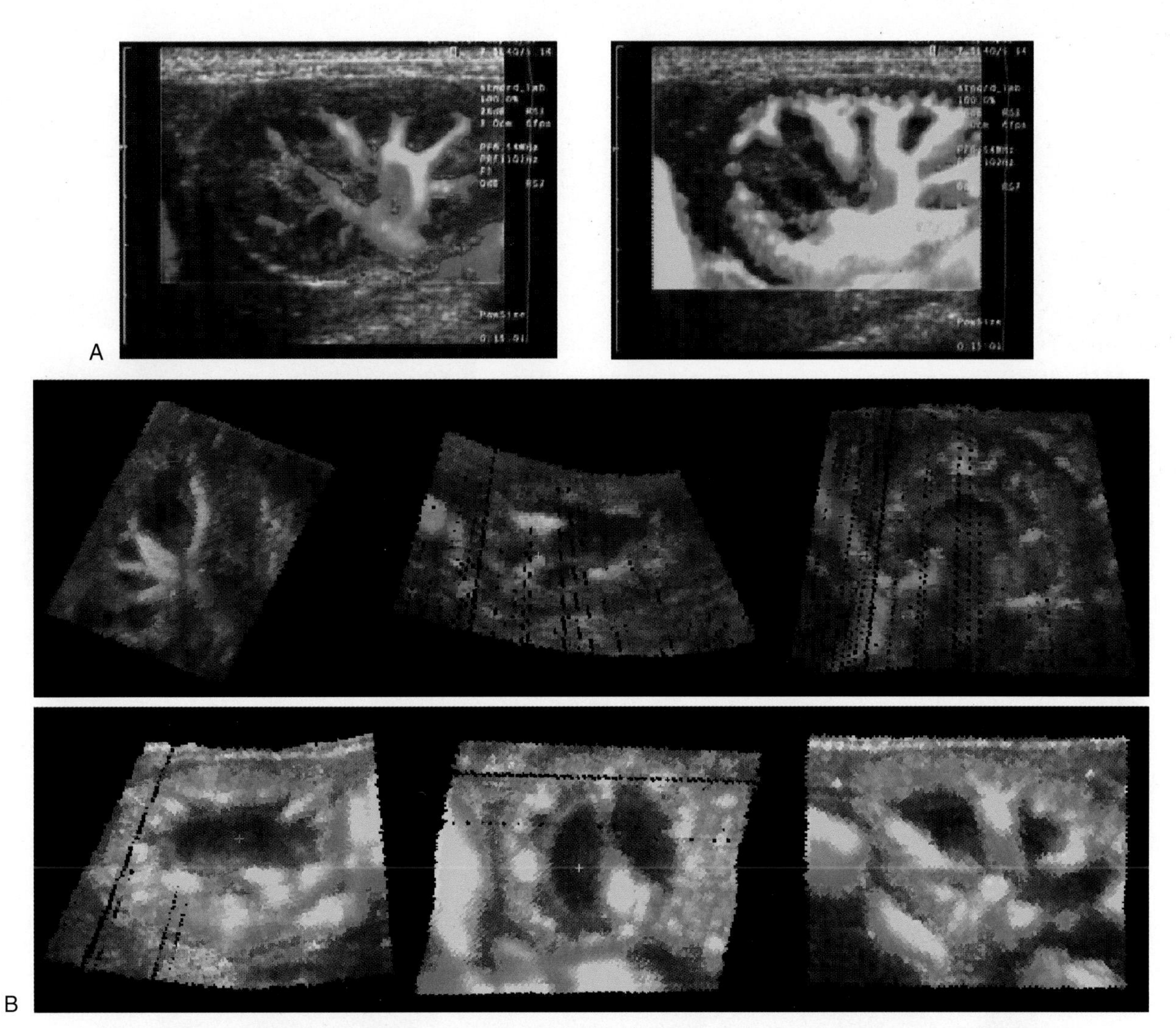

Figure 15–1 3DUS power Doppler study of a dog kidney imaged using contrast material (Imagent, Alliance Pharmaceuticals, San Diego, CA). A: Two 2DUS images from a free-hand acquisition with position sensing. The left image is without contrast material and the right is with contrast material. B: Two panels showing three orthogonal slices through the 3DUS volumes from A. The upper panel is without contrast; the lower is with contrast. 3DUS data provide clear visualization of the entire kidney and an appreciation of the 3D geometry of the organ. Although the kidney is well visualized without contrast material, the smaller vessels and parenchyma are much more clearly seen after contrast. See black and white version of figure in text.

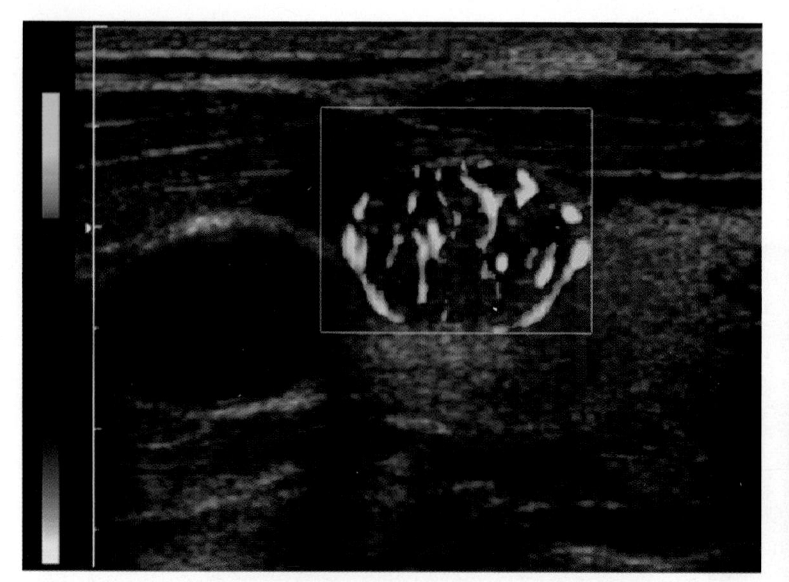

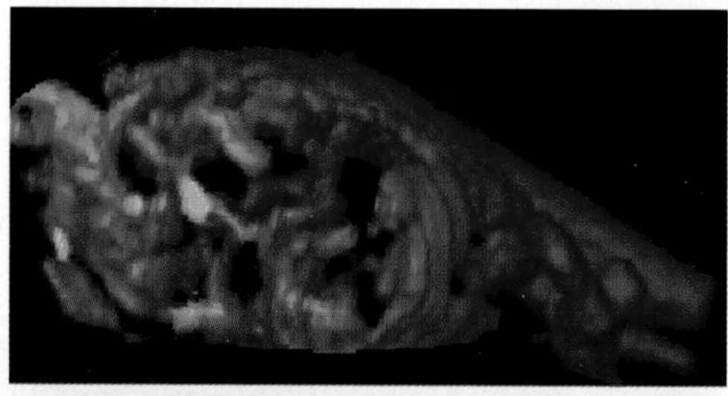

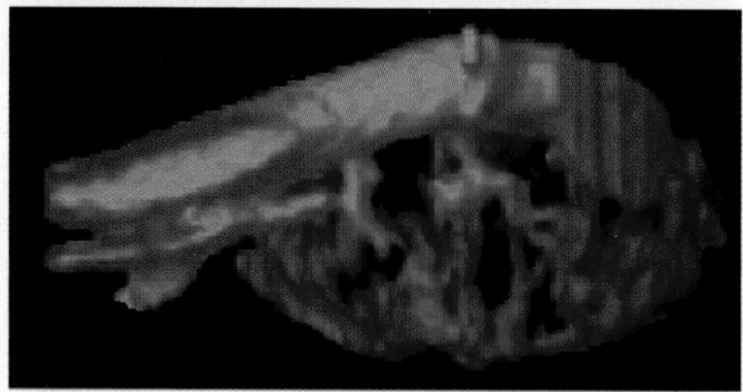

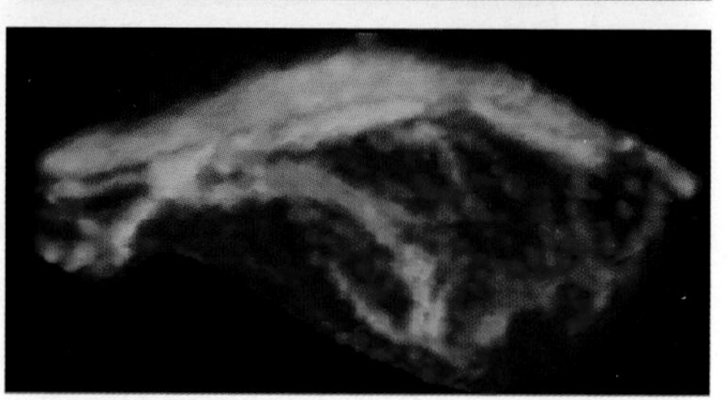

Figure 15–9. 3DUS power Doppler study of thyroid adenoma. The upper image is a 2DUS power Doppler image. The complex vasculature is more clearly appreciated in the 3DUS-rendered images shown below than in the 2DUS power Doppler image. (Image courtesy of Siemens Ultrasound, Issaquah, WA.) See black and white version of figure in text.

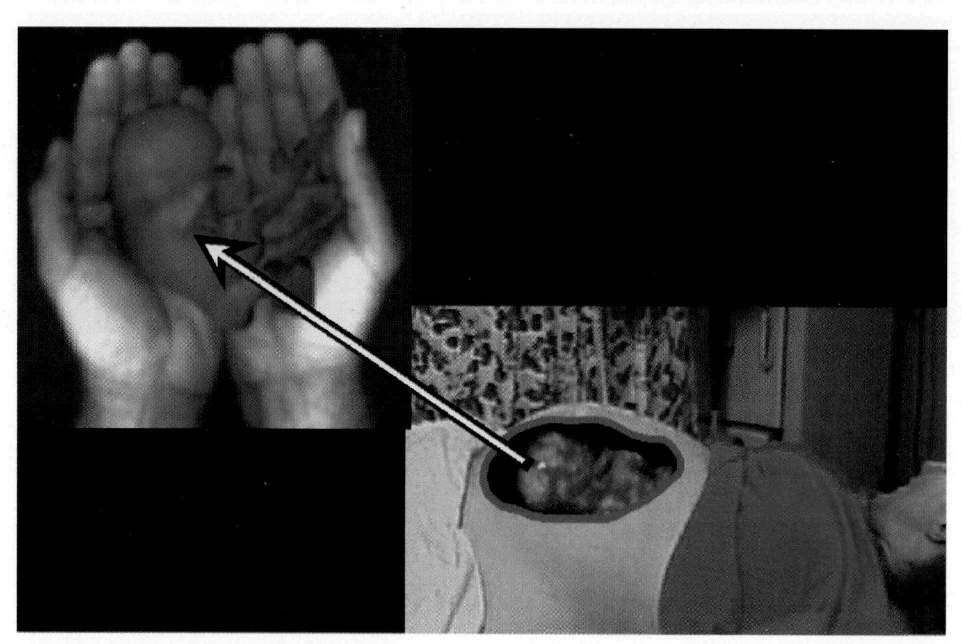

Figure 16–1. Future volume imaging in ultrasound, and other modalities, will offer real-time interactive review and manipulation of patient data. Instead of viewing patient data on a computer or scanner console, the sonographer or physician will directly view the internal patient anatomy and organs in an intuitively straightforward manner that will allow physicians to feel as if they are holding a model of the organ in their hands and allow them to ``see'' the organ or fetus as it actually is. Direct viewing of internal anatomy would facilitate more rapid comprehension of the patient's condition and provide improved feedback for interventional procedures. See black and white version of figure in text.

Ectopic Renal Tissue

Congenital anomalies of the ureteral bud are rare. The most common is a duplicated collecting system; renal agenesis, crossed fused ectopia, horseshoe kidney, pelvic kidney, or supernumerary kidney are all rare (5). Theoretically, 3DUS might provide a more thorough understanding of the anatomic appearances and relationships of some of these unusual conditions.

3DUS LIMITATIONS

Obtaining satisfactory 3D images of the kidneys presents significant challenges. Getting good 2DUS images depends on the patient's body habitus, the type of ultrasound equipment used, and the experience of the operator (5). Kidneys are paired, retroperitoneal structures that typically lie high in the retroperitoneum, where scanning access is difficult. A significant portion of the kidney is covered laterally and posteriorly by ribs. Anteriorly, bowel gas blocks the passage of sound waves—the problem being more prominent on the left. The orientation and location of the kidneys is variable, as is the amount of overlying abdominal fat that degrades the image (5). Despite these challenges, it is usually possible to see both kidneys by gently compressing the transducer on the skin, scanning from a variety of perspectives, altering the patient's position, and varying the patient's respiratory pattern. In addition to these challenges, voluntary and involuntary motion must be contended with during 3DUS acquisitions. Both the kidneys themselves and structures adjacent to them, such as liver and bowel, also move with respiration. There may be extreme difficulty in obtaining 3DUS images that cover the entire kidney.

HOW TO DO IT
Practical Clinical Tips for the Kidney

- Free-hand scanning with a magnetic localizing device is the optimal method of obtaining 3D renal images. Mechanical tilt scanning also can produce some good images, and, less commonly, so can mechanical linear scanning.
- It is important to have the patient relaxed and practice breath-holding prior to obtaining the scan data. If the patient is relaxed and one has a good window, try to get multiple data sets before reconstructing the images.
- Spend some time ensuring the 2DUS B-mode image and Doppler images are as good as they can be prior to scanning.
- Use a sufficient amount of warm coupling gel to ensure good skin contact.
- In general, to overcome the problems discussed earlier, one tries to obtain the entire 3D data set during one breath-hold, which can be extremely challenging. Because the typical adult kidney is 11 cm long, 2.5 cm anteroposterior, and 5 cm wide, more than 150 images are typically obtained. If one cannot get the entire kidney, then try and ensure that the area of interest is included in the early part of the scan.
- Experiment with different scanning windows, patient positions, and breath-hold positions to get the best images.
- Multiplanar imaging is the technique of choice for displaying data. Color and power Doppler data sets may provide additional information with volume rendering.
- In evaluating data, sometimes it is useful to line up the data set with the kidney long axis.

3DUS OF THE URETER

Miniaturization of ultrasound transducers has enabled them to be mounted on catheters and inserted in various tubular structures in the body, including the ureter, bladder, and urethra. Exams are usually performed in conjunction with endoscopic procedures. 3DUS shows excellent potential for clinical utility by providing visual cues to help the examiner understand which portion of the structure is being evaluated. 3DUS also permits accurate measurements to be obtained. Initial reports with 3DUS show potential clinical utility in assessment of strictures, stones, tumors, periureteral disease and in surgical planning and guidance. Excellent-quality 3DUS images of the ureter and urethra can be obtained by inserting an imaging catheter beyond the area of interest and withdrawing it through the suspicious location at a predetermined, steady rate to create a volume of data.

Conventional transcutaneous real-time 2DUS usually provides satisfactory images of the proximal ureter and the ureterovesical junction (UVJ) (5). Stones in these regions are readily appreciated, as are proximal ureterectasis, large retroperitoneal masses, and fluid collections (5,6). However, retrograde pyelography and ureteroscopy are currently the imaging tests of choice for assessing ureteric tumors, strictures, and most mid-ureteric pathology, as these diseases are rarely appreciated adequately with 2DUS (3). Initial clinical reports using 3D intraureteric ultrasound (3D-IUUS) are promising (25).

Miniaturization of ultrasound transducers has enabled them to be mounted on catheters and inserted in various tubular structures in the body, including the ureter (25–27). These transducers are typically radial mechanical scanners with scanning frequencies between 12.5 and 20 MHz (27). They show excellent ureteric wall detail and have sufficient penetration to allow periureteric structures to be seen easily. The transducers are typically mounted on the distal end of 3.5 and 6.2 French diameter guide wires (27). These readily fit in the normal ureter and can pass through all but the tightest of strictures. They can be introduced with a cystoscope, ureteroscope, or nephroscope.

Imaging with these devices has been reported to be clinically useful in the management of ureteric strictures, tumors, and stones (27). 3D-IUUS can stage tumors accurately by determining their size and the depth of mural invasion. It can also provide biopsy guidance and planning assistance for planned therapy. Several centers are now doing endoluminal pyeloplasties for infundibular stenoses and pelviureteric junction obstructions. 3D-IUUS helps the operator during this procedure by demonstrating periureteral vessels, which frequently cross the pelviureteric junction. Without 3D-IUUS the operator might otherwise inadvertently cut into them during a pyelotomy. The commonest reason for performing ureteral canalizations are for the assessment of stone disease and strictures, and 3D-IUUS readily demonstrates strictures and intraluminal stones. There is rarely a major clinical benefit in showing the stones except when they are located submucosally. Then 3D-IUUS is tremendously helpful, as these may not be seen at ureteroscopy (27).

CLINICAL BENEFITS OF 3D-IUUS

Adding 3DUS capability to IUUS greatly expands clinical utility (25). Excellent 3DUS images of the ureter can be obtained by coupling the IUUS catheter to a mechanical pullback device that withdraws the catheter through the ureter at a predetermined, steady rate (Figure 8-4). Multiple 2D images and the rate of withdrawal are collected by the computer system, which allows accurate reconstruction of the ureter in three dimensions. Withdrawing the catheter slowly and steadily by hand while 2D images are stored allows a faster and less cumbersome exam but does not allow accurate measurement. Multiplanar viewing appears the best display method for evaluating the data (25).

Liu et al. report that there are numerous potential applications for 3D endoluminal ultrasound images in urology, including accurate staging of tumor infiltration in the ureter, accurate visualization of changes in wall thickness and luminal patency in patients with strictures, guidance for biopsies and endoureteral interventions, and detection of intramural and periurethral disease (25).

With standard IUUS it is impossible to measure the longitudinal length of a stricture or a tumor accurately, as images are only obtained perpendicular to the vessel wall. 3D-IUUS data sets, cut in a longitudinal plane, provide this longitudinal dimension. This obviates the need to develop an IUUS transducer that images in the longitudinal plane,

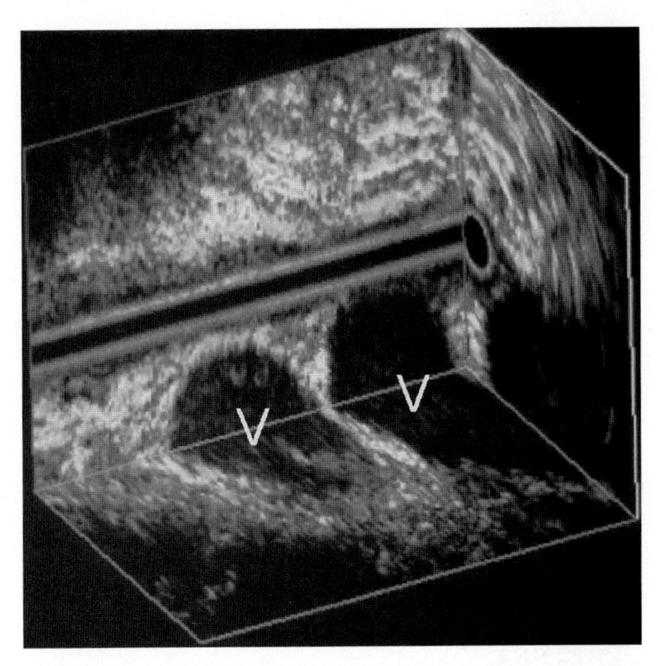

Figure 8-4. 3DUS reconstruction of a normal ureter showing the renal vessels crossing adjacent to the normal ureteric wall close to the hilum of the kidney. The long tubular structure is the catheter within the ureter. V, vessels. (Image courtesy of Dr. Ji-Bin Liu.)

which would be both expensive and technically challenging (27). Liu et al. also report that strictures can be more thoroughly understood and more accurately measured using 3D-IUUS (25).

Liu also claims that 3D-IUUS gives valuable information about anatomic and pathologic abnormalities and is of particular value in imaging the upper urinary tract (25). 3D endoluminal ultrasound can show obstructions in this segment, which can then be treated by endopyelotomy. Not only are strictures in the ureter visualized better than with 2D cross-sectional images, but this technique also gives a good overview of the surrounding anatomy by rotating the 3D model in real time. The 3D perspective is valuable in giving a more thorough understanding of the surrounding vessels, particularly when their courses cross. This can be difficult to image using 2DUS, without placing the transducer in different positions.

Endoluminal transducers are able to provide far better images of tumors arising within the lumen than when the transducer is outside a tubular structure. It may be possible to provide earlier detection and accurate staging of tumors using this technique. Some features of tumors, such as the presence and degree of obstruction, are enhanced by 3DUS, as is more accurate staging and the choice of the best site for biopsy (25).

3DUS Limitations

As in other organs, involuntary movement caused by respiration may degrade the image. In addition, because the ureter lies in close proximity to arterial vessels there can also be considerable motion artifact (25). One of the major problems in imaging the ureter lies in deciding exactly what portion of the ureter should be imaged. Liu et al. recommend recording 2D images of the entire ureter on videotape first and then selecting the segment of interest from this for 3D reconstruction. Even using this method, it may be difficult to pinpoint the segment of the ureter the images came from, and real-time assessment may be desirable, possibly with radiographic guidance (25).

HOW TO DO IT
Practical Clinical Tips for the Ureter

- Mechanical pullback produces better 3D-IUUS images than free-hand pullbacks.
- Try to keep the patient relaxed, as tachycardia may degrade the images.
- It is essential to work in close cooperation with the endoscopist, ensuring that one has a clear, concise understanding of the clinically important issues in each case.
- As it is sometimes very difficult to know exactly where one is in the ureter, fluoroscoping the catheter may help.
- It is a good idea initially to do a long, slow, complete pullback of the catheter through the whole ureter and record it on videotape. The tape should then be reviewed to assess areas of potential disease. A 3D data set should be obtained from each suspicious area. (It is currently impossible on most systems to do high-resolution 3DUS evaluations of the entire ureter in one data set, as the computer memory required would be excessive.)
- Take the longest possible data set through areas of abnormality.
- Multiplanar imaging is the display technique of choice.

3DUS OF THE URETHRA

Most urethral assessments are performed with cystoscopes or contrast studies, either retrograde or anterograde cystourethrograms (3). The recent improvement in ultrasound transducer technology has allowed better imaging of both the male and female urethra with transabdominal and transrectal ultrasound, and of the female urethra with translabial and transvaginal scanning (5,28,29). Reported promising clinical uses include the assessment of urinary incontinence and the detection of urethral diverticula in women (28–31). More recently, miniaturized transducers have been assessed in a variety of intracorporeal tubular structures (27). These techniques are currently in their infancy, with their main role in the urethra likely to be as an adjunct to standard endoscopy exam or for guidance during surgery (30).

Possible clinical utility has been reported in the assessment of epithelial overgrowth on urethral stents, inserted for benign prostatic hypertrophy, strictures or detrusor–sphincter dyssynergia. It may also have utility for the assessments of strictures and in the guidance of intraoperative collagen injection used to treat some types of incontinence (30).

CLINICAL BENEFITS OF 3DUS

3DUS has the potential to standardize ultrasound imaging from any of the perspectives (Figure 8-5). It also may speed up the examination, which would be desirable from the patient's perspective. Having the data collected in a virtual cube allows angular measurements that are recorded during incontinence assessment and can be examined at a later time.

Limited 3DUS evaluation of the urethra work has been done. In the largest reported series to date, Ng et al. evaluated a transrectal technique on 23 patients with various urethral conditions and considered the data they obtained to be a major advance in imaging (31). They believed it would allow better urethral stents to be designed and would facilitate understanding urodynamics. From a technical viewpoint, it is relatively easy to obtain high-quality 3D images of the urethra transrectally, transperineally, or transvaginally. Whether this will prove to be clinically useful is still unknown. Adding 3D capability to these small-catheter transducers has been successful in other parts of the body (27), is relatively easy, and may produce clinically desirable results.

3DUS Limitations

The fact that some of the current assessments for incontinence are done dynamically may limit the utility of 3DUS in the short term.

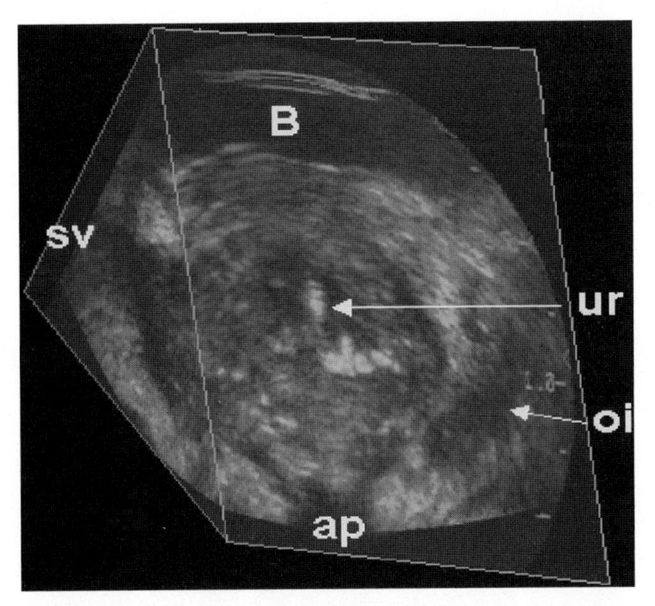

Figure 8-5. Coronal oblique view of the prostatic urethra. Interactive multiplanar slicing allows the prostate to be sliced at an oblique angle. On this image it is possible to see the internal and the external sphincters at the same time. In addition, the relationship of the ureter to the transition zone calcification is clearly shown. B, bladder; sv, seminal vesicles; ap, apex; oi, obturator internis; ur, urethra.

HOW TO DO IT
Practical Clinical Tips for the Urethra

- Mechanical pullback produces better 3D images than free-hand pullbacks.
- It is essential to work in close cooperation with the endoscopist, ensuring that one has a clear, concise understanding of the clinically important issues in each case.
- Never force the catheter; it may damage the urethra.
- Take the longest possible data set through areas of abnormality.
- Multiplanar imaging is the display technique of choice.

3DUS OF THE BLADDER

3DUS has been shown to measure the volume of urine in the bladder more accurately than 2DUS, which is clinically important in assessing a variety of structural and functional diseases of the bladder. Intraluminal 3DUS may improve the staging of transitional cell neoplasms of the bladder.

It is relatively easy to obtain good-quality conventional transcutaneous ultrasound images of the bladder when it is moderately filled. Then it is close to the transducer, with few overlying structures that impede visibility. The filled bladder has excellent acoustic properties, and intraluminal, mural, and perivesical pathologies are readily appreciated. The trigone region of the bladder and the distal ureters can also be seen in greater detail on both transrectal and transvaginal scanning. Intraluminal radial sector scanners introduced with cystoscopes have been shown to be useful in assessing the depth of invasion of transitional cell bladder tumors. Scanning is usually performed to assess the presence of intraluminal or mural disease, disorders of lower urinary tract function, or involvement of the bladder from other pathologic processes such as Crohn's disease.

CLINICAL BENEFITS OF 3DUS

In many diseases that interfere with normal bladder emptying, accurate assessment of the amount of urine in the bladder is necessary to decide on the correct diagnosis and appropriate treatment plan (32). This may be particularly important in accurately measuring residual volume, which has conventionally required catheterization (Figure 8-6) (33). This can be performed either by using standard ultrasound (3) or by passing a catheter into the bladder. Standard ultrasound methods use a formula that multiplies the bladder length by

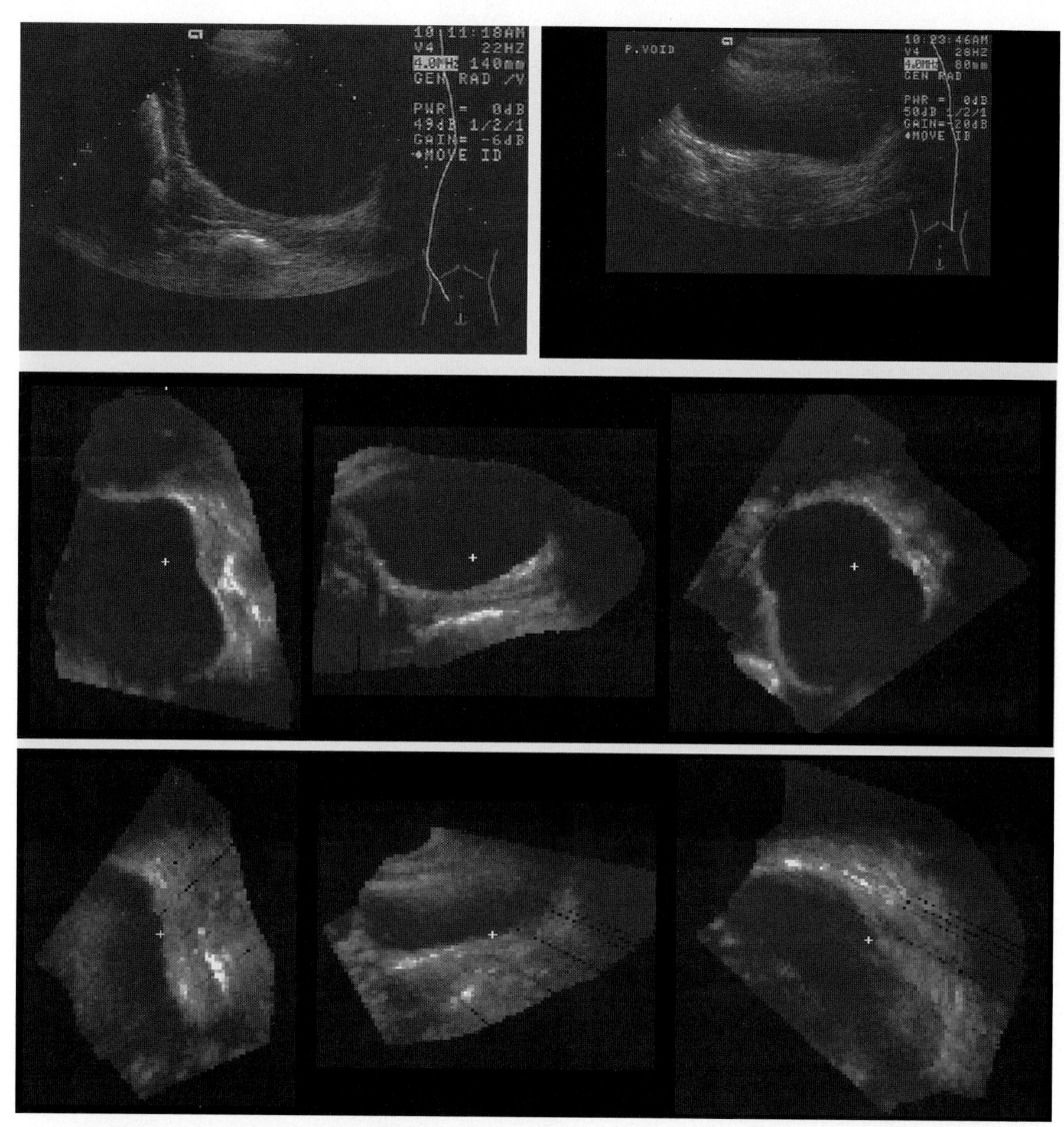

Figure 8-6. 3DUS study of the bladder showing pre- and postvoid images to measure residual urine volume. The upper panel shows the pre- and postvoid 2DUS images. The middle panel shows three orthogonal slices through the 3DUS data for the prevoid bladder volume measurement. The lower panel shows three-orthogonal slices through the 3DUS data for the postvoid bladder volume measurement. The field of view for the pre- and postvoid was different to accommodate changes in the bladder volume, which also was reflected in the the centimeter/voxel scaling for the 3DUS data. Measurement of residual urine volume is readily accomplished.

the width, the height, and a conversion factor of 0.5236 and that is accurate to within 10% to 20% (34). In one study comparing conventional ultrasound with 3DUS, the absolute error was 4.9% for measurement using 3DUS and 27.5% for the conventional method (32).

Staging of bladder neoplasia is very important in ensuring that the appropriate therapy is instituted and in determining the prognosis (35). Tumors invading the muscle layer are treated with radical cystectomy, whereas those confined to the mucosa and submucosa are treated with local resection. Conventional ultrasound faces two challenges in accurately assessing these lesions (30). First, up to 45% of lesions measuring less than 5 mm are missed by conventional scanning; second, 2DUS is inaccurate in assessing if the lesion is confined to the superficial layers (30). Intravesical ultrasound has a 70% accuracy in detecting tumor invasion of the muscular layer (30). 3D catheter-based intravesical ultrasound may improve on this performance.

3DUS Limitations

Although obtaining scanning access to the bladder is rarely a challenge, ensuring that the entire bladder is obtained in an image is more difficult. This is because the volume can be in excess of 700 ml and can extend upward as far as the umbilicus. Because of the large area to be covered, it is better to use a free-hand technique rather than a mechanical motion. Determining the edges of the bladder can be challenging, especially with large bladders. Catheter-based ultrasound assessment is invasive.

HOW TO DO IT
Practical Clinical Tips for the Bladder

- For standard transcutaneous scanning, using free-hand magnetic localizers is the technique of choice.
- Copious amounts of warm gel and a little direct pressure on the bladder are desirable.
- It is crucial to ensure that the extreme periphery of the bladder is imaged, as this is the area most prone to artifacts.
- Multiplanar imaging is the display technique of choice.
- For 3D intravesical scanning it is crucial to make sure the entire mural lesion is covered. It is probably wise to obtain at least five good data sets, and from different perspectives if possible.

3DUS OF THE PROSTATE

The prostate is the GU organ that has been studied the most widely with 3DUS. There is considerable evidence that 3DUS methods are superior to 2DUS methods for evaluating prostate disease. 3DUS prostate volume estimations have been more accurate and more reproducible than those made with conventional transrectal ultrasound (TRUS). 3DUS also has significant potential to improve the planning, performance, and follow-up of minimally invasive procedures of the prostate. 3DUS also could help standardize TRUS exams and assist in the diagnosis and staging of prostate cancer. Technically, it is relatively easy to obtain good-quality 3DUS images.

Transrectal ultrasound (TRUS) has a much more important and better understood role in the diagnosis and management of prostate disease (36,37) than when it was introduced by Watanabe over 25 years ago (38,39). Improvements in ultrasound instrumentation (40), refinements in scanning and biopsy technique (36), and a better understanding of prostate zonal anatomy (41,42) and disease (43–45) have improved the clinical utility. It is now the imaging modality most commonly used to measure the prostate, to detect prostate

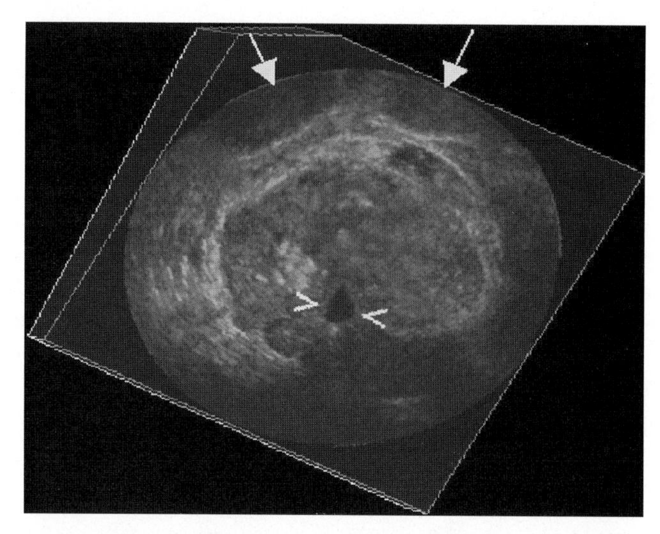

Figure 8-7. Ejaculatory duct cyst. Coronal perspective shows an ejaculatory duct cyst (*arrowheads*). *Arrows,* pubic bones. Some calcification is noted in the left transition zone.

cancer and other prostate pathologies, to guide biopsies and other interventional procedures, and to follow up medical interventions for prostate diseases (36,37). 3DUS has proven utility in improving the ability of TRUS to provide accurate measurements (46–48) and has considerable potential in improving interventional techniques (36,37,49–52) and post-treatment assessments (52–54). It may also have a role in improved cancer detection (55,56) and cancer staging. Structures such as ejaculatory duct cysts can be imaged in planes not generally available with 2DUS (Figure 8-7). Technically, 3DUS images of the prostate are relatively easy to obtain (57). Data sets are relatively small and images tend to be high quality with few artifacts (57). Unfortunately, side-firing transducers, although they produce the best 3DUS images, are becoming less popular. End-fire transducers have gained in popularity, because of the ease with which they can perform biopsies (58).

CLINICAL BENEFITS OF 3DUS

Improved Measurement Accuracy

Accurate and repeatable prostate measurements using 3DUS assist in determining the optimal management of patients with elevated serum prostate-specific antigen (PSA) (Figure 8-8) (59). In patients with biopsy-proven prostate cancer, they assist in staging the disease by determining the volume of cancer present (47,48). Also, they provide objective proof of the utility of some medical interventions in the prostate (52–54).

Unfortunately, conventional 2DUS assessments of prostate volumes are inaccurate, with errors in the 10% to 20% range (48,60–62). This lessens both the popularity and the clinical utility of this technique. Despite this, two measuring techniques are commonly employed (48,63): (a) an ellipsoid volume technique (prostate height × prostate width × prostate length × 0.5236) and (b) an ellipsoid outlining technique in which the prostate outline is manually traced on the ultrasound machine screen in two orthogonal planes.

Volume calculation formulas are built into the software on several ultrasound machines to compute these results. Some other formulas also have been advocated for assessing the volume of the prostate rapidly. These include the sphere volume calculation ($0.5236 \times [\text{height} \times \text{width} \times \text{length}/3]^3$) and the quick sphere volume calculation ($0.5236 \times [\text{height} \times \text{width} /2]^3$) (34). All these calculations are less accurate than planimetry—a slow technique in which multiple parallel prostate images are obtained a few millimeters apart. Measurement requires that the ultrasound transducer be placed in a mechanical stepping

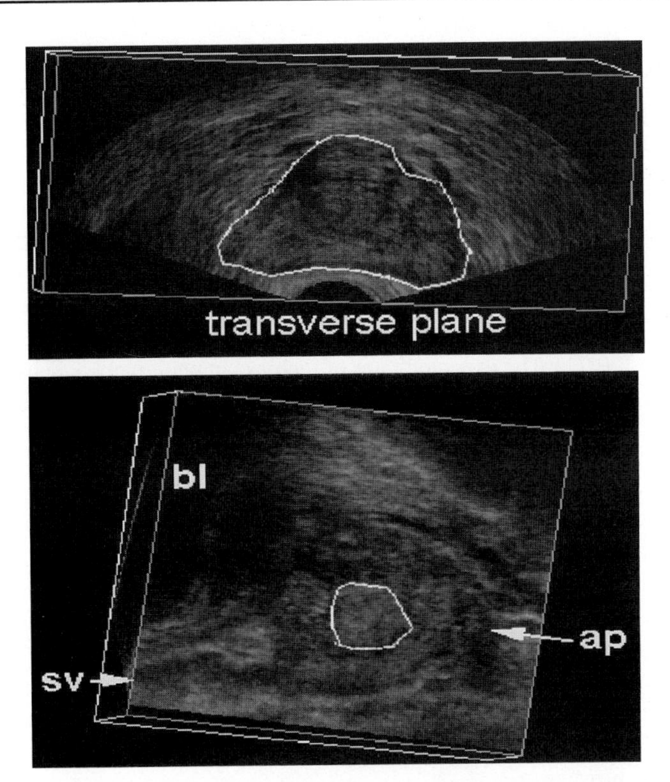

Figure 8-8. 3DUS image showing a prostate being outlined during the measuring process. While 3D volume measurement is more accurate and reproducible, it also may be more tedious than conventional methods, depending on the method or algorithm used to identify or segment the region of interest. bl, bladder; sv, seminal vesicles; ap, apex.

device attached to the patient's couch (48,63). Still images are captured 4–5 mm apart with the area of each slice being calculated, and subsequently summed to give the true volume (48,63).

Several authors have shown that 3DUS is better than 2DUS at measuring the volume of the prostate (60–62) and several other solid organs (13,14). Elliot showed 3DUS was almost perfect in estimating the true volume of resected cadaver prostates (61), and Gilja showed similar results in assessing porcine kidneys and stomach (13). Tong reports that the prostate volume can be measured with greater than 92% accuracy, using a mechanical motorized device coupled with a side-firing transducer (57). These high levels of accuracy and precision have also been shown with side-firing transducers that have been coupled with pullback motion (62). However, like planimetry, these techniques are time-consuming and labor intensive, as the contour of each slice of prostate has to be manually outlined. Aarnink et al. reported considerable success in measuring prostate volumes accurately using a technique that segments multiple 2DUS images of the prostate (64,65). This allows the edges of the prostate to be determined rapidly with minimum operator input. The Pearson correlation coefficient of the automated technique correlated with the true volume of the prostate was 0.92. They believe that this technique will be useful in the sequential assessment of prostates, as it removes the variability introduced by individual subjectivity in determining the edge of the prostate. This will allow an increase in the reliability of multicenter trials (63). Combining this software with fast 3D volume acquisitions is likely to assist significantly in the clinical management of these patients.

The incidence of age-adjusted prostate cancer worldwide is increasing by 3% annually (66). This increase can be partially attributed to the increased screening of the population with serum PSA tests (58). TRUS findings are routinely combined with digital rectal examinations (DRE) and serum PSA tests to improve the diagnosis and clinical outcome of patients suspected of having significant prostate cancer (37). Mildly elevated serum

PSA levels are nonspecific, with less than 25% of those with levels between 4 and 10 mg/ml actually having prostate cancer (67). In general, men with larger transition zones due to benign prostatic hyperplasia (BPH) will have higher levels of serum PSA (68). Obviating the need for biopsies in some of these men is highly desirable, as biopsies are painful, expensive, and carry a small risk of infection (37,66). Rather than performing biopsies on all patients with elevated PSA levels, several authors have advised that a patient's PSA can be best interpreted clinically when it is combined with an accurate assessment of the volume of the prostate (46–48). Kalish and Aarnink have suggested it is even better to correlate the serum PSA with the volume of the transition zone (68,69). Several studies suggest that 3DUS will have utility in both identifying and measuring these zones accurately (70,71). The expected rates of increase in benign prostate hyperplasia and the serum PSA in older male populations are starting to be defined using automated techniques (68).

Improved Measurement Repeatability

3DUS measurements (see Figure 8-8) are much more repeatable than corresponding 2DUS assessments (13,72). Accurate and repeatable 3D volume measurements probably will help assess the clinical utility of the plethora of new surgical, interventional, and medical treatments for both benign and malignant prostate diseases (49,52,73,74). 3D-TRUS will provide valuable objective diagnostic information and postinterventional follow-up in patients with benign prostate disease (37,54,73). It also could be used to measure both gland size and the size of any transurethral resection of the prostate (TURP) type defect within the prostate.

Potential Role in Improving Interventional Procedures

Ultrasound-guided prostate biopsy is a well-established and clinically valuable technique, and ultrasound is gradually being accepted as the imaging modality of choice for guiding other prostate interventions (36,37). 3DUS has been shown to be clinically useful in prostate brachytherapy (50,51), laser therapy (52), and cryosurgery procedures (49). It is likely that it will be useful also for a variety of other procedures. Flannigan worried that, to date, many of the newer interventional therapies have been employed without adequately relating dosimetry to prostate volume (34). 3D-TRUS may provide the information necessary for more accurate dosimetry to be obtained (see Chapter 14). To date, clinical utility for 3DUS-guided prostate biopsy has been limited.

Role in Improving Cancer Detection

Prostate cancer most often presents as a focal, peripherally located, subcapsular hypoechoic area on B-mode imaging (36,40). The appearance is nonspecific, and the differential diagnosis of this ultrasound appearance includes prostatitis, granulomatous prostatitis, tuberculous prostatitis, prostatic intraepithelial neoplasia, infarcts, and cysts (36). The positive predictive value of a hypoechoic nodule for cancer has been reported to be between 18% and 60% (75–78). Occasionally, cancers are reported to be hyperechoic (36,37), and about 25% are isoechoic (37,79). Other sonographic findings in prostate cancer include gland asymmetry, especially in the peripheral zone; hypervascularity on Doppler exams; and capsular distortion with loss of the normal differentiation between the central and peripheral zones (37). Currently, approximately 25% of tumors are not clearly seen as abnormal with 2DUS, although color Doppler and power Doppler may increase that diagnostic accuracy slightly (80). As TRUS has such a low specificity, all suspicious lesions are biopsied to obtain an accurate diagnosis (36,75). The risk of diagnosing clinically insignificant cancers with TRUS is thought to be less than 5% (36,81).

To date, the increased number of viewing perspectives afforded by 3DUS has not been shown to have altered either the sensitivity or the specificity of ultrasound in diagnosing cancer on the basis of its B-mode appearance. However, there is progress in using computer

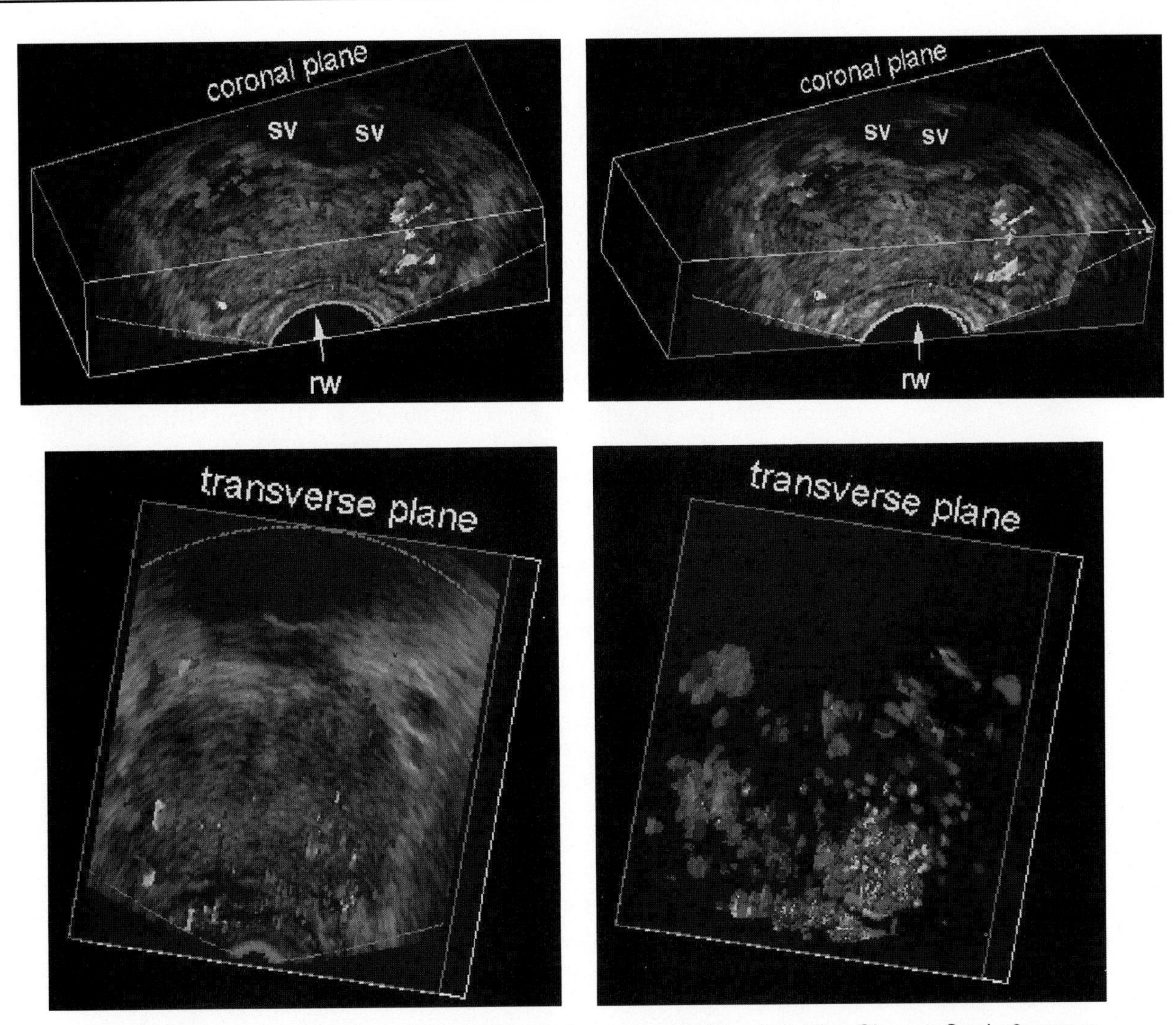

Figure 8-9. Pre- and postultrasound contrast views of a prostate containing Gleason Grade 6 cancer at the right base. Precontrast image shows vascular prominence at the right base where the cancer is located. On an equivalent image obtained several minutes following injection an increased Doppler signal is seen from vessels in both the cancer and the remainder of the prostate. A thin-section transverse plane B mode and color Doppler image of the same prostate obtained slightly more caudally in the prostate. Note the slight decrease in echogenicity and the increased vascularity of the cancer at the right base. Equivalent projected image with the B-mode removed and the color Doppler signal summed. Note the vascularity is markedly increased in the region of the tumor. rw, rectal wall, sv, seminal vesicles. (Echogen Contrast courtesy of Sonus Pharmaceuticals.) See color version of figure.

algorithms to detect specific B-mode sonographic patterns in transrectal ultrasound images (82–86). The AUDAX software reports a 84.8% sensitivity and 87.5% specificity in detecting prostate cancer and a 90.6% sensitivity and 64.2% specificity in detecting prostatitis using automated techniques (82).

3DUS power Doppler may prove helpful in improving cancer diagnosis. Lees reported that contrast-enhanced 3DUS power Doppler detected 27 of 28 cancers (56) and had a site-specific accuracy of 84% (56). Subsequently, a single case report showed 3D power Doppler detected a prostate cancer without contrast enhancement (55). Although both reports are encouraging, more experience with the technique will be required before any conclusions can be drawn (Figure 8-9).

Potential Role in Improving Cancer Staging

In determining the prognosis and optimal treatment of prostate cancer, accurate staging is required, as pathologic staging is the single most powerful prognostic variable (87,88).

Prostate cancer was classically staged by the Jewitt/Whitmore classification, but the TNM classification is now becoming more popular because of its ability to integrate clinical, imaging, and pathologic findings (89). In general, organ-confined disease is curable (about 90% 5-year disease-free survival), whereas extraglandular spread carries a much worse prognosis (89). It is crucial to assess the stage of the cancer, as the prognosis and management are very closely dependent on them (88).

To date, no technique does a good job of preoperative staging (88), with more than half of the patients thought to have organ-confined disease being shown to have positive margins at radical prostatectomy (89). The most recent progress from an imaging perspective has involved endorectal MRI and ultrasound staging biopsies. Neither technique has been evaluated in a large series, though accuracies of between 77% and 82% have been reported with endorectal MRI in small series (87). In the only large study performed, no statistical difference was found between the abilities of TRUS and MRI in staging prostate cancer based on seeing direct spread—both did poorly (77% MRI accuracy versus 66% for TRUS) (87,90). 2DUS has been disappointing in improving cancer staging, based on either 2DUS appearances of direct invasion or measurement of the lesion size (77,87). As current sensitivities are between 40% and 90%, and specificities are 46% to 90% for direct visualization of the cancer spreading out of the prostate, it is not cost-effective to stage prostate cancer with ultrasound (87,90).

The theory behind using prostate cancer volume for staging is based on McNeal's reports that tumors must be at least 1 ml in volume before they can spread from the prostate (91). To date, TRUS has largely failed in the direct volumetric assessment of the lesions, as it tends to significantly underestimate the volume of the prostate cancers detected (46).

The clinical utility of 3D-TRUS in staging prostate cancers has not yet been assessed. It might improve staging by allowing the prostate to be visualized from a more advantageous position or by combining more accurate prostate measurement with some other clinical parameters. With B-mode assessments the operator attempts to assess spread into the seminal vesicles (Figure 8-10), into the neurovascular bundles, and into and beyond the prostate capsule. Sometimes the optimal image plane for assessing spread is unavailable. Using 3DUS, images can be captured in different orientations and reconstructed in the plane of interest (92). In the prostate, the coronal plane often assists in assessing whether

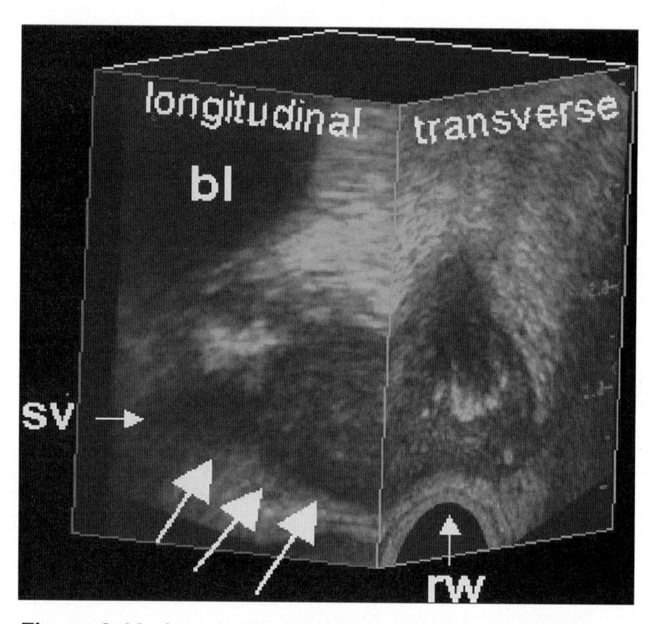

Figure 8-10. Longitudinal and transverse views through a prostate showing a hypoechoic tumor (*arrow*) extending into the seminal vesicles (SV). bl, bladder; rw, rectal wall.

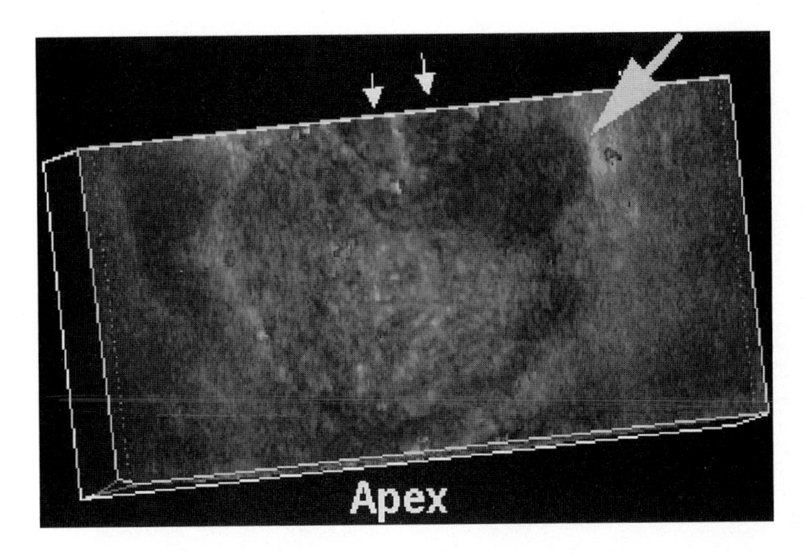

Figure 8-11. Coronal view of left base prostate cancer. An extensive hypoechoic cancer that extends along the left lateral aspect of the prostate toward the apex is clearly shown (*large arrows*). Ejaculatory ducts are noted (*small arrows*). The tumor clearly extends outside of the prostate. The tumor is mildly more vascular than the remainder of the prostate. See color version of figure.

the disease has spread out of the prostate at the base (Figures 8-11). 3D intraluminal scanning of the prostate via the urethra also may assist in evaluating prostate tissue, normal and cancerous, adjacent to the urethra (Figure 8-12).

D'Amico et al. recently defined a new clinical entity called the calculated prostate cancer volume, which they found more useful than PSA in predicting both the actual cancer volume in resected prostate specimens and pathologic stage T3 in patients with clinical stage T1 and T2 disease (i.e., organ-confined disease) (93). This calculated prostate volume depends on the prostate volume measured at ultrasonography, the pathologic Gleason score, and the PSA. If these findings are verified in other studies, then there will be an even more compelling reason to assess the prostate volume accurately.

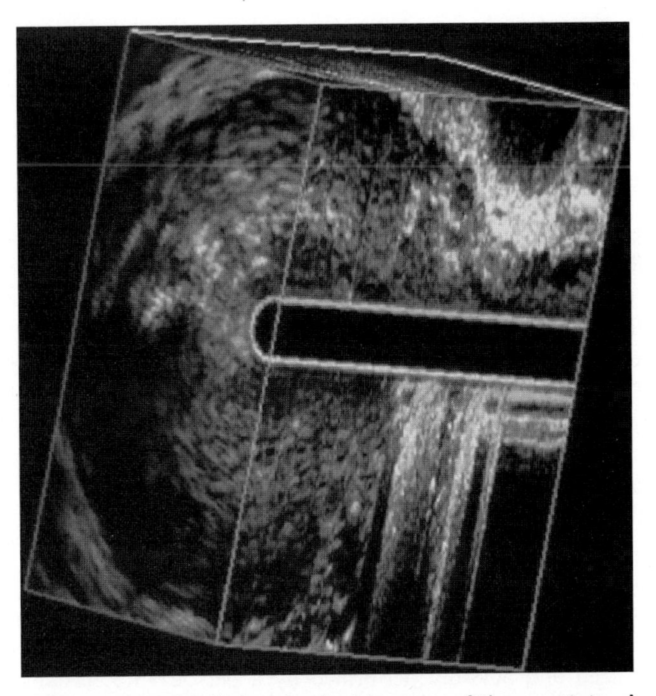

Figure 8-12. 3DUS intraluminal images of the prostate. A combined longitudinal and transverse perspective of the prostate is provided by a catheter that has been inserted through the urethra. This provides excellent detail of the area of the prostate adjacent to the imaging crystal. (Image courtesy of Ji-Bin Liu.)

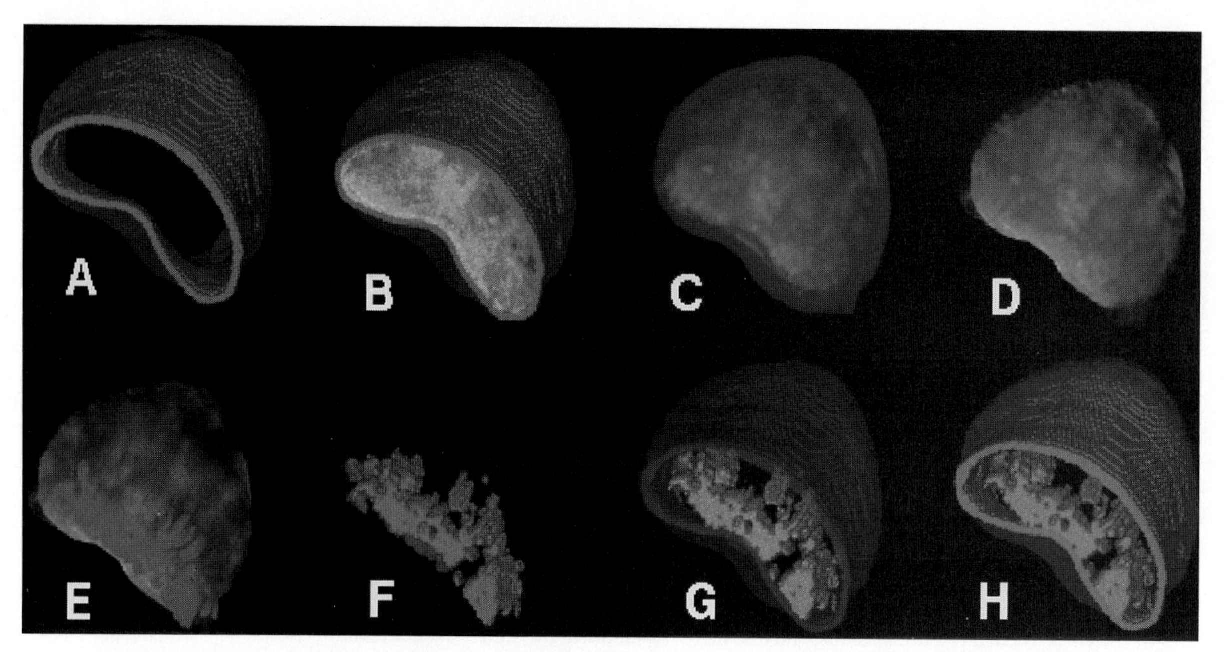

Figure 8-13. Eight 3D renderings of a biopsy proven cancerous prostate viewed from the apex. (**A**) Opaque capsule; (**B**) parenchyma within capsule; (**C**) transparent parenchyma with capsule; (**D**) transparent parenchyma without capsule; (**E**) parenchyma with cancerous lesions coded in red; (**F**) cancerous lesions alone; (**G**) cancerous lesions with transparent capsule; and (**H**) cancerous lesions with opaque capsule. The analysis was automatic after identification of capsule boundaries. (With permission from Ref. 94.) See color version of figure.

Feleppa et al. recently have presented data demonstrating differentiation of cancerous from noncancerous tissues based on spectrum analysis (Figure 8-13) (94). Their work suggests that more sophisticated algorithms may assist in identifying cancer based on features in the acoustic signature of different tissues.

Potential Role of Standardizing 3D-TRUS Examinations

The number of patients referred for prostate ultrasound exams has grown dramatically over the last decade (58) and will continue to grow. The recent rise in life expectancy in North America has been accompanied by a similar increase in the prevalence of prostate disease, as both benign and malignant prostate diseases affect older men (58). Currently, there are more than 22 million men over the age of 50 in the United States, and this number is expected to double before 2010 (95). There is also an increased public awareness of prostate disease and many new diagnostic tests that are likely to continue the trend of increasing numbers of patients being referred for diagnostic ultrasound assessment and ultrasound-guided biopsy.

Current methods of performing TRUS exams are expensive, inefficient, and generally performed by a physician who is required to mentally integrate multiple 2DUS images into a 3D representation of prostate anatomy (92)—a process that is highly operator dependent, complicated, and sometimes inaccurate. 3DUS has the potential to provide a more standardized, less expensive exam (63) because data acquisition, reconstruction, and display all can be standardized and possibly performed by a person with less intensive training. Accurate 3DUS data sets can be obtained as long as there is no motion of the patient or the transducer during data acquisition. The physician can subsequently review the images either locally or remotely via a PACS network to assess the gland, decide whether a biopsy is appropriate, decide the optimal location for the biopsy, and plan the procedure in a more timely, efficient manner (64,70,82). Other potential advantages include automated gland volume estimations (64) and the application of image characterization programs (82–85).

3DUS Limitations

Disadvantages to using this type of 3DUS assessment of the prostate include the loss of the ability to palpate suspicious areas in the prostate with the transducer, which sometimes aids diagnosis in standard TRUS exams. Some artifacts may be worse with 3DUS. For example, shadowing may block visibility more than with 2DUS, where the operator can take steps to compensate for areas of decreased visibility either by applying more pressure or by altering the viewing perspective (61). As a result, optimizing the scan position before acquiring a data volume is essential. Scanning with end-fire transducers is more prone to motion artifacts. Unfortunately, most ultrasound manufacturers are tending toward the end-firing transducer, as this is the most commonly used transducer to guide biopsies, and the vast majority of people currently presenting for transrectal ultrasound are being biopsied.

HOW TO DO IT
Practical Clinical Tips for the Prostate

Transrectal Transducers

- Excellent-quality 3DUS images can be reconstructed by collecting between 200 and 400 2DUS images over 12–20 sec. The best data sets are obtained using a side-firing transrectal probe with either pullback or side-to-side rotation.
- 3DUS images also can be obtained with an end-firing transducer, although imaging is more susceptible to motion artifacts. The extreme posterolateral aspects of the prostate also may not be included in the data sets.
- Potential fixes for the problem with the end-fire transducers include the standoff, which we believe would have to be more solid than fluidlike. The other method would be to obtain an exceedingly long convex crystal.
- It is crucial that there is as little motion as possible during scanning. We suggest that the examiner keep his or her arms close to the body for stability and ensure that the cable doesn't drag against the mechanical motion.
- Ask the patient to stay as still as possible. If there is still motion, ask the patient to hold his breath.
- The coronal perspective is often helpful during interventional procedures and in assessing direct spread of cancers.
- 3D-IUUS catheters also can be used to image the prostate with a 9-MHz transducer, providing a penetration depth of up to 4 cm, which provides full visualization of the prostate with excellent detail in the transducer near field.
  - Reconstructed 3D-TRUS images can serve as an intraoperative tool for assessing tumor size, location, and radioactive seed placement.
- 3D-TRUS also is appropriate for other minimally invasive treatments for both benign and malignant prostatic disease.

3DUS OF THE SCROTUM, PENIS, AND SEMINAL VESICLES

Ultrasound is the imaging modality of choice for evaluating scrotal disease (96). It detects intra- and extra-testicular disease in exquisite detail, and occasionally requires other imaging modalities such as MRI. Both 2DUS B-mode and color Doppler information can usually characterize masses as being intra- or extra-testicular. They also can show whether the blood flow is excessive or compromised.

Though it is possible to get excellent 3DUS images of the seminal vesicles, scrotal

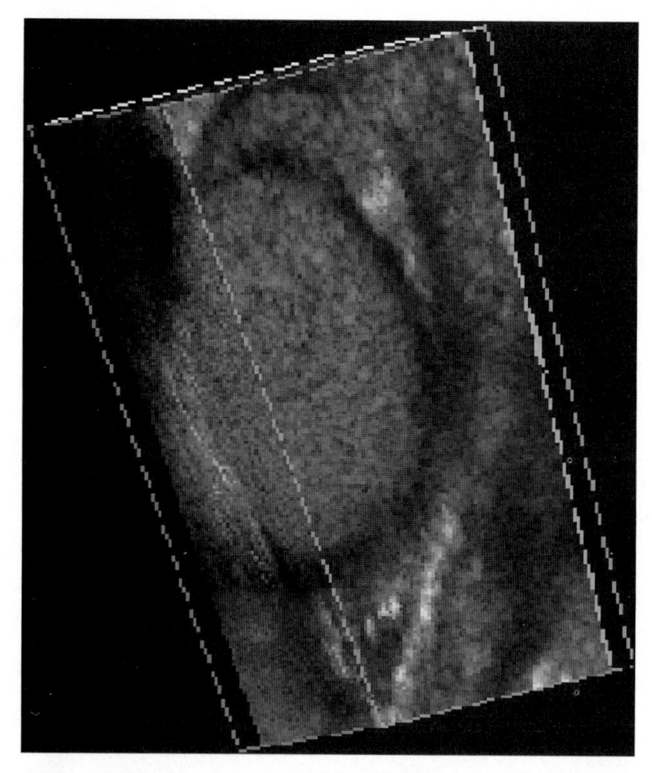

Figure 8-14. Reconstructed longitudinal view of the testes and scrotum. The relationship between the epididymal head and the body of the testes is clearly shown. The clinical utility of this however is unclear.

contents, and penis, to date no compelling clinical reason for performing these scans has been shown. The potential exists for 3DUS to assist in displaying the relationship of the epididymis to the testes (Figure 8-14), particularly when there is enlargement and/or increased flow in inflammatory states. At times it is difficult to differentiate varicoceles from inflammation and 3DUS may assist in understanding the true location of vessels in the scrotum. 3DUS also has potential to better comprehend the mediastinum of the testes. At times it can be difficult to differentiate normal tubular ectasia from testicular tumor, and 3DUS should assist in this differentiation (97). 3DUS also may be important in evaluating complex hydroceles and assist in guidance when percutaneous drainage is appropriate. 3DUS also can assist in assessing invasion of the seminal vesicles in prostate cancer and can be used to guide staging biopsies of the seminal vesicles in select cases of prostate cancer (see earlier).

Clinical benefits of 3DUS

To date, no definite indications or benefits for 3DUS have been identified. Although it might be interesting to more clearly define and measure intratesticular tumors, such measurements are unlikely to affect management.

3DUS Limitations

Currently, 3D imaging adds to the length of time to perform all three types of scans. This might have a negative effect in cases of suspected torsion.

HOW TO DO IT
Practical Clinical Tips for the Scrotum, Penis, and Seminal Vesicles

- Linear transducers are most appropriate for scrotal and penile scanning.
- A generous amount of warm gel is necessary.
- Multiplanar imaging is a valuable display technique.
- Seminal vesicle scanning tips are the same as those for prostate ultrasound.

REFERENCES

1. Bree RL. Hematuria. In: Bluth EI, Arger PH, Hertzberg BS, Middleton WD, eds. *RSNA special course in ultrasound 1996*. Oak Brook: RSNA Publications, 1996:211–216.
2. Cronan JJ. Acute Urinary tract obstruction in the adult. In: McClennan BL, ed. *RSNA categorical course in genitourinary radiology 1994*. Oak Brook: RSNA Publications, 1994:103–108.
3. Dunnick NR, McCallum RW, Sandler CM. Examination techniques. In: Dunnick NR, McCallum RW, Sandler CM, eds. *Textbook of uroradiology*. Baltimore: Williams & Wilkins, 1991:41–69.
4. Lee RA, Kane RA, Lantz EJ, Charboneau JW. Intraoperative and laparoscopic ultrasound of the abdomen. In: Rumack CM, Wilson SR, Charboneau JW, eds. *Diagnostic ultrasound*, 2nd ed. St. Louis: Mosby, 1997: 671–699.
5. Thurston W, Wilson SR. The urinary tract. In: Rumack CM, Wilson SR, Charboneau JW, eds. *Diagnostic ultrasound*, 2nd ed. St. Louis: Mosby, 1997:329–397.
6. Downey DB. The retroperitoneum and great vessels. In: Rumack CM, Wilson SR, Charboneau JW, eds. *Diagnostic ultrasound*, 2nd ed. St. Louis: Mosby, 1997:453–486.
7. Rumack CM. Pediatric abdominal masses. In: Bluth EI, Arger PH, Hertzberg BS, Middleton WD, eds. *RSNA special course in ultrasound 1996*. Oak Brook: RSNA Publications, 1996:87–90.
8. Bosniak MA. Problematic renal masses. In: McClennan BL, ed. *RSNA categorical course in genitourinary radiology 1994*. Oak Brook: RSNA Publications, 1994:183–191.
9. Licht MR, Novick AC. Nephron sparing surgery for renal cell carcinoma. *J Urol* 1993;149:1–7.
10. Enislidis G, Wagner A, Ploder O, Ewers R. Computed intraoperative navigation guidance—a preliminary report on a new technique. *Br J Oral Maxillofac Surg* 1997;35:271–274.
11. Hata N, Dohi T, Iseki H, Takakura K. Development of a frameless and armless stereotactic neuronavigation. *Neurosurgery* 1997;41:608–613.
12. Rankin RN, Fenster A, Downey DB, Munk PL, Levin MF, Vellet AD. Three-dimensional sonographic reconstruction: techniques and diagnostic applications. *AJR* 1993;161:695–702.
13. Gilja OH, Thune N, Matre K, Hausken T, Odegaard S, Berstad A. *In vitro* evaluation of three-dimensional ultrasonography in volume estimation of abdominal organs. *Ultrasound Med Biol* 1994;20:157–165.
14. Gilja OH, Smievoll AI, Thune N, et al. *In vivo* comparison of 3D ultrasonography and magnetic resonance imaging in volume estimation of human kidneys. *Ultrasound Med Biol* 1995;21:25–32.
15. Hughes SW, D'Arcy TJD, Maxwell DJ, et al. Volume estimation from multiplanar 2D ultrasound images using a remote electromagnetic position and orientation. *Ultrasound Med Biol* 1996;22:561–572.
16. Hughes SW, D'Arcy TJ, Maxwell DJ, Saunders JE, Chinn S, Sheppard RJ. The accuracy of a new system for estimating organ volume using ultrasound. *Physiol Meas* 1997;18:73–84.
17. Cronan JJ. Renal Failure. In: Bluth EI, Arger PH, Hertzberg BS, Middleton WD, eds. *RSNA special course in ultrasound 1996*. Oak Brook: RSNA Publications, 1996:195–200.
18. Pruthi RS, Angell SK, Dubo F, Meiguerian P, Shortliff LD. The use of renal parenchymal area in children with high grade vesicoureteral reflux. *J Urol* 1997;158:1232–1235.
19. Absy M, Metreweli C, Matthews C, Al Khadar A. Changes in transplanted kidney volume measured by ultrasound. *Br J Radiol* 1987;60:525–529.
20. Kelcz F, Pozniak MA, Pirsch JD, Oberly TD. Pyramidal appearance and resistive index: insensitive and nonspecific sonographic indicators of renal transplant rejection. *AJR* 1990;155:531–535.
21. Grant EG, Perrella RR. Wishing won't make it so: duplex sonography in the evaluation of renal transplant dysfunction. *AJR* 1990;155:538–539.
22. Merz E, Weber G, Bahlmann F, Macchiella D. [3DUS in prenatal diagnosis]. *Gynakol Geburtshilfliche Rundsch* 1995;35(suppl 1):118–121.
23. Davidson AJ. A Systematic Approach to the radiological diagnosis of parenchymal disease of the kidney. In: Davidson AJ, ed. *Radiology of the kidney*. Philadelphia: WB Saunders, 1985:113–124.
24. Stanley RJ. Benign Renal Neoplasms. In: McClennan BL, ed. *RSNA categorical course in genitourinary radiology 1994*. Oak Brook: RSNA Publication, 1994:193–202.
25. Liu JB, Miller LS, Bagley DH, Bonn J, Forsberg F, Goldberg BB. Three-dimensional endoluminal ultrasound. In: Liu JB, Goldberg BB, eds. *Endoluminal ultrasound: vascular and nonvascular applications*. St. Louis: Mosby, 1997:325–346.
26. Bagley DH, Liu JB, Goldberg BB. Use of endoluminal ultrasound of the ureter. *Semin Urol* 1992;10: 194–198.
27. Bagley DH, Liu JB. Upper Urinary Tract. In: Liu JB, Goldberg BB, eds. *Endoluminal ultrasound: vascular and nonvascular applications*. St. Louis: Mosby, 1997:81–122.
28. Jolic V, Gilja I. Vaginal vs. transabdominal ultrasonography in the evaluation of female urinary tract anatomy, stress urinary incontinence and pelvic organs static disturbances. *Zentralbl Gynakol* 1997;119:483–491.

29. Kuo HC. Transrectal sonography of the female urethra in incontinence and frequency-urgency syndrome. *J Ultrasound Med* 1996;15:363–370.
30. Chancellor MB, Liu JB. Lower Urinary Tract. In: Liu JB, Goldberg BB, eds. *Endoluminal ultrasound: vascular and nonvascular applications.* St. Louis, MO: Mosby, 1997:123–146.
31. Ng KJ, Gardener JE, Rickards D, Lees WR, Milroy EJ. Three-dimensional imaging of the prostatic urethra—an exciting new tool. *Br J Urol* 1994;74:604–608.
32. Riccabona M, Nelson TR, Pretorius DH, Davidson TE. *In vivo* three-dimensional sonographic measurement of organ volume: validation in the urinary bladder. *J Ultrasound Med* 1996;15:627–632.
33. Simforoosh N, Dadkhah F, Hosseini SY, Asgari MA, Nasseri A, Safarinejad MR. Accuracy of residual urine measurement in men: comparison between real-time ultrasonography and catheterization. *J Urol* 1997; 158:59–61.
34. Flannigan GM. Editorial: imaging, stone disease. *Curr Opin Urol* 1997;7:U10–U12.
35. Sakr WA, Grignon DJ, Crissman JD. Histopathology of transitional cell carcinoma of the urinary bladder. *Curr Opin Urol* 1997;7:287–292.
36. Clements R. The changing role of transrectal ultrasound in the diagnosis of prostate cancer. *Clin Radiol* 1996;51:671–676.
37. El Din KE, De la Rosette JJ. Transrectal ultrasonography of the prostate. *Br J Urol* 1996;78:2–9.
38. Watanabe H, Kaiho H, Tanaka M, Tersawa Y. Diagnostic application of ultrasonography to the prostate. *Invest Urol* 1971;8:548–549.
39. Watanabe H. History and applications of transrectal sonography of the prostate. *Urol Clin North Am* 1989; 16:617–622.
40. Lee F, Torp-Petersen ST, Siders DB, Littrup PJ, McLeary RD. Transrectal ultrasound in the diagnosis and staging of prostate carcinoma. *Radiology* 1989;170:609–615.
41. McNeal JE. Regional morphology and pathology of the prostate. *Am J Clin Pathol* 1968;49:347–357.
42. McNeal JE. The zonal anatomy of the prostate. *Prostate* 1981;2:35–49.
43. McNeal JE, Redwine EA, Freiha FS, Stamey TA. Zonal distribution of prostatic adenocarcinoma: correlation with histologic pattern and direction of spread. *Am J Surg Pathol* 1988;12:897–906.
44. Reissigl A, Pointner J, Strasser H, Ennemoser O, Klocker H, Bartsch G. Frequency and clinical significance of transition zone cancer in prostate cancer screening. *Prostate* 1997;30:130–135.
45. Stamey TA, McNeal JE, Freiha FS, Redwine E. Morphometric and clinical studies on 68 consecutive radical prostatectomies. *J Urol* 1988;139:1235–1241.
46. Lee F, Littrup PJ, Christensen LL, et al. Predicted prostate specific antigen results using transrectal ultrasound gland volume: differentiation of benign prostate hypertrophy from prostate cancer. *Cancer* 1992;70:211–220.
47. Littrup PJ, Kane RA, Williams CR, et al. Determination of prostate volume with transrectal US for cancer screening. Part 1. Comparison with prostate-specific antigen studies. *Radiology* 1991;178:537–542.
48. Littrup PJ, Williams CR, Egglin TK, Kane RA. Determination of prostate volume with transrectal US for cancer screening. Part 2. Accuracy of *in vitro* and *in vivo* techniques. *Radiology* 1991;179:49–53.
49. Chin JL, Downey DB, Onik G, Fenster A. Three-dimensional prostate ultrasound and its application to cryosurgery. *Tech Urol* 1996;2:187–193.
50. Stock RG, Stone NN, Wesson MF, DeWyngaert JK. A modified technique allowing interactive ultrasound-guided three-dimensional transperineal prostate implantation. *Int J Radiat Oncol Biol Phys* 1995;32:219–225.
51. Stone NN, Stock RG. Brachytherapy for prostate cancer: real-time three-dimensional interactive seed implantation. *Tech Urol* 1995;1:72–80.
52. Strasser H, Janetschek G, Horninger W, Bartsch G. Three-dimensional sonographic guidance for interstitial laser therapy in benign prostatic hyperplasia. *J Endourol* 1995;9:497–501.
53. Richard WD, Grimmell CK, Bedigan K, Frank KJ. A method for three-dimensional prostate imaging using transrectal ultrasound. *Comput Med Imaging Graph* 1993;17:73–79.
54. Richard WD, Keen CG. Automated texture-based segmentation of ultrasound images of the prostate. *Comput Med Imaging Graph* 1996;20:131–140.
55. Downey DB, Fenster A. Three-dimensional power Doppler detection of prostatic cancer [letter]. *AJR* 1995; 165:741.
56. Lees WR, Balen F, Allen CM, Kessel D. Diagnostic value of 3D tumor blood flow maps derived from Doppler data. *Radiology* 1994;193(P):335–336.
57. Tong S, Downey DB, Cardinal HN, Fenster A. A three-dimensional ultrasound prostate imaging system. *Ultrasound Med Biol* 1996;22:735–746.
58. Bree RL. The Prostate. In: Rumack CM, Wilson SR, Charboneau JW, eds. *The prostate*, 2nd ed. St. Louis: Mosby, 1997:399–429.
59. Littrup PJ, Kane RA, Mettlin CJ, et al. Cost-effective prostate cancer detection. Reduction of low-yield biopsies. Investigators of the American Cancer Society National Prostate Cancer Detection Project [see comments]. *Cancer* 1994;74:3146–3158.
60. Basset O, Bimenez G, Mestas JL, Cathignol D, Devonec M. Volume measurement by ultrasonic transverse or sagittal cross-sectional scanning. *Ultrasound Med Biol* 1991;17:291–296.
61. Elliot TL, Downey DB, Tong S, Mclean CA, Fenster A. Accuracy of prostate volume measurements *in vitro* using three-dimensional ultrasound. *Acad Radiol* 1996;3:401–406.
62. Sehgal CM, Broderick GA, Whittington R, Gorniak RJ, Arger PH. Three-dimensional US and volumetric assessment of the prostate. *Radiology* 1994;192:274–278.
63. Aarnink RG, Giesen RJ, De la Rosette JJ, Huynen AL, Debruyne FM, Wijkstra H. Planimetric volumetry of the prostate: how accurate is it? *Physiol Meas* 1995;16:141–150.
64. Aarnink RG, Huynen AL, Giesen RJ, De la Rosette JJ, Debruyne FM, Wijkstra H. Automated prostate volume determination with ultrasonographic imaging [see comments]. *J Urol* 1995;153:1549–1554.
65. Aarnink RG, De la Rosette JJ, Huynen AL, Giesen RJ, Debruyne FM, Wijkstra H. Standardized assessment to enhance the diagnostic value of prostate volume. Part I. Morphometry in patients with lower urinary tract symptoms. *Prostate* 1996;29:317–326.
66. Brewster SF. Antimicrobial prophylaxis for transrectal prostatic biopsy. *Curr Opin Urol* 1997;7:57–60.

67. Catalona JW, Smith DS, Ratliff TL, et al. Measurement of prostate-specific antigen in the serum as a screening test for prostate cancer. *N Engl J Med* 1991;324:1156–1161.
68. Aarnink RG, De la Rosette JJ, Huynen AL, Giesen RJ, Debruyne FM, Wijkstra H. Standardized assessment to enhance the diagnostic value of prostate volume. Part II. Correlation with prostate-specific antigen levels. *Prostate* 1996;29:327–333.
69. Kalish J, Cooner WH, Graham SD Jr. Serum PSA adjusted for volume of transition zone (PSAT) is more accurate than PSA adjusted for total gland volume (PSAD) in detecting adenocarcinoma of the prostate. *Urology* 1994;43:601–606.
70. Chen CH, Lee YH, Yang WH, Chang CM, Sun YM. Segmentation and reconstruction of prostate from transrectal ultrasound imaging. *Biomed Eng Applic Basis Communications* 1995;150:395–399.
71. Strasser H, Janetschek G, Reissigl A, Bartsch G. Prostate zones in three-dimensional transrectal ultrasound. *Urology* 1996;47:485–490.
72. Tong S, Cardinal HN, Downey DB, Fenster A. Inter- and intra-observer variability and reliability of prostate volume measurement via 2D and 3D ultrasound imaging. *Ultrasound Med Biol* 1998;24:673–681.
73. Madersbacher S, Djavan B, Marberger M. Minimally invasive treatment for benign prostatic hyperplasia. *Curr Opin Urol* 1998;8:17–26.
74. McConnell JC. Benign prostatic hyperplasia. *Curr Opin Urol* 1998;8:1–3.
75. Alexander AA. Editorial to color Doppler image the prostate or not: that is the question. *Radiology* 1995; 195:11–13.
76. Lee F, Torp-Petersen S, Littrup PJ, et al. Hypoechoic lesions of the prostate: clinical relevance of tumor size, digital rectal examination and prostate-specific antigen. *Radiology* 1989;170:29–32.
77. Rifkin MD, Sudakoff GS, Alexander AA. Prostate: techniques, results and potential applications of color Doppler US scanning. *Radiology* 1993;186:509–513.
78. Spencer JA, Alexander AA, Gomella L, Matteucci T, Goldberg BB. Clinical and US findings in prostate cancer: patients with normal prostate-specific antigen levels [see comments]. *Radiology* 1993;189:389–393.
79. Shinohara K, Wheeler TM, Scardino PT. The appearance of prostate cancer on transrectal ultrasound: correlation of imaging and pathological examinations. *J Urol* 1989;142:76–82.
80. Downey DB. Power Doppler in prostate cancer. *Curr Opin Urol* 1997;7:93–99.
81. Terris MK, McNeal JE, Stamey TA. Detection of clinically significant prostate cancer by transrectal ultrasound-guided systematic biopsy. *J Urol* 1992;148:829–832.
82. De la Rosette JJ, Giesen RJ, Huynen AL, Aarnink RG, Debruyne FM, Wijkstra H. Computerized analysis of transrectal ultrasonography images in the detection of prostate carcinoma. *Br J Urol* 1995;75:485–491.
83. De la Rosette JJ, Giesen RJ, Huynen AL, et al. Automated analysis and interpretation of transrectal ultrasonography images in patients with prostatitis. *Eur Urol* 1995;27:47–53.
84. Giesen RJ, Huynen AL, Aarnink RG, et al. Computer analysis of transrectal ultrasound images of the prostate for the detection of carcinoma: a prospective study in radical prostatectomy specimens. *J Urol* 1995;154: 1397–1400.
85. Giesen RJ, Huynen AL, Aarnink RG, De la Rosette JJ, Debruyne FM, Wijkstra H. Construction and application of hierarchical decision tree for classification of ultrasonographic prostate images. *Med Biol Eng Comput* 1996;34:1.
86. Hendrix AJM, Wijkstra H, Maes RM, et al. Audex medical. A new system for digital processing and analysis of ultrasonographic images of the prostate. *Scand J Urol Nephrol Suppl* 1991;137:95–100.
87. Jager GJ. Magnetic resonance imaging in prostate cancer. *Curr Opin Urol* 1997;7:88–92.
88. Naito K, Yoshihiro S, Oba K, Yamamoto M. New assay for staging and prognostic information in prostate cancer. *Curr Opin Urol* 1997;7:263–267.
89. Prescott S. Editorial: prostate. *Curr Opin Urol* 1997;7:U61–U62.
90. Rifkin MD, Zerhouni EA, Gatsonis CA, et al. Comparison of magnetic resonance imaging and ultrasonography in staging early prostate cancer—results of a multi-institutional cooperative trial. *N Engl J Med* 1990; 323:621–625.
91. McNeal JE, Kindrachuk RA, Freiha FS, Bostwick DG, Redwine EA, Stamey TA. Patterns of progression in prostate cancer. *Lancet* 1986;1:60–63.
92. Fenster A, Downey DB. 3DUS imaging: A review. *IEEE Eng Med Biol* 1996;15:41–51.
93. D'Amico AV, Chang H, Holupka E, et al. Calculated prostate cancer volume: the optimal predictor of actual cancer volume and pathologic stage. *Urology* 1997;49:385–391.
94. Feleppa EJ, Fair WR, Kalisz A, Larchian W, Liu T, Reuter V, Rosado A. Spectrum analysis and three-dimensional imaging for prostate evaluation. *Mol Urol* 1997;1:109–116.
95. Chapple CR. Alpha blocker update—prostate specific blockade. *Curr Opin Urol* 1997;7:8–14.
96. Horstman WG, Middleton WD. Testicular and scrotal imaging. In: McClennan BL, ed. *RSNA categorical course in genitourinary radiology 1994.* Oak Brook: RSNA Publications, 1994:159–174.
97. Tartar VM, Trambert MA, Balsara ZN, Matterey RF. Tubular extasia of the testicle: sonographic and MR imaging appearance. *AJR* 1993;160:539–542.
98. Downey DB, Fenster A, Three-dimensional ultrasound: a maturing technology. *Ultrasound Q* 1998;14: 25–40.

9

Abdomen

OVERVIEW

This chapter presents an overview of the current clinical utility and limitations of three-dimensional ultrasound (3DUS) imaging in the abdomen, including the liver and spleen, the pancreas and lymph nodes, the gallbladder and biliary system, and the gastrointestinal tract. The genitourinary tract is reviewed in Chapter 8 and vascular imaging is discussed in Chapter 10. Potential future benefits of the technology are also discussed.

KEY CONCEPTS

- Data sets from the abdomen tend to be more prone to artifacts.
- Free-hand scanning is a good way to collect the data.
- Volume rendering of color and power Doppler data produces elegant images, although its potential clinical utility remains to be established.
- 3DUS can improve the measuring capabilities of endoluminal scanning.
- 3DUS can provide a more panoramic rendering of endoluminal data.
- 3DUS can measure structures more accurately and reproducibly with simplified and standardized acquisition protocols.

3DUS OF THE LIVER

Obtaining good three-dimensional ultrasound (3DUS) data sets of the entire liver may required imaging from several different positions to obtain an adequate exam because the organ is too large to be seen from any single perspective and because the liver can be partially obscured by the ribs, lungs, and bowel gas. Motion artifacts from respiration can pose a problem, although vascular and cardiac motion also can interfere with scan quality. Despite these issues, several studies indicate that 3DUS increased volumetric measurement accuracy and repeatability can be of clinical benefit in the management of some liver diseases, both prognostically and to gauge the volume of remaining tissue after resection, or for deciding the best size for donor organs. Volume-rendered 3DUS power Doppler images of the liver and spleen show the blood vessels to good effect, though their clinical utility remains unclear.

At 1500 grams the liver is the largest solid organ in the body. It is affected frequently by both local and systemic diseases (1). Ultrasound plays an extremely important role in evaluating liver diseases worldwide (1). In developed countries computed tomography (CT), magnetic resonance imaging (MRI), and ultrasound all have both complementary and competing roles in assessing the liver (2,3). Two-dimensional ultrasound (2DUS) often is the initial liver imaging test performed, because it is more portable, more flexible, and less expensive, and because it provides dynamic information about blood flow and organ motion (1). 2DUS also is the modality of choice for guiding many abdominal interventional procedures, because the imaging is in real time (4). The limitations of 2DUS include a lesser ability to give an overall panoramic view of large structures, such as the liver, and a sensitivity to artifacts, because of degradation by gas, bone, and to a lesser extent, fat. Until recently, 2DUS also lacked the ability to show enhancement using contrast agents, although this is changing with the recent arrival of effective ultrasound contrast materials (5).

CLINICAL BENEFITS OF 3DUS

Improved Understanding of Complex Anatomy

Liver scanning is hampered by patient motion and limited scanning access (1). The liver is partially obscured by bowel gas, the ribs, and the lungs. Usually, it is necessary to image the patient in different positions at different phases of respiration. Typically, one must scan from at least three different scanning positions to obtain an adequate exam, because the organ is too large to be seen from any single perspective. The examiner must visualize exactly which portions of the liver have been imaged from each transducer location to be sure the entire organ has been evaluated. This can be challenging. A good 3DUS system should document exactly which portions of the liver were evaluated, and therefore facilitate a more thorough exam. Preliminary results by Wagner et al. are very encouraging. In 28% of 93 patients with chronic liver disease, 3DUS was superior to 2DUS. Improved assessment of the extent and localization of liver tumors was seen in nine of 20 cases (45%). In addition, improved visualization of vascular thrombi were seen in 2 of 5 patients with Budd–Chiari syndrome (Figure 9-1)(6).

Over the last decade hepatic surgeons have altered their perspective of patients with primary liver cancers and metastatic colon cancers. They no longer view them as end-stage victims but as possible long-term survivors, as up to 25% of cases are potentially resectable for cure (2). This has led to an increasing number of benign and malignant liver masses being treated surgically (3). Resections can be focal wedge resections, in which small superficial lesions are shelled out. Alternatively, more formal liver resections based on Couinaud's segmental liver anatomy can be performed (2,7). These operations include lobectomies, segmentectomies, and trisegmentectomies. Preoperatively, the surgeons need to know not only which of Couinaud's segments are involved, but also how close the lesions are to the edge of each segment (Figure 9-2). In addition, there is considerable anatomic variation in these segments, especially in the right lobe (8). Currently, the ultrasound operator initially has to check for anatomic variations and then determine the relationship of each edge of the lesion to the nearest anatomic border of the segment where it is located (Figures 9-2, 9-3) (9). This process is time-consuming and extremely operator dependent. Interacting with a 3DUS data set on a computer workstation would be faster and produce less variable results. It is also clear that 3D planning displays help surgeons plan their surgeries better (8,10–14).

Volume Measurements of the Liver

Several studies indicate that 3DUS's increased volumetric measurement accuracy and repeatability is of clinical benefit in the management of some liver diseases (15). Hughes

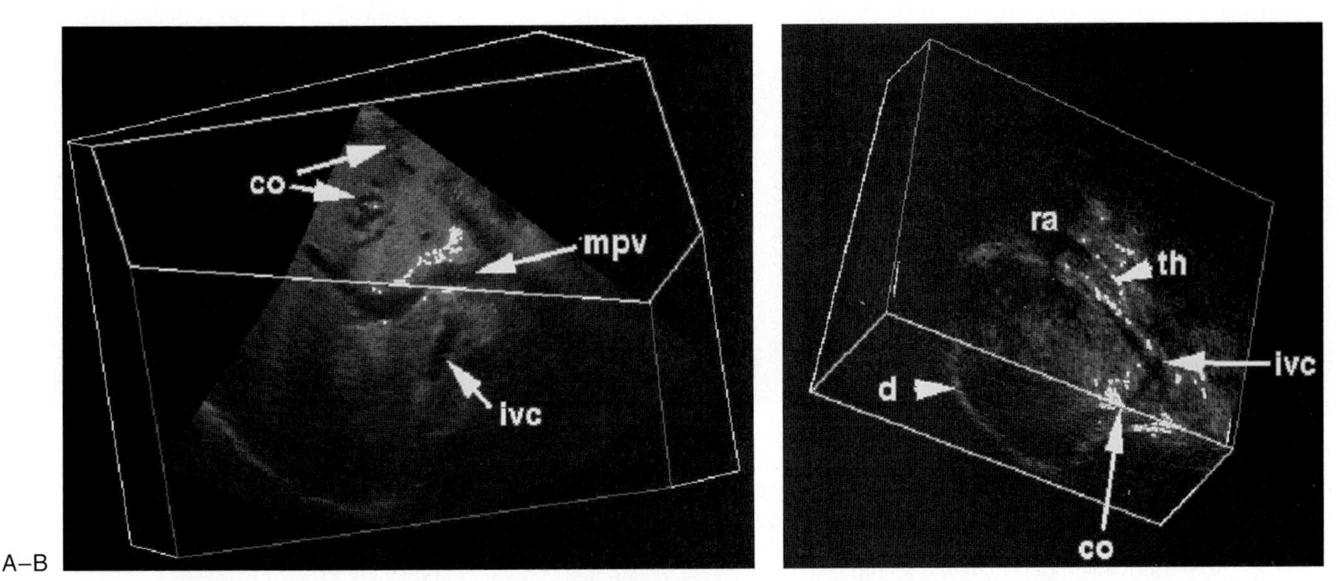

Figure 9-1. Budd–Chiari syndrome. **A:** By combining multiple oblique sections into the volume a tortuous collateral (co) that extends from the left main portal vein to the falciform ligament can be traced. MPV, main portal vein; ivc, inferior vena cava. **B:** This coronal perspective through the liver showing a thrombus (th) in the upper portion of the inferior vena cava (ivc) just below the right atrium (ra). d, diaphragm; co, collateral vein.

suggests that sequential volumetric assays of the fetal liver and other organs are more accurate and repeatable than standard ultrasound measurements and advocates using the technology in screening cases of suspected intrauterine growth retardation (16). Liess showed that 3DUS provided more accurate and repeatable measurements of focal liver lesions than either 2DUS or CT in sequential studies (15). Wolf et al. have shown that 3DUS not only is more accurate than 2DUS and comparable to three-dimensional dynamic computed tomography (3D-DCT) for measurement of liver volume but also is faster than 3D-DCT (17). This is extremely important information, as it is crucial to have an accurate

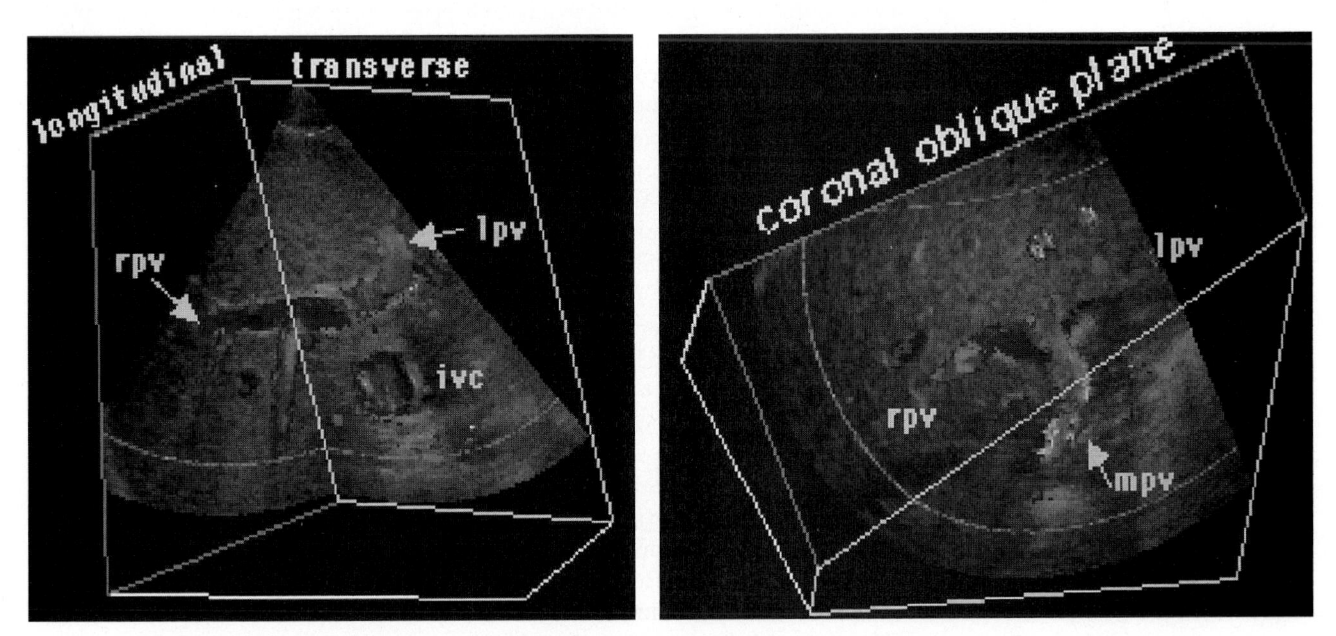

Figure 9-2. Normal portal venous anatomy. The complex anatomy of the portal veins can be viewed from a variety of different perspectives by interactively slicing into the volume cube. rpv, right portal vein; mpv, main portal vein; lpv, left portal vein; ivc, inferior vena cava. See color version of figure.

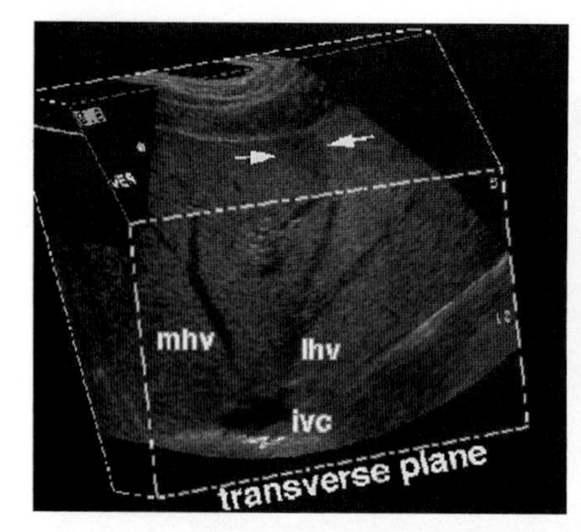

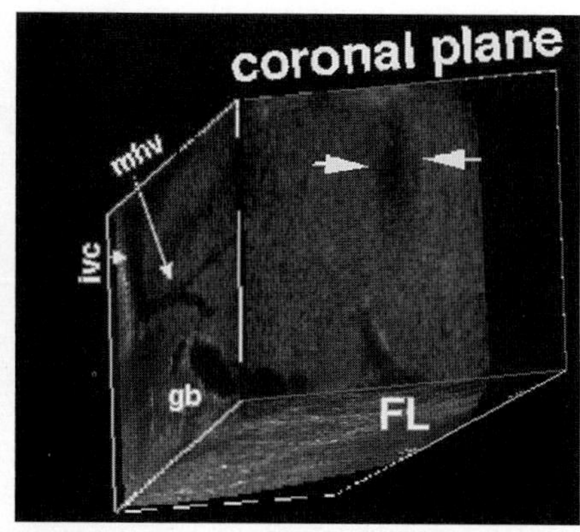

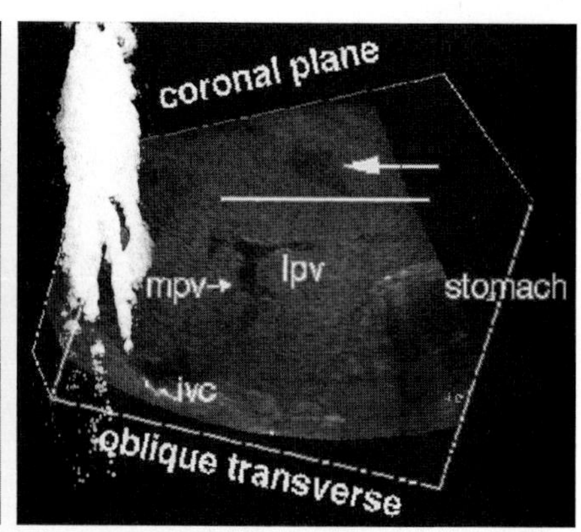

Figure 9-3. Resolving liver abscess in a 37-year-old man. The hypoechoic area anteriorly in the left lobe (*arrows*) is clearly seen on standard transverse image. Its exact relationship to the left hepatic vein can be seen on an appropriately angled oblique plane. The coronal plane provides an interesting perspective on this lesion. Its relationship to the falciform ligament (FL) is well seen. ivc, inferior vena cava; mhv, middle hepatic vein; gb, gallbladder. This view shows the relationship of the lesion to the left portal vein (lpv). By using a combination of these views, it is possible to precisely locate the lesion in the appropriate segment of the liver and determine its crucial relationships. mpv, main portal vein; ivc, inferior vena cava.

measurement of the volumes of liver lesions that are to be treated with minimally invasive therapies. The amount of therapy given to the patient is usually directly proportional to the volume of the lesion. For example, with alcohol ablation of hepatocellular carcinomas in cirrhotic patients, the dose of alcohol injected into the lesion is directly proportional to the volume of the tumor (18).

Yamashita commented that accurate classification of the histologic subtype of hepatocellular cancer (HCC) is important in deciding the optimal, minimally invasive therapy (19). He suggests that quantification of the amount of vascularity in these hepatocellular carcinomas also may influence which method is the optimal treatment (19). He noted that the hypervascular lesions that are seen most commonly in the frank type of hepatocellular cancer generally have a good response to transcatheter arterioembolization (TAE). Sclerosing, poorly differentiated, or undifferentiated hepatocellular carcinomas tend to be poorly vascularized and do poorly following transcatheter arterial embolization. These, he suggested, would be better treated with alcohol ablation. Whether the accurate quantification of blood flow within the tumor using 3DUS can obviate the need of biopsy needs to be studied in a prospective manner.

Knowing the true volume of the entire liver is clinically important sometimes. Both native and transplant liver tissue have the ability to regenerate, and imaging often is used to monitor this regeneration process following surgery (20). Prior to performing a hepatic resection it is essential to assess what volume of residual liver will remain following the procedure. Although hepatic regeneration will occur in most cases, resecting too much liver tissue can result in hepatic failure and death (21). Caldwell argues that it is important to measure the volume of the liver accurately in patients with advanced liver disease who are being considered for orthotopic liver transplantation (22). This information can be used to gauge the appropriate size of donor organs and may have prognostic value. He noted that lower mean liver volumes were observed in Child–Pugh class C patients as opposed to those with Child–Pugh class AB.

To date, MRI and volumetric CT are the modalities most frequently employed for these assessments (20,22). 3DUS has the potential to replace CT and MRI in this role, if the technical problems in acquiring high-quality data sets can be overcome. This would be desirable, as assessment could be done more frequently, since ultrasound is more readily available, uses no ionizing radiation, and is less expensive.

Volumetric and functional blood flow parameters are different in a transplant liver compared with a normal liver (23). More accurate assessment of these parameters may

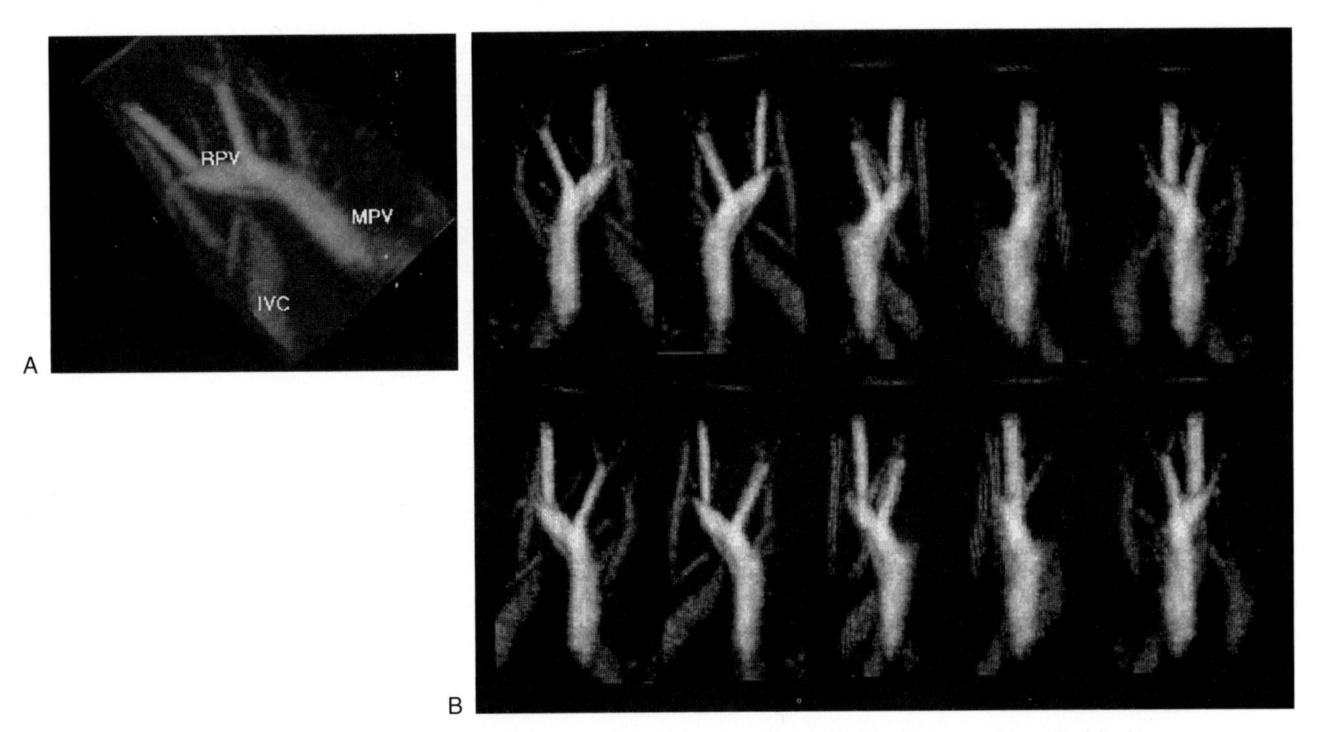

Figure 9-4. Maximum-intensity projection (MIP) images of the portal venous system. **A:** Maximum-intensity projection-rendered image shows the relationship of the main portal vein (MPV) to the right main portal vein branch (RPV) and the inferior vena cava (IVC). Data are unenhanced power Doppler data acquisitions from a healthy volunteer. **B:** A series of these MIP images can be rendered from different perspectives and combined into a video loop. It is possible to produce a rotational effect by displaying the images rapidly. This can increase understanding of the 3D relationships of the vessels.

be possible with 3DUS, with or without contrast agents. However, the actual clinical benefits of achieving this currently are unknown.

3DUS power Doppler of the portal veins produces relatively clear images of the portal venous system when viewed in volume-rendered mode (Figure 9-4) (24,25). There remains tremendous potential for this technique in assessing blood flow patterns within the liver, although this is as yet unproven. In addition, assessing the patency of portal systemic shunts and transjugular-intrahepatic-portosystemic shunt (TIPS) procedures may be more elegantly displayed in three dimensions, especially with 3DUS power rendering. The role of contrast enhancement of liver tumors using three dimensions holds tremendous promise, although this is still untested.

Improved Guidance for Interventional Procedures

Although it is relatively easy to obtain biopsies from most lesions located caudally in the liver (26), obtaining satisfactory biopsies of lesions high up in the right lobe often proves extremely difficult. This is especially true in patients who are obese or who have difficulty maintaining even respirations. In these patients, fast 3DUS guidance may prove useful. In addition, the number and types of interventional procedures in the liver are increasing, and having appropriate treatment planning and guidance software coupled with the 3DUS would be extremely useful. (See Chapter 14.)

3DUS Limitations

It is extremely difficult to obtain accurate 3DUS data sets of the entire liver in most patients because of the previously described size and scanning access problems. The sheer size of the data sets necessary to show reasonable detail in the liver is prohibitive. Often one is left with a 3DUS scan of a portion of the liver. Motion artifacts due to respiration also can be a considerable problem. Vascular motion from the aorta also can interfere with the ability to get good-quality scans. Cardiac motion most frequently degrades imaging of the left lobe of the liver. As 3DUS data usually are obtained from one particular perspective, the overall quality of the images may not be as good as that from a conventional 2DUS scan that is optimized for a particular image plane. With conventional 2DUS the operator frequently makes small adjustments in patient positioning to improve the image, which is impossible with 3D scans. This is why the 2DUS prescan is essential to optimize scan quality before acquiring the volume data.

3DUS OF THE SPLEEN

The spleen typically measures 12 cm × 7 cm × 3.5 cm and has an average weight of 150 grams. As with the liver, scanning access is limited by ribs, bowel gas, and lungs (27). Diffuse splenomegaly is the most commonly diagnosed abnormality within the spleen. Focal splenic abnormalities include cysts, benign and malignant tumors, and calcifications. Post-traumatic changes within the spleen include tears and fragmentation and the usual accompaniment either of a localized or a diffuse bleed. Accessory splenunculi are extremely common.

Clinical Benefit of 3DUS

The diagnosis of splenomegaly is most frequently made visually because exact volume measurements determined by 2DUS are limited predominantly by the variable, irregular contour of the spleen. 3DUS has the potential to make this diagnosis more accurate (27,28).

De Odorico et al. calculated splenic volumes using a volume transducer (Medison, Korea) in 52 healthy volunteers (28). Splenic volumes were calculated automatically after serial slices each had a manually drawn region of interest assigned around the spleen.

Figure 9-5. 3DUS power Doppler maximum-intensity projection (MIP) images of a normal spleen acquired using a free-hand scan without position sensing. (Image courtesy of ATL.)

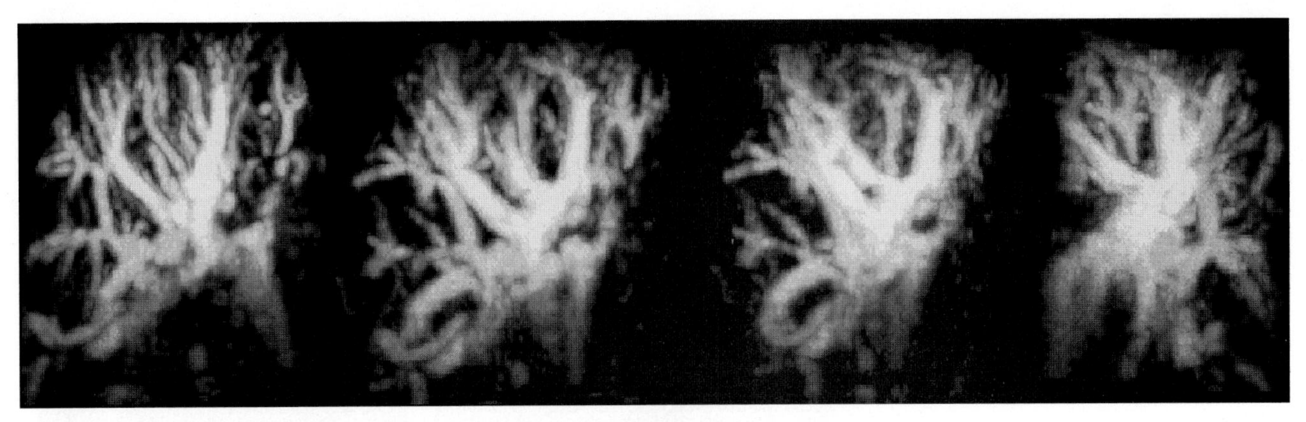

Figure 9-6. 3DUS power Doppler MIP projections of the vasculature of a normal spleen shown from different perspectives. Multiple such images can be displayed rotating in a cineloop, which can increase understanding of the 3D nature of the spleen.

3DUS measurements of splenic volume were consistently smaller than 2DUS estimations and were within ±2% of *in vitro* water displacement volume measurements.

The exact clinical benefit of increased volumetric measurement accuracy remains unclear. It is well known that the spleen enlarges because of a large number of clinical conditions, including infection, portal hypertension, leukemia, and lymphoma. There is, however, often little correlation between the degree of splenomegaly and the severity of the disease. Similarly, the spleen size may never return to normal even after the disease that caused the enlargement has disappeared. One advantage and contribution of 3DUS may be in preventing individuals with long, thin spleens being labeled as having splenomegaly.

High-quality 3DUS power Doppler images of the spleen are relatively easily obtained (24). When they are volume rendered without the B-mode data, they show the intrasplenic vascularity to good effect (Figures 9-5, 9-6), although the clinical value has not been evaluated at this time. This type of evaluation may assist in evaluating splenic infarcts and trauma.

3DUS Limitations

Care must be taken to ensure that the entire spleen is included in the study. Artifacts can arise from motion and from shadowing from overlying structures.

HOW TO DO IT
Practical Clinical Tips for the Liver and Spleen

- Free-hand scanning with a magnetic localizer is the best method of scanning the liver/spleen.
- Mechanical positioners may produce good images of small segments of the liver.
- When possible, it is desirable to image entire volume during one breath-hold.
- Ensure that the liver/spleen area of interest is centered in the scan field.
- Multiplanar imaging is the best display method for assessing B-mode structures.
- Volume rendering is also useful for assessing portal vessels in the liver and spleen.
- In the presence of ascites, interesting renderings of cirrhotic livers are possible.

3DUS OF THE PANCREAS

Although ultrasound remains the most widely available and least expensive method of visualizing the pancreas, there is considerable controversy over its exact role relative to

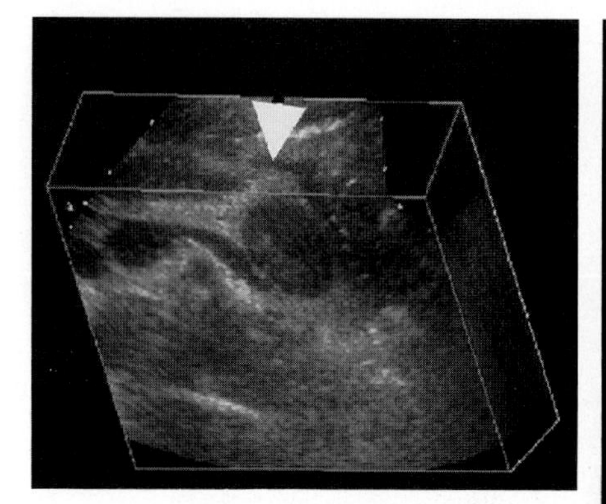

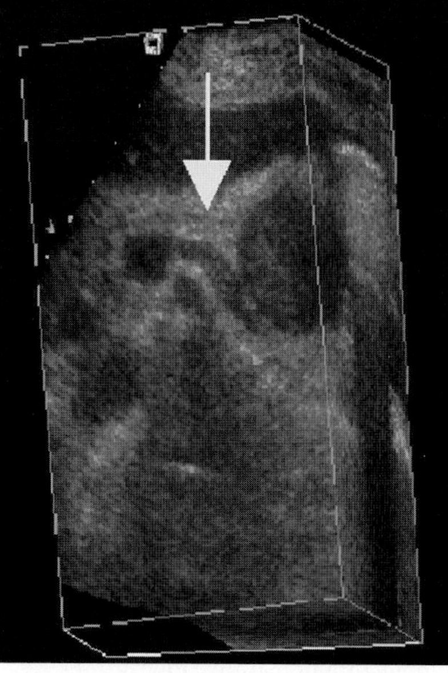

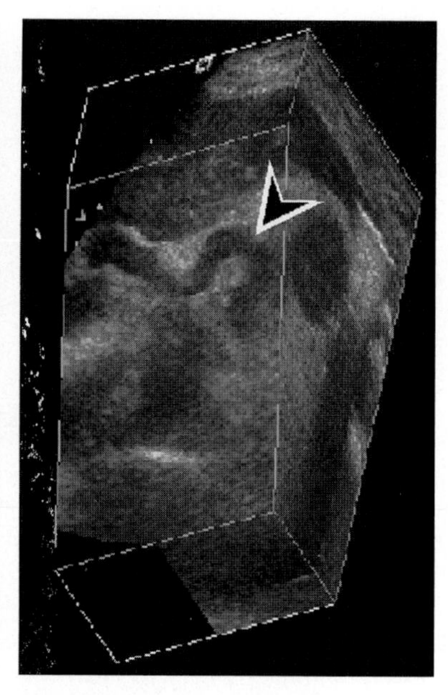

Figure 9-7. Sharply demarcated pancreatic cancer. The coronal perspective shows a small amount of normal pancreatic tissue (*arrowhead*) surrounding the hypoechoic mass. This confirmed that the lesion was truly in the pancreas. This perspective showed the mass to lie within 2 mm of the splenic vein. A transverse oblique image shows the pancreatic duct (*arrow*) can be traced right up to abut the hypoechoic mass, again confirming that the lesion was in the pancreas. The relationship of the mass to the tortuous splenic artery (*black arrowhead*) is clearly seen.

the other imaging modalities such as computed tomography, magnetic resonance imaging, endoscopic retrograde cholangio pancreatography (ERCP), and more recently, endoscopic ultrasound (29). Despite this, pancreatic disease is frequently evaluated with ultrasound in most institutions.

In scanning the pancreas the sonographer faces challenges in visualizing the entire organ, detecting disease within it, and deciding the etiology and extent once the disease has been definitively diagnosed (27). Most pancreases can be seen with ultrasound, although the tail is frequently obscured (Figure 9-7) (29). Lesions seen in or about the pancreas usually can be subdivided into cystic or solid categories. Ultrasound rarely can diagnose the exact cause of a solid intrapancreatic abnormality, and biopsy is usually necessary to distinguish between the two most common disorders—pancreatic adenocarcinoma and focal pancreatitis. Ultrasound has a well-established role in guiding these biopsies, and in drainage procedures of pancreatic and peripancreatic collections.

Transduodenal and transgastric endoscopic sonography of pancreatic cancer is a relatively new technique that is increasing in popularity (30). Although invasive, it produces excellent pictures of the entire pancreas in more than 95% of patients (30). The technique has been most frequently used to assess the pancreas for the presence of small tumors such as those seen in multiple endocrinadenopathies and Zollinger–Ellison syndromes. Clinical utility has also been reported in staging pancreatic cancer (30). In addition, tiny catheter-based transducers have been introduced for separate endoscopes with frequencies between 15 and 20 MHz, compared with 7.5–12.5 MHz for the endoscopic transducers. These have less utility in the pancreas because they can rarely image through the bowel wall (30).

Clinical Benefits of 3DUS

3DUS has little chance of detecting pancreatic lesions not already seen with 2DUS, but it may have a role in assessing the proximity of detected lesions to vessels (see Figure 9-7), which may have a benefit in deciding on appropriate therapy (e.g., whether to operate on a pancreatic cancer). 3DUS also may be helpful in deciding the extent of peripancreatic cystic disease due to either pseudocyst formation or tumors and may be helpful in measuring intrapancreatic lesions more accurately, which may be clinically useful in following the pseudocyst.

Whether adding 3DUS capability to the endoscopic or intraluminal device increases its accuracy is currently unknown. However, if it proves similar to the 3D intraluminal imaging elsewhere in the body (31), it will be helpful in assessing the exact extent of masses. Selner et al. commented on the importance of tumor volume in the prognosis of radically periumpullary cancers. He found that the larger the cancers, the worse the prognosis (32). 3DUS should provide more accurate measurements.

3DUS Limitations

Achieving good-quality scans, especially of the pancreatic tail, may be very difficult because of intervening bowel gas with transcutaneous scanning. Endoscopic and catheter-based 3DUS imaging have limitations similar to those discussed later for bowel imaging.

HOW TO DO IT
Practical Clinical Tips for the Pancreas

- Free-hand magnetic scanning will probably give the best images, although mechanical tilting scans may also provide good data for transcutaneous scans.
- Having a fasting patient and pressing firmly on the abdomen will be beneficial.

3DUS OF THE LYMPH NODES

Traditionally, intra-abdominal nodal pathology has been examined with CT, and it is likely for the foreseeable future that CT will remain the imaging modality of choice (33). This is because bowel gas frequently obscures intra-abdominal nodes, especially in the para-aortic area. Normal lymph nodes are seen with ultrasound in 72% of healthy subjects in the peripancreatic area and the liver hilum regions, and enlarged nodes can usually be appreciated in all but the most obese subjects (34). Enlarged para-aortic nodes may also be seen at sonography, although this is less frequent (33).

Clinical Benefits of 3DUS

Some data suggest that accurate measurement of lymph nodes may prove helpful in diagnosing some diseases (34) and that serial measurements may help monitor the activity in other diseases (33,35,36). Because 3DUS measurements are more accurate than standard ultrasound, it may have a role in the sequential assessment of nodal disease (see Figure 9-8).

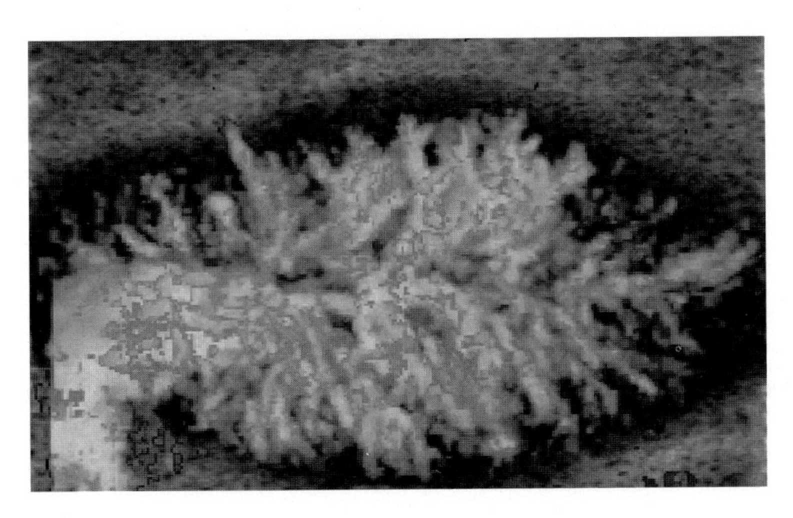

Figure 9-8. 3DUS power Doppler study of an enlarged lymph node. This image helps to appreciate the 3D anatomy of the node. (Image courtesy of GE Ultrasound) See color version of figure.

Dietrich et al. have shown that using good ultrasound technique, large lymph nodes can be seen in the hepatoduodenal ligament in almost all patients with Hep B and Hep C (34). They also reported that enlargement of the perihepatic lymph nodes in patients with chronic hepatitis C is predictive for the presence of severe inflammatory disease within the liver (35). Monitoring these nodes can help evaluate the patient's response to treatment.

Hashimoto et al. found 3DCT measurement of volume of retroperitoneal lymph node to be extremely useful in the follow-up of patients with testicular cancers (35). Liszka and his colleagues reported similar findings in patients with head and neck region cancers (36). They concluded that the efficacy of radiotherapy could be judged by determining the changes in volume of lymph nodes in the neck (37). However, it is important to understand the limitations of techniques that rely solely on node size to diagnose malignancy. Hawnaur found that MR had a sensitivity of 75% and a specificity of 88% in predicting metastatic lymphadenopathy when a node size of more than 1.5 cm was the only criterion used for suspected malignancy (38).

3DUS Limitations

As with pancreatic imaging, it is often difficult to get good-quality data sets because of overlying bowel gas.

How to Do It
Practical Clinical Tips for the Lymph Nodes

- Free-hand magnetic scanning will probably give the best images, although mechanical tilting scans may also provide good data for transcutaneous scans.
- Having a fasting patient and pressing firmly on the abdomen will be beneficial.

3DUS OF THE GALLBLADDER AND BILIARY SYSTEM

The gallbladder may be one of the most difficult organs to visualize satisfactorily using 3DUS techniques. Very little research has been carried out in this area, and conventional 2DUS scanning is excellent for detecting gallstones and intra- and extra-hepatic biliary dilation. The increased accuracy of volume measurements afforded by 3DUS may be useful in tests of gallbladder function.

Conventional 2DUS has a high sensitivity and specificity for detecting gallstones in addition to intrahepatic and extra-hepatic biliary dilation (39). The fact that it is a speedy, safe, flexible, and portable exam that does not involve ionizing radiation and that can be performed without contrast material makes it universally accepted as the imaging modality of choice for gallbladder and biliary disease (39). The gallbladder is most frequently assessed in patients with jaundice, abnormal liver function, pancreatitis, and chronic cholecystitis. It is also possible to evaluate adjacent structures during the examination.

Unlike many other ultrasound exams the dynamics involved in data acquisition are often extremely important in making the diagnosis (39) and these nuances may be lost in 3DUS acquisition. During a standard 2DUS exam the gallbladder is typically imaged with the patient in a left lateral decubitus position. Subtle variations in the patient's position and breathing pattern are often necessary to detect sludge, gallbladder wall thickening and tiny stones in the neck of the gallbladder. 3DUS may offer assistance in appreciating the

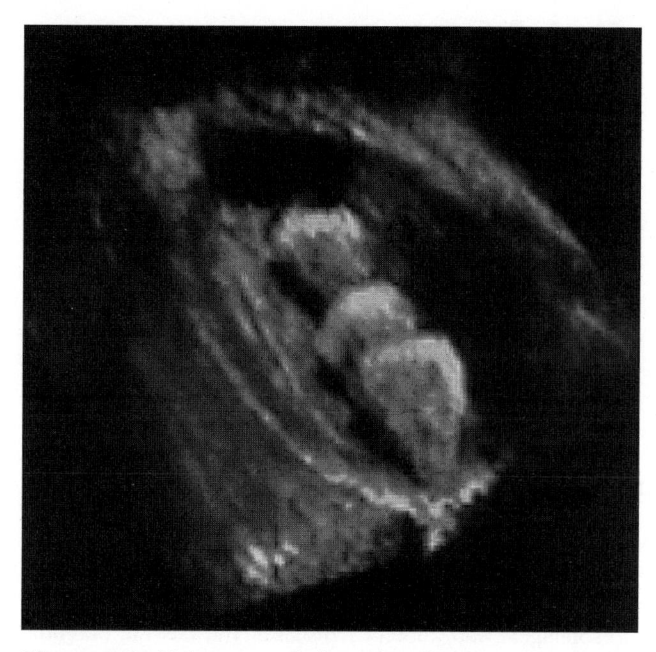

Figure 9-9. Volume rendering of gallstones. This image helps one appreciate the 3D nature of the stones. (Image courtesy of ATL.) See color version of figure.

size and structure of gallstones (Figure 9-9) and surrounding fluid in or near the wall. Preliminary data suggest that 3DUS assisted in identifying and localizing the correct number of gallstones within the biliary tree in 8 of 16 patients (6), especially in cases of cystic duct and prepapillary stones. Changes in patient position and respiration may also be needed to trace the entire length of the common duct. Significant variations in the amount of pressure placed on the probe may be necessary to visualize the distal end of the duct. 3DUS also may be helpful in evaluating the extent of biliary tract alterations in primary sclerosing cholangitis (6).

Clinical Benefits of 3DUS

3DUS can provide a more accurate measurement of gallbladder volume than standard 2DUS methods (39,40). The clinical relevance of the slightly increased accuracy in this assessment is not clear, but it may have utility in improving currently used tests of gallbladder function. 3DUS images of the gallbladder and dilated biliary tree have facilitated visualization of the dilated biliary system (40).

Knowing the exact location of the gallbladder is important prior to surgery or radiologic interventions. Although extremely rare, anomalies in gallbladder location, such as a left-sided gallbladder or an intrahepatic gallbladder, may make laparoscopic cholecystectomy and other procedures extremely challenging. These anomalies might be better appreciated in 3DUS exams, when a more panoramic view of the gallbladder and surrounding structures is presented. Getting a clear perspective of the pericholecystic region also may be desirable in trying to assess the appropriate surgical approach for gallbladder masses, either carcinoma or metastatic disease.

3DUS Limitations

3DUS of the gallbladder may prove exceptionally challenging. Frequently, gas artifacts can be present around the gallbladder that mimic gallstones. During a typical examination the examiner may go to extreme lengths to displace this artifact and obtain good pictures. In addition, the gallbladder can be subjected to significant artifacts, such as side-lobe artifacts or separation artifacts (39), and if these are present it may severely confuse interpretation of the 3DUS images.

HOW TO DO IT
Practical Clinical Tips for the Gallbladder and Biliary System

- Patient position and phase of respiration can produce considerable gallbladder movement.
- All 3DUS examinations should be obtained during a single breath-hold.
- Either free-hand or mechanical tilt or translation can produce good-quality images.

3DUS OF THE GASTROINTESTINAL TRACT

Initial studies on the gastrointestinal tract suggest that adding 3DUS capability will probably be of more use in intraluminal studies, rather than the transcutaneous scans. Early results have shown that 3D endosonography provides clinically useful information in patients with esophageal and anorectal diseases and has the potential to allow more accurate staging and guidance during biopsies. It can also be extremely helpful in assessing the overall volume of tumors and for measuring distances for surgical planning.

2DUS can evaluate the GI tract either from the traditional transcutaneous perspective with linear or curved array transducers or from an intraluminal perspective using either a higher-frequency (7.5–20 MHz) transducer mounted on either an endoscope or a catheter (41). Initial studies suggest that adding 3DUS capability will likely be more useful in intraluminal studies than the transcutaneous scans (Figures 9-10, 9-11, 9-12).

Transcutaneous sonography has been reported to be clinically useful in assessing focal mass lesions, diffuse lesions of the bowel wall (e.g., inflammatory bowel disease), and intraluminal pathologies (e.g., intussusception and radio-opaque foreign bodies).

Intraluminal ultrasound produces exquisite images of the different layers of the bowel, but over a much shorter segment, and has been shown to have good clinical utility in assessing the depth of involvement of specific pathologic processes. It can also detect

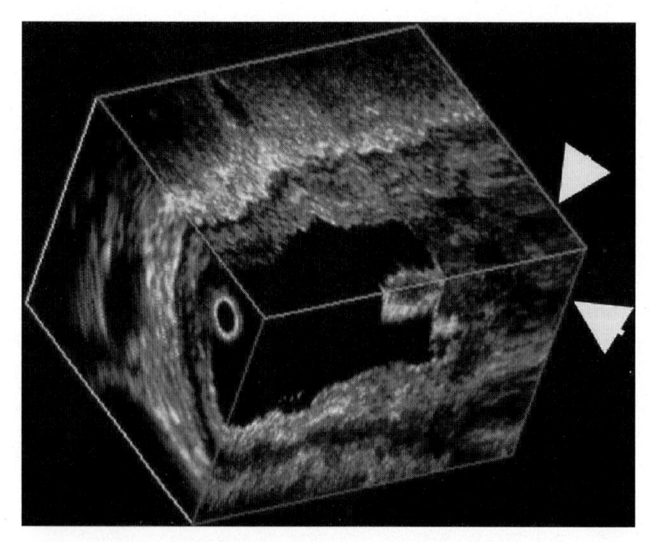

Figure 9-10. Intraluminal 3DUS view of esophageal cancer. The left side of the image shows a normal bowel wall appearance. Toward the right side of the image, the lumen is grossly abnormal and detail on the wall is obscured (*arrowheads*) by the esophageal tumor. (Image courtesy of Dr. Ji-Bin Liu.)

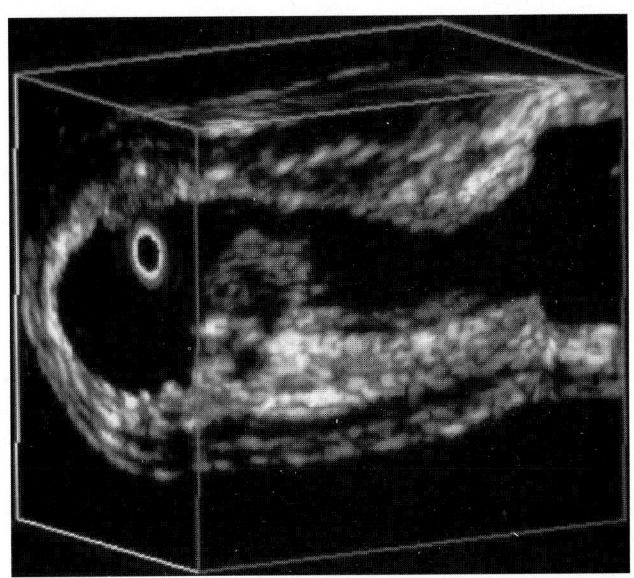

Figure 9-11. Pullback 3DUS image of the normal esophagus. Extreme care must be taken when slicing into these images to ensure that supposed mural-based lesions are truly lesions. The apparent filling defects seen in this image are due to slicing through a mucosal fold and do not represent a true mass. (Image courtesy of Dr. Ji-Bin Liu).

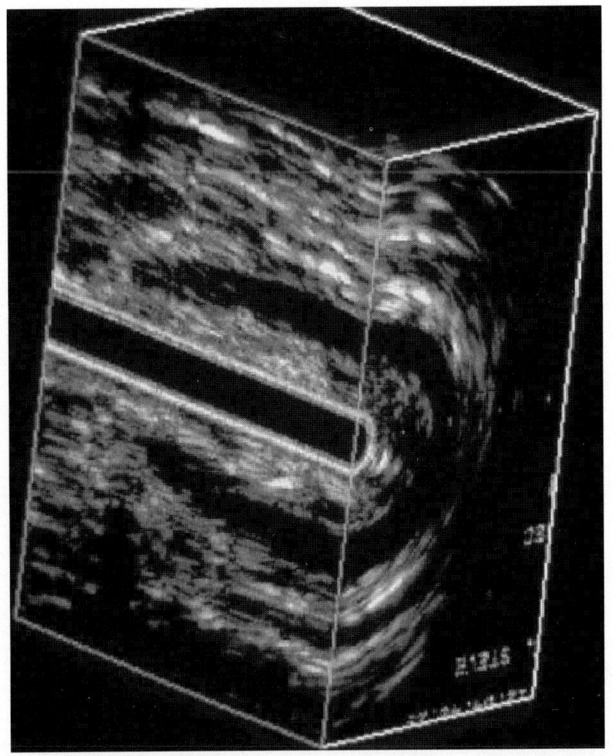

Figure 9-12. 3DUS study of a normal anal sphincter using an endoluminal pullback technique. (Image courtesy of Dr. Ji-Bin Liu.)

intraluminal pathology and has a limited role in assessing disease just outside the serosa (30,31).

Endosonography has rapidly evolved into the imaging modality of choice for assessing the early stages of colon and rectal tumors and recurrent cancer at the site of the surgical anastomosis (42). In the upper GI tract the main role of endosonography is in assessing the depth of invasion of stomach and esophageal cancers, and determining the local lymph node status (30). It also determines the etiology of submucosal lesions and some other diffuse bowel wall processes (30).

Clinical Benefits of 3DUS

Early results have shown that 3D endosonography provides clinically useful information in patients with esophageal (see Figures 9-10 and 9-11) and anorectal (Figures 9-12 and 9-13) diseases (43–45). These initial reports suggest that adding 3D capability has the potential to allow more accurate cancer staging (44,45). By allowing the examiner to evaluate the data from different perspectives 3DUS may permit a more thorough understanding of the disease (43) and guide biopsies more accurately (45).

3DUS can be extremely helpful in assessing overall tumor volume and for measuring some distances (e.g., the distance from a rectal tumor to the external sphincter, which may dictate how and when surgery is planned).

Hunerbein and Schlag assessed 3DUS endosonography in examining 100 patients with rectal cancer (43). Their overall conclusion was that the technique allowed visualization of the lesions in different planes, which would have been impossible with conventional 2DUS imaging. Improvements in the diagnostic information supplied by the 3DUS data were sufficient to recommend further studies. For example, the assessment of infiltration depth was 88%, compared with 82% for conventional 2DUS, and the determination of lymph node involvement was 79%, compared with 74% using 2DUS techniques.

Ivanov and Diavoc investigated the use of 3D transrectal ultrasound (TRUS) in 18 patients and considered it a useful addition to conventional TRUS (44). They thought it

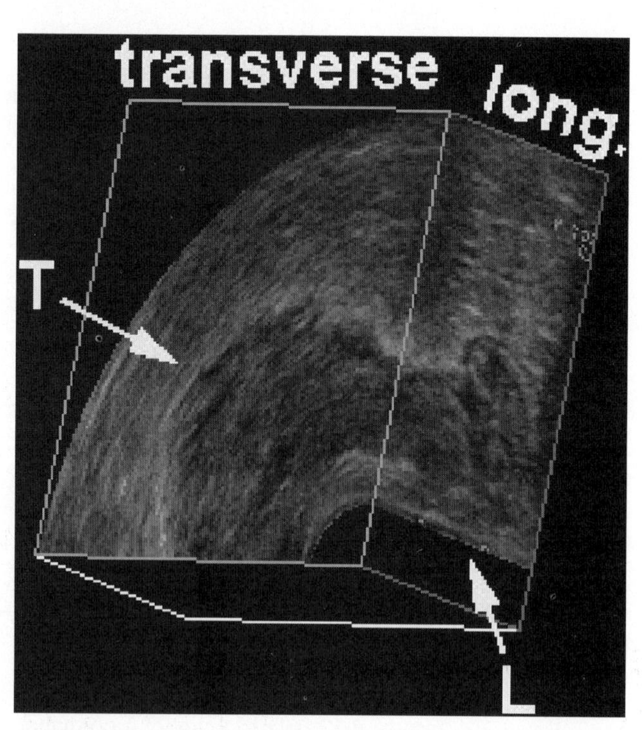

Figure 9-13. 3DUS study showing rectal cancer that has spread through the rectal wall in a 56-year-old man. T, tumor; L, bowel lumen.

would be helpful in the staging of rectal carcinoma and in the treatment planning for rectal surgery.

Stuhldreier et al. used the method of linking the axially rotating transducer to a mechanical pulling device to create 3D images of the pelvic floor and rectal wall in children to diagnose continence problems (46). They found this technique worked well even with small children and described the advantages over conventional ultrasound imaging. These included the ability to reconstruct sections of interest and to volume-render organs and perform volumetry.

Kallimanis et al. discussed the utility of 3D endosonography of the esophagus, showing that it was technically possible and that it had good potential to improve cancer staging (45).

3DUS Limitations

Adding 3D capability to transcutaneous ultrasound has not been reported to date. As transcutaneous bowel ultrasound is always challenging and frequently frustrating, because of the problems with imaging artifacts, by adding 3D capability the resulting images may become even more confusing.

Conventional intraluminal ultrasonography is sometimes impossible because of obstructions of the lumen by tumors. Although useful information of these obstructions can be achieved using 3D images reconstructed in different planes, the quality is rarely as good as with nonobstructing lesions (43).

HOW TO DO IT
Practical Clinical Tips for the Gastrointestinal System

- Have a well-prepped bowel that facilitates good viewing.
- Mechanical pullback produces better 3DUS images than free-hand pullback.
- Work in close cooperation with the endoscopist, to have a clear, concise understanding of the clinically important issues in each case.
- Fluoroscopy may assist in localizing the catheter because it sometimes can be difficult to know exactly where one is in the bowel.
- High-resolution 3DUS evaluation of entire bowel in one volume is impossible on most 3DUS systems:
  - First, video-record a long, complete pullback of the catheter through the whole bowel.
  - Review the tape to assess areas of potential disease.
  - Obtain a complete 3DUS data set from each suspicious area.
- Multiplanar image review is the display technique of choice.

REFERENCES

1. Withers CE, Wilson SR. The liver. In: Rumack CM, Wilson SR, Charboneau JW, eds. *Diagnostic ultrasound,* 2nd ed. St. Louis: Mosby, 1997:87–154.
2. Ferrucci JT. Liver tumor imaging: current concepts. *AJR* 1990;155:473–484.
3. Ohahsi I, Ina H, Okada Y, et al. Segmental anatomy of the liver under the right diaphragmatic dome: evaluation with axial CT. *Radiology* 1996;200:779–783.
4. Lee RA, Kane RA, Lantz EJ, Charboneau JW. Intraoperative and laparoscopic ultrasound of the abdomen. In: Rumack CM, Wilson SR, Charboneau JW, eds. *Diagnostic ultrasound,* 2nd ed. St. Louis: Mosby, 1997: 671–699.
5. Cosgrove D. Why do we need contrast agents for ultrasound? *Clin Radiol* 1996;51(Suppl 1):1–4.
6. Wagner S, Gebel M, Bleck JS, Manns MP, Clinical application of three-dimensional sonography in hepatobiliary disease. *Bildgebung* 1994;61(2):104–109.
7. Hughes KS, Simon R, Songhorabodi S, et al. Resection of the liver for colorectal carcinoma metastases: a multi-institutional study of patterns of recurrence. *Surgery* 1986;100:278–284.
8. Van Leeuween MS, Noordzij J, Hennipman A, Fledberg MA. Planning of liver surgery using three dimensional imaging techniques. *Eur J Cancer* 1995;31A:1212–1215.

9. Smith DF, Downey DB, Spouge AR, Soney SJ. Sonographic demonstration of Couinaud's liver segments. *J Ultrasound Med* 1998 (*in press*).
10. Enislidis G, Wagner A, Ploder O, Ewers R. Computed intraoperative navigation guidance—a preliminary report on a new technique. *Br J Oral Maxillofac Surg* 1997;35:271–274.
11. Gil R, von Birgelen C, Prati F, di Mario C, Ligthart J, Serruys PW. Usefulness of three-dimensional reconstruction for interpretation and quantitative analysis of intracoronary ultrasound during stent deployment. *Am J Cardiol* 1996;77:761–764.
12. Hata N, Dohi T, Iseki H, Takakura K. Development of a frameless and armless stereotactic neuronavigation. *Neurosurgery* 1997;41:608–613.
13. Olivier A, Alonso-Vanegas M, Comeau R, Peters TM. Image-guided surgery of epilepsy. *Neurosurg Clin N Am* 1996;7:229–243.
14. Prati F, di Mario C, Gil R, et al. Usefulness of on-line three-dimensional reconstruction of intracoronary. *Am J Cardiol* 1996;77:455–461.
15. Liess H, Roth C, Umgelter A, Zoller WG. Improvements in volumetric quantification of circumscribed hepatic lesions by three-dimensional sonography. *Z Gastroenterol* 1994;32:488–492.
16. Hughes SW, D'Arcy TJ, Maxwell DJ, Saunders JE. *In vitro* estimation of foetal liver volume using ultrasound, x-ray computed tomography and magnetic resonance imaging. *Physiol Meas* 1997;18:401–410.
17. Wolf GK, Lang H, Prokop M, Schreiber M, Zoller WG, Volume measurements of localized hepatic lesions using three-dimensional sonography in comparison with three-dimensional computed tomography. *Eur J Med Res* 1998;3:157–164.
18. Alexander DG, Unger EC, Seeger SJ, Karmann S, Krupinski EA. Estimation of volumes of distribution and intratumoral ethanol concentrations by computed tomography scanning after percutaneous ethanol injection. *Acad Radiol* 1996;3:49–56.
19. Yamashita Y, Matsukawa T, Arakawa A, Hatanaka Y, Urata J, Takahashi M. US-guided liver biopsy: predicting the effect of interventional treatment of hepatocellular carcinoma. *Radiology* 1995;196:799–804.
20. Chari RS, Baker ME, Sue SR, Meyers WC. Regeneration of a transplanted liver after right hepatic lobectomy. *Liver Transpl Surg* 1996;2:233–234.
21. Hino I, Tamai T, Satoh K, Takashima H, Ohkawa M, Tanabe M. Index for predicting post-operative residual liver function by pre-operative dynamic liver SPET. *Nucl Med Commun* 1997;18:1040–1048.
22. Caldwell SH, de Lange EE, Gaffey MJ, et al. Accuracy and significance of pretransplant liver volume measured by magnetic resonance imaging. *Liver Transpl Surg* 1996;2:438–442.
23. Henderson JM, Mackay GJ, Kutner MH, Noe B. Volumetric and functional liver blood flow are both increased in the human transplanted liver. *J Hepatol* 1993;17:204–207.
24. Downey DB, Fenster A. Vascular imaging with a three-dimensional power Doppler system. *AJR* 1995;165: 665–668.
25. Ritchie CJ, Edwards WS, Mack LA, Cyr DR, Kim Y. Three-dimensional ultrasonic angiography using power-mode Doppler. *Ultrasound Med Biol* 1996;22:277–286.
26. Dodd GD. Biopsy or drainage of a mass: Role of sonography. In: Bluth EI, Arger PH, Hertzberg BS, Middleton WD, eds. *RSNA special course in ultrasound 1996.* Oak Brook: RSNA Publication, 1996:325–330.
27. Atri M, Finnegan PW. The Pancreas. In: Rumack CM, Wilson SR, Charboneau JW, eds. *Diagnostic ultrasound,* 2nd ed. St. Louis: Mosby, 1997:225–278.
28. De Odorico I, Spaulding KA, Pretorius DH, Lev-Toaff AS, Bailey TB, Nelson TR, Normal splenic volumes estimated using 3-dimensional ultrasound, *Radiology* 1998 (*in press*).
29. Liu JB, Schiano TD, Miller LS. Upper gastrointestinal tract. In: Liu JB, Goldberg BB, eds. *Endoluminal ultrasound: vascular and nonvascular applications.* St.Louis: Mosby, 1997:147–199.
30. Alexander AA, Miller LS, Schiano TD, Liu JB. Lower gastrointestinal tract. In: Liu JB, Goldberg BB, eds. *Endoluminal ultrasound: vascular and nonvascular applications.* St. Louis: Mosby, 1997:201–227.
31. Sellner F, Machacek E. The importance of tumour volume in the prognosis of radically treated periampullary carcinomas. *Eur J Surg* 1993;159:95–100.
32. Downey DB. The retroperitoneum and great vessels. In: Rumack CM, Wilson SR, Charboneau JW, eds. *Diagnostic ultrasound*, 2nd ed. St. Louis: Mosby, 1997:453–486.
33. Dietrich CF, Gottschalk R, Herrmann G, Caspary WF, Zeuzem S. [Sonographic detection of lymph nodes in the hepatoduodenal ligament]. *Dtsch Med Wochenschr* 1997;122:1269–1274.
34. Dietrich CF, Lee JH, Herrmann G, et al. Enlargement of perihepatic lymph nodes in relation to liver histology and viremia in patients with chronic hepatitis C. *Hepatology* 1997;26:467–472.
35. Hashimoto S, Isaka S, Okano T, Shimazaki J. [Three dimensional image analysis on retroperitoneal lymph node metastasis of testicular cancer]. *Nippon Hinyokika Gakkai Zasshi* 1994;85:1388–1394.
36. Liszka G, Thalacker U, Somogyi A, Nemeth G. [Volume changes to the neck lymph node metastases in head-neck tumors. The evaluation of radiotherapeutic treatment success.] *Strahlenther Onkol* 1997;173: 428–430.
37. Hawnaur JM, Johnson RJ, Buckley CH, Tindall V, Isherwood I. Staging, volume estimation and assessment of nodal status in carcinoma of the cervix: comparison of magnetic resonance imaging with surgical findings. *Clin Radiol* 1994;49:443–452.
38. Laing FC. The gall bladder and bile ducts. In: Rumack CM, Wilson SR, Charboneau JW, eds. *Diagnostic ultrasound,* 2nd ed. St. Louis: Mosby, 1997:175–223.
39. Pauletzki J, Sackmann M, Holl J, Paumgartner G. Evaluation of gallbladder volume and emptying with a novel three-dimensional ultrasound system: comparison with the sum-of-cylinders and the ellipsoid methods. *J Clin Ultrasound* 1996;24:277–285.
40. Fine D, Perring S, Herbetko J, Hacking CN, Fleming JS, Dewbury KC. Three-dimensional (3D) ultrasound imaging of the gallbladder and dilated biliary tree: reconstruction from real-time B-scans. *Br J Radiol* 1991; 64:1056–1057.
41. Thoeni RF. Colorectal cancer. Radiologic staging. *Radiol Clin North Am* 1997;35:457–485.
42. Hunerbein M, Schlag PM. Three-dimensional endosonography for staging of rectal cancer. *Ann Surg* 1997; 225:432–438.

43. Hunerbein M, Dohmoto M, Haensch W, Schlag PM. Evaluation and biopsy of recurrent rectal cancer using three-dimensional endosonography. *Dis Colon Rectum* 1996;39:1373–1378.
44. Ivanov KD, Diavoc CD. Three-dimensional endoluminal ultrasound: new staging technique in patients with rectal cancer. *Dis Colon Rectum* 1997;40:47–50.
45. Kallimanis G, Garra BS, Tio TL, et al. The feasibility of three-dimensional endoscopic ultrasonography: a preliminary report. *Gastrointest Endosc* 1995;41:235–239.
46. Stuhldreier G, Kirschner HJ, Astfalk W, Schweizer P, Huppert PE, Grunert T. Three-dimensional endosonography of the pelvic floor: an additional diagnostic tool in surgery for continence problems in children. *Eur J Pediatr Surg* 1997;7:97–102.

10

Vascular System

OVERVIEW

This chapter provides an overview of the applications of three-dimensional ultrasound (3DUS) imaging methods to the evaluation of vascular anatomy and disease. Both external and intraluminal techniques are discussed with respect to examination of the major peripheral and abdominal arteries, the coronary arteries, and the upper- and lower-limb venous systems. Other chapters also discuss imaging of 3DUS vascular anatomy. Current and future clinical benefits and limitations are considered.

KEY CONCEPTS

- Evaluation of vascular anatomy using 2DUS and 3DUS is increasing in popularity:
  - Vascular evaluation by external imaging—transcutaneous.
  - Vascular evaluation by intravascular imaging—catheters.
- 3DUS vascular imaging utilizes several modes to image vessels:
  - B-mode imaging.
  - Color and power Doppler imaging.
  - Microbubble contrast materials.
  - Harmonic imaging modes.
- Technical challenges to be overcome include:
  - Shadowing.
  - Positional accuracy for intravascular studies.
  - Mapping of flow patterns.
  - Cardiac and respiratory gating for physiologic synchronization.
- Several volume data acquisition methods useful:
  - Slow, motorized pullback gated technique best for 3D-IVUS imaging.
  - Free-hand technique best for transcutaneous assessment of long vessel segments.
  - Motorized linear translation useful for selected short-vessel segments.

KEY CONCEPTS, *continued*

- Several interactive display methods valuable for optimizing display of vessel morphology:
  - Multiplanar slices.
  - Volume rendering—maximum-intensity projection.
- 3DUS vascular imaging provides several clinical benefits:
  - Measuring and analyzing the volume and composition of atherosclerotic plaque.
  - Deciding optimal treatment methods for patients.
  - Providing information useful for planning, monitoring, and evaluating therapy.
  - Doppler (color/power) imaging produces excellent qualitative information.

Imaging of blood flow and vessel anatomy is one of the most important areas of medical imaging. The prevalence and severity of vascular disease increases with age, and an increasing number of vascular ultrasound exams will be done as the population over the age of 65 continues to increase (1). Other diagnostic imaging tests (e.g., contrast angiography, magnetic resonance angiography [MRA], electron beam tomography [ECT], computed tomography [CT], and synchrotron imaging) generally only permit the patent lumen to be imaged. Because of its noninvasiveness, relatively low cost, and advanced technology and contrast materials, ultrasound imaging has taken on an important role in vascular imaging. Ultrasound is now providing important information about a variety of vascular pathologies (2), and the number of vascular exams is increasing dramatically (1).

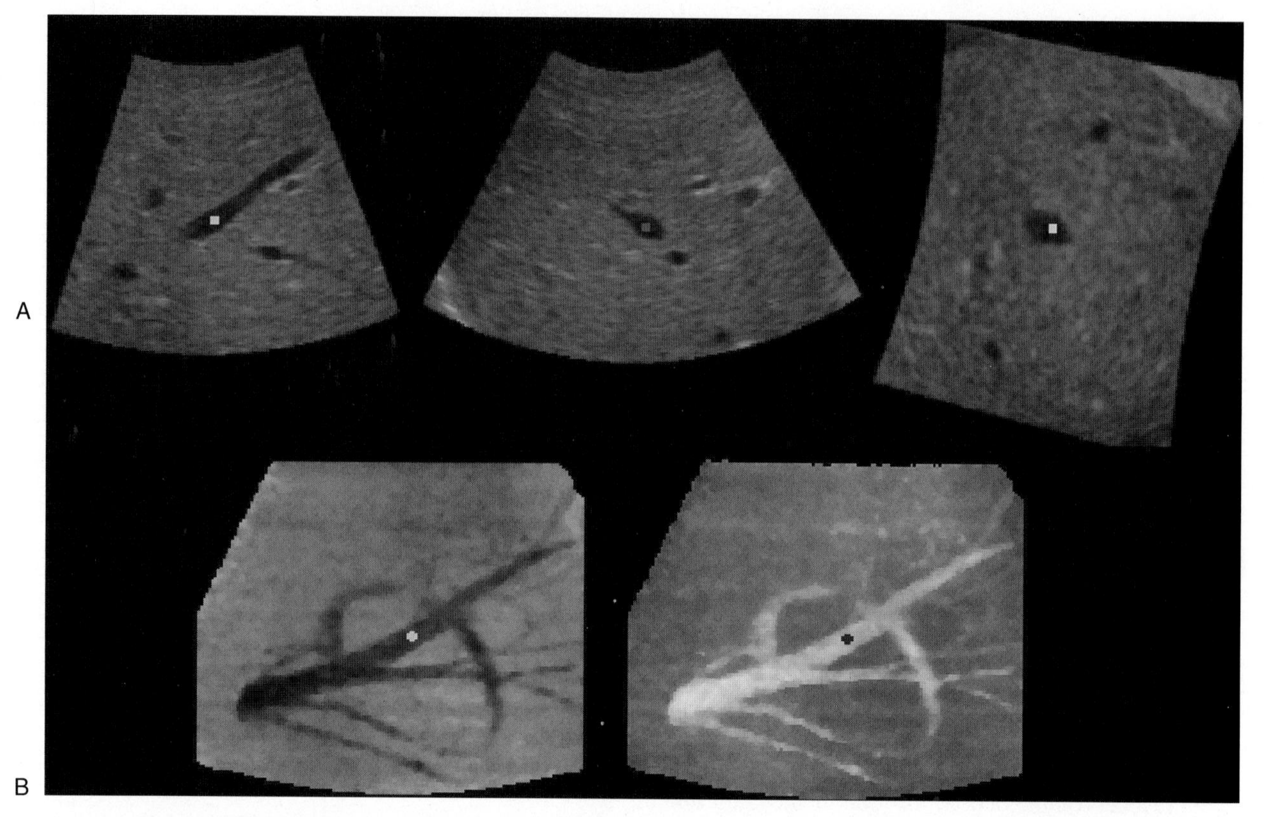

Figure 10-1. 3DUS acquisition of the liver. **A:** Three orthogonal slices through the volume data. The dot represents the point of intersection between the planes. **B:** Two-volume rendered images from the volume. Both images are the same but with the gray-scale inverted with respect to each other. The vessels are rendered using a minimum-intensity projection that emphasizes the anechoic properties of the vessels. The vessel continuity is clearly appreciated from the display to a much greater extent than possible in the 2DUS images.

Ultrasound techniques allow the contour of the vascular lumen and wall to be evaluated, often permitting accurate measurement and quantification of plaque. Color and power Doppler ultrasound examinations also play an increasingly important role in the evaluation of vascular diseases. Many vascular surgeons plan, carry out, and follow up on surgery by relying solely on ultrasound rather than using it in conjunction with angiography (3–5).

3DUS methods show much promise to provide clear pictures of vascular anatomy and blood flow dynamics, particularly with the advent of contrast materials and harmonic imaging (6–10). 3DUS vascular images may be viewed in a multiplanar or rendered format (Figure 10-1). Whether they are acquired using B-mode or power/color Doppler techniques, each method offers advantages. Vessels can be readily recognized as they course through a volume. As a result, 3DUS is a rapidly evolving technology that overcomes some of the current limitations of conventional 2DUS vascular ultrasound.

Evaluation of blood flow and vessel anatomy using 3DUS falls into two broad categories that overlap to some extent: (a) external imaging using conventional imaging equipment and methods; (b) intravascular imaging using specialized catheters. Although the ultimate clinical value of 3DUS remains to be established, reports, though anecdotal, have been encouraging. This chapter will provide an overview of the applications of 3DUS imaging methods in the evaluation of vascular anatomy and disease and discuss where 3DUS may add clinical value.

VASCULAR EVALUATION BY EXTERNAL IMAGING

External 3DUS imaging of blood flow and vessel anatomy using a transcutaneous approach with conventional imaging equipment has shown promise, particularly in imaging the carotid vessels (Figure 10-2) (11–21). 3DUS imaging may assist in understanding vascular anatomy, particularly in identifying the relationship of specific vessels (see Figure

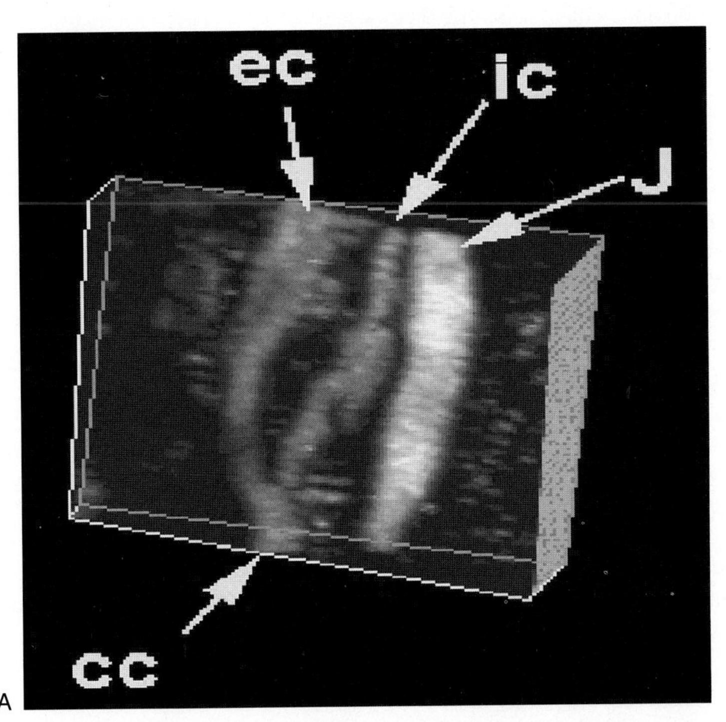

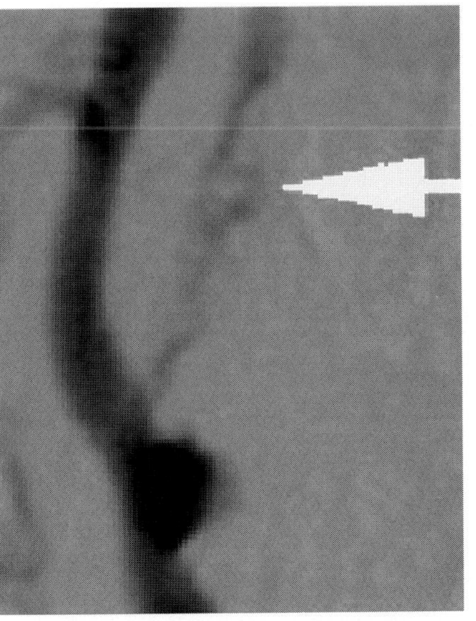

Figure 10-2. A: A maximum-intensity projection of a mechanical scanned common carotid and internal carotid artery showing decrease in the size of the lumen of the internal carotid proximally. ic, internal carotid artery; ec, external carotid artery; cc, common carotid artery; J, internal jugular vein. Although the image is not as crisp as the accompanying angiogram, alterations in blood flow can be appreciated. (Image courtesy of Life Imaging Systems, Inc.) **B:** Digital subtraction angiogram (DSA) of the same case showing significant impairment of flow in the region of the internal carotid artery (*arrow*).

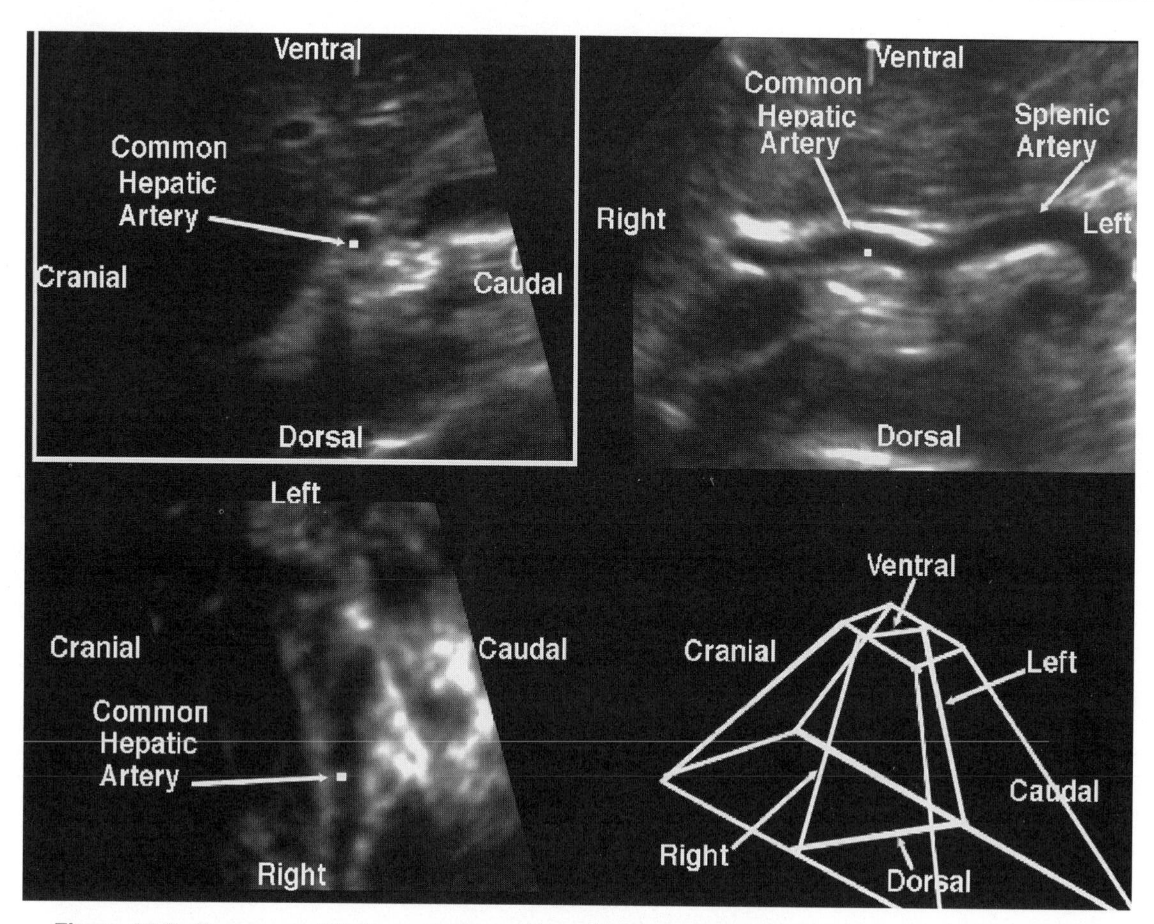

Figure 10-3. Transverse 3DUS acquisition of the celiac axis. The small white cursor in each image is located within the common hepatic artery. The lower right hand image shows the orientation of the upper left image in the data volume. (Image courtesy of ref. 22.)

10-2). Evaluation of the celiac axis by Spaulding et al. showed that 3DUS offered useful advantages compared with 2DUS (Figure 10-3) (22). 3DUS provided more data from within a volume than were obtained in any single 2DUS plane. 3DUS data permitted adjusting the position and orientation of planes interactively to optimize the appearance of an individual vessel in three orthogonal planes. As a result, the spatial relationship of a vessel to either other adjacent vessels or an organ could be readily determined. It is important to note that a coronal plane was available for viewing that is not available or possible with 2DUS. Vascular variants were identified in two of the subjects using 3DUS techniques.

Conventionally, all vascular ultrasound examinations produce a 2DUS B-mode image of the long axis of the vessel that usually is unhelpful for accurately assessing luminal diameters as they rarely represent the true midline of the vessel (Figure 10-4) (23,24). To obtain the best view of the true diameter of the vessel, it is best to image in a transaxial plane perpendicular to the artery axis (23). Interactive 3DUS display assists in evaluation of vessel morphology and stenosis by permitting the view of the vessel to be rotated to optimize the display.

Plaque Characterization and Volume

3DUS methods have demonstrated the potential for transabdominal assessment of arterial plaque (12,21,25). Evidence is mounting that particular subtypes of plaques have different natural histories and should be treated differently (26). Identification of atherosclerotic plaque structure and composition, based on differences in their acoustic properties, requires high-frequency transducers in close proximity to the area of interest (26).

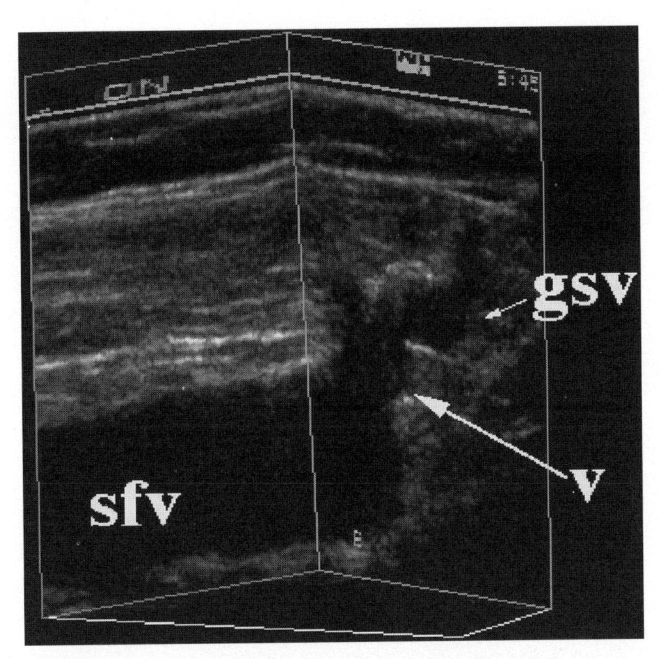

Figure 10-4. Multiplanar reconstruction of the junction of the greater saphenous vein and common femoral vein. The volume image can be sliced into to show the relationship between the two vessels. (v, valve; sfv, superficial femoral vein; gsv, greater saphenous vein.) The true vessel diameter can be calculated from these images. (Image courtesy of Life Imaging Systems, Inc., London, Ontario, Canada.)

There is increasing clinical interest in using high-frequency ultrasound to characterize plaque composition (26–33).

Improved plaque characterization may help in assessing prognosis for development of emboli. Conventionally, plaque is assessed for ulceration, irregularity, and intraplaque hemorrhage and for lipid and proteinaceous material, all of which probably increase the risk of emboli (34–36). Ultrasound assessment of intraplaque hemorrhage in the carotid bulb shows a sensitivity of from 72% to 94% and a specificity of from 75% to 90% (23,37).

Quantitation of plaque volume using 2D B-mode images has been tried (3), with 3DUS methods showing considerable promise to quantify plaque volume more accurately in the internal carotid artery (12–14,38) (Figures 10-5, 10-6, and 10-7). The mean variability (MV) in carotid plaque volume measurement was assessed at follow-up using two different 3D techniques: the watershed algorithm (MV = 5.3%) and the threshold procedure (MV = 8.9%) (13). These techniques have been used successfully to serially assess patients at high risk for plaque progression (38). Similar results have been found by Ge, who reported that his group found plaque in 54% of 812 coronary artery segments that were reported to be normal on angiography (1).

External 3DUS imaging of abdominal aortic wall plaque also has shown favorable results (21). Webber et al. found that 3DUS methods were superior to angiography in a small series but provided no incremental advantage over 2DUS (21). Although these reports are preliminary, they are encouraging but require further work to establish clinical utility.

Aneurysms

Accurate determination of the exact size of an aneurysm is important because the risk of rupture directly correlates with the absolute diameter and the rate of growth in diameter (39). Several authors have emphasized the necessity of measuring aneurysms at their

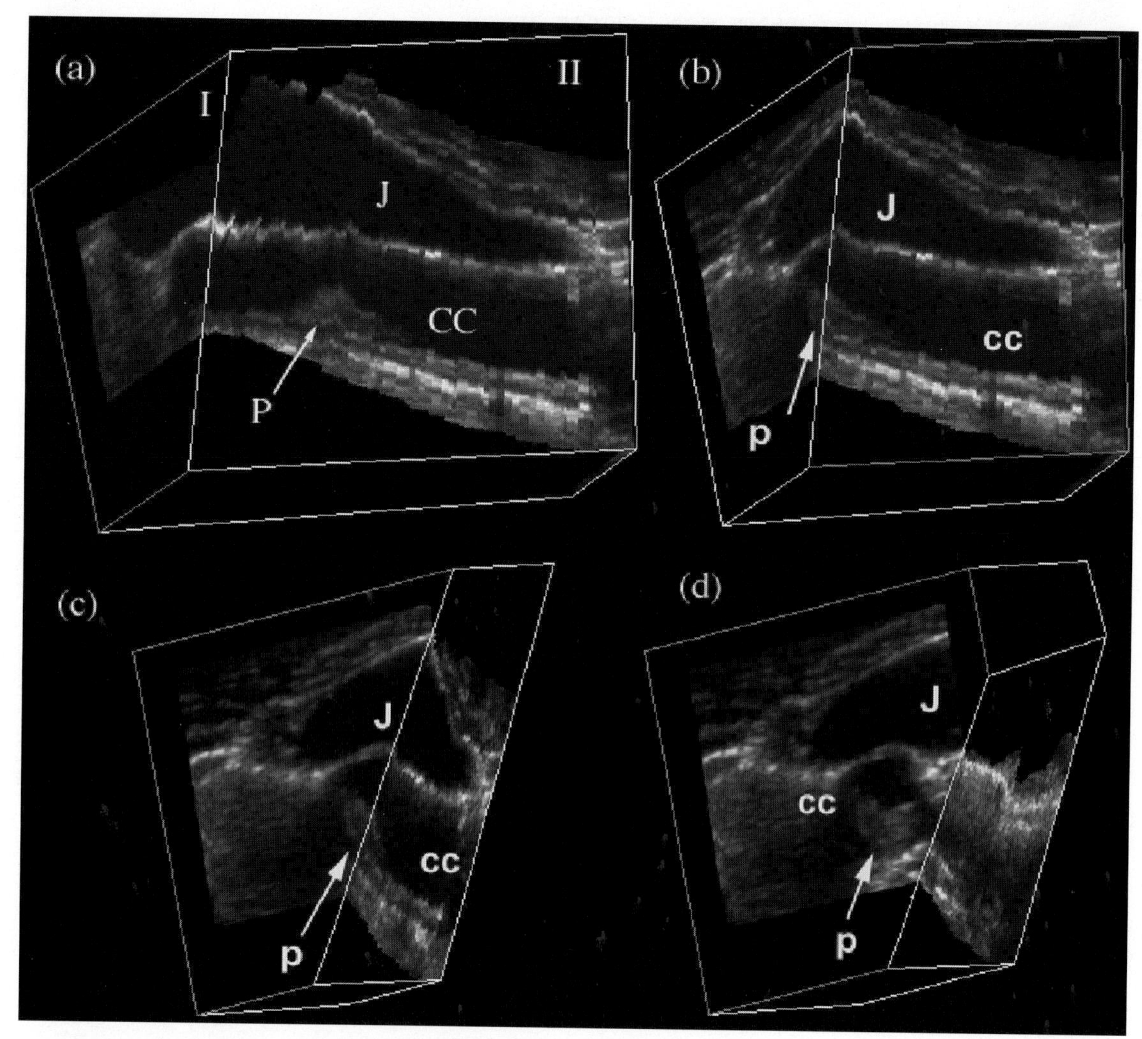

Figure 10-5. Collage of four images showing 3D representation of the common carotid (cc) and internal jugular vein (J). Volume was obtained by transcutaneously scanning with a linear array transducer using a magnetic position sensor. Both the plaque (P) and the patent lumen are well shown. (Image courtesy of S. Sherebrin.)

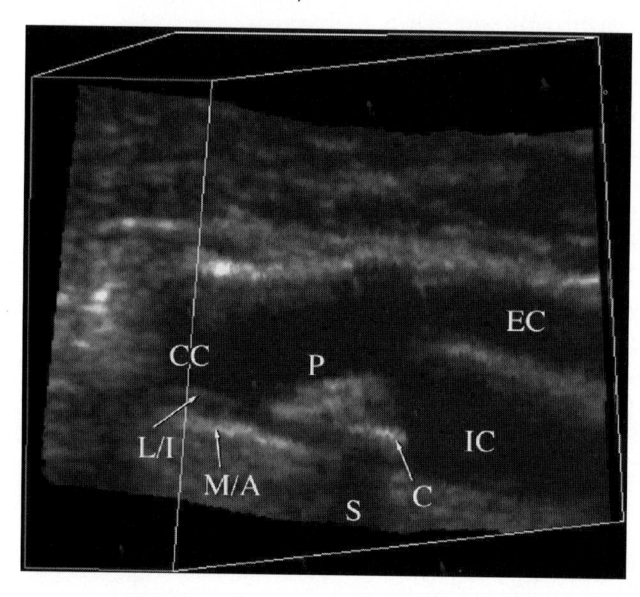

Figure 10-6. 3DUS image of an internal carotid plaque showing details of the wall and plaque structure. 3DUS data were acquired using free-hand scanning with position sensing. CC, common carotid; C, calcification in plaque; S, shadowing; P, plaque; IC, internal carotid; EC, external carotid; L/I, lumen intima interface; m/a, media/adventitia interface. (Image courtesy of S. Sherebrin.)

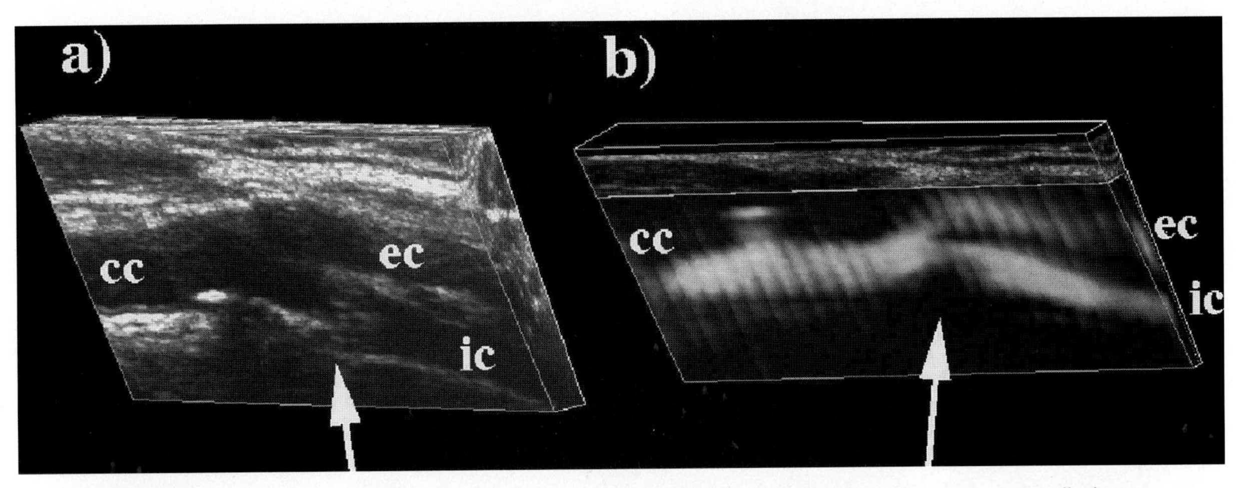

Figure 10-7. 3DUS study of the carotid artery. **a:** Longitudinal perspective of a posterior wall plaque that protrudes into the lumen (*arrow*). Note the luminal surfaces of the plaque are poorly defined. ec, external carotid; ic, internal carotid; cc, common carotid. **b:** Similar perspective but with only power Doppler data. An angiographic-like image is produced. Note the lumen wall interface is better defined. *Arrow*, plaque; ec, external carotid; ic, internal carotid; cc, common carotid.

widest dimension and ensuring that the sound beam is perpendicular to the wall (37,39). Although it has yet to be proved in clinical practice, 3DUS offers the ability to view the aneurysm at its greatest diameter and optimize the cross-section measurements through the maximum diameter orthogonal to the wall.

Venous Studies

Thus far the clinical utility of 3DUS for assessing veins for either DVT or chronic venous insufficiency has not been demonstrated. Because compression and other physiologic maneuvers make up an important part of both the upper- and lower-limb exams, 3DUS may have limited applicability in the near future. 3DUS may have a role in showing the extent of collateral veins or possible extent of other pathology, such as a Baker's cyst. 3DUS also might be beneficial for follow-up studies assessing the size of DVTs.

Vascular Rendering Using Color and Power Doppler

Color and power Doppler can improve visualization of vascular anatomy by showing the extent and geometry of the vessel (11,17,18,40,41). Preliminary 3DUS work has evaluated a variety of body areas, including the carotid arteries (Figures 10-8, 10-9) (15), the portal venous system (40), placentas (Figure 10-10) (41,42), spleen (40), native kidneys (Figure 10-11) (40), transplanted kidneys (41), heart (43), and intracranial blood vessels (Figure 10-12) (44). Some techniques allow the B-mode data to be displayed with Doppler, which provides a more thorough understanding of the pathology. Generally, 3DUS vascular data are best displayed using maximum-intensity projection techniques, which produce images similar to those produced by angiography (see Figures 10-7, 10-10) and multiplanar slicing (see Figures 10-8, 10-9). Displays of both techniques together assist in localization. The quality of the images compares favorably with magnetic resonance angiography (MRA). To date, 3DUS power/color Doppler images have not been compared objectively with angiography, MRA, or conventional ultrasound.

3DUS vascular imaging benefits by incorporating recent advances in ultrasound contrast imaging and harmonic imaging (6–9,39,45–50). This is because ultrasound contrast material and harmonic imaging combined increase the conspicuity of signals from larger vessels

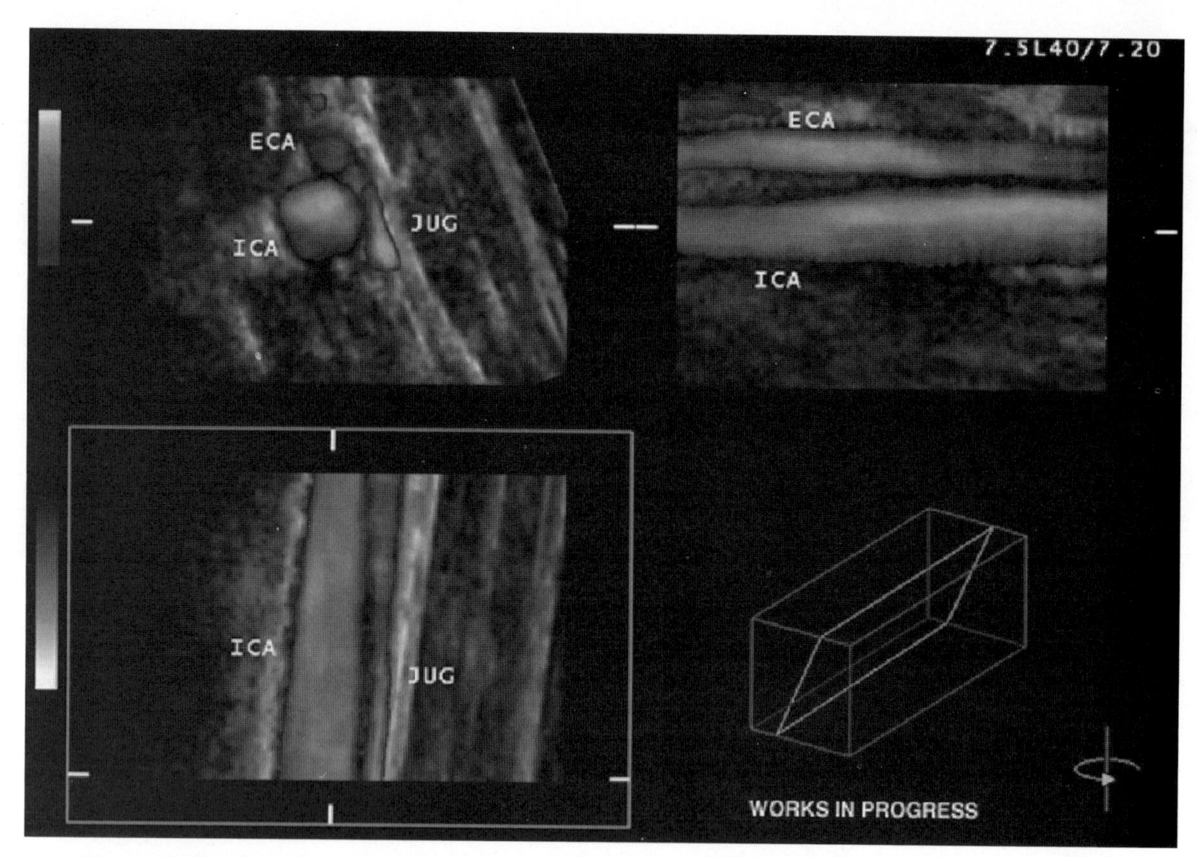

Figure 10-8. 3DUS power Doppler study of the carotid artery obtained using a free-hand technique. Doppler data helps define the edges of the vessels. (Image courtesy of Siemens Ultrasound, Issaquah, WA.) See color version of figure.

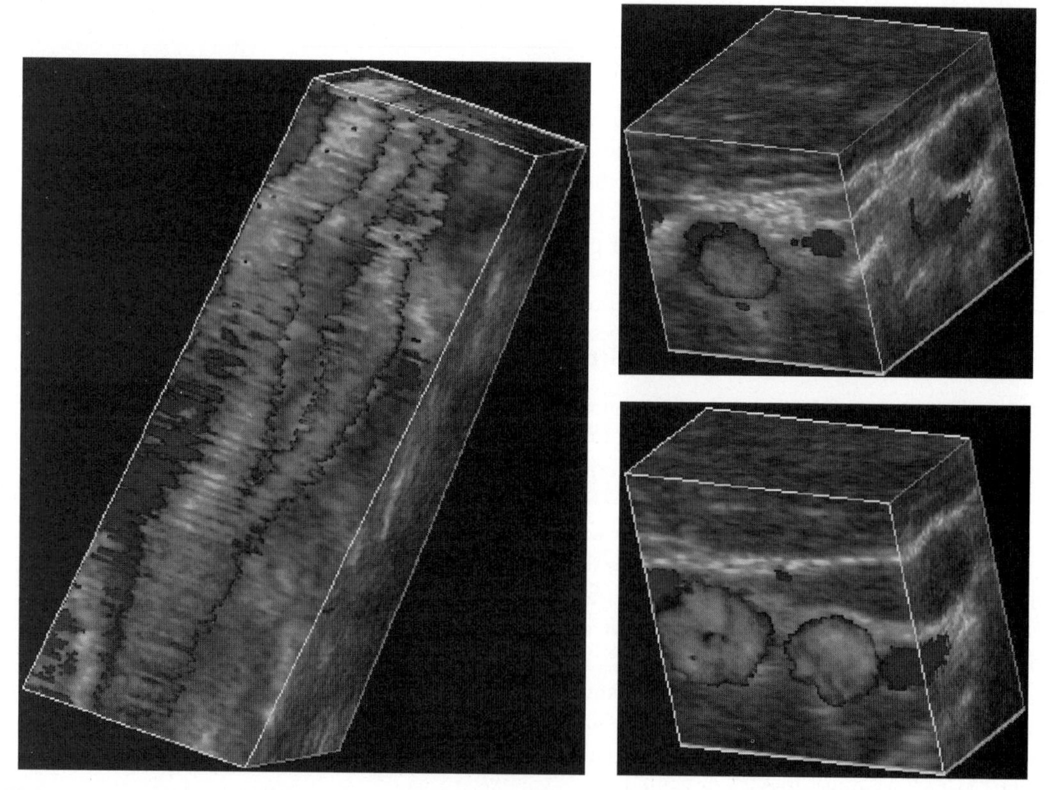

Figure 10-9. 3DUS acquisition of a normal volunteer's carotid artery combining color Doppler and B-mode data. The entire course of the carotid artery can be followed. Note the slight irregularity of the boundary of the color caused by artifact. Image was obtained following cardiac gating. (Image courtesy of Life Imaging Systems, Inc., London, Ontario, Canada.) See color version of figure.

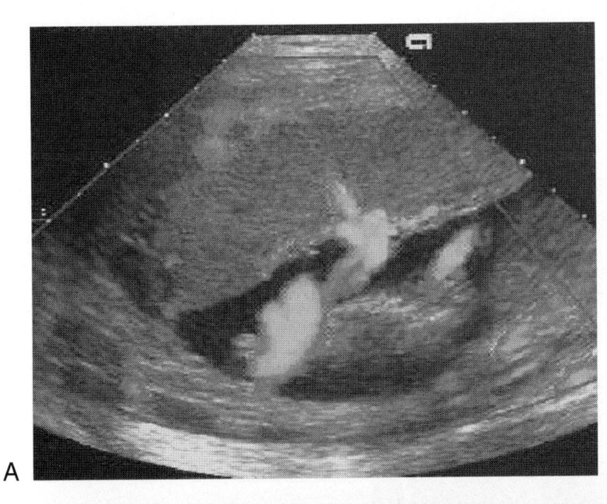

A

Figure 10-10. 3DUS power Doppler study of the placental circulation using free-hand scanning with position sensing. **A:** One frame from the 2DUS acquisition showing the point of umbilical cord insertion. **B:** Three-volume rendered images without the B-mode data showing the vasculature of the placental circulation at progressively higher magnifications. (*Arrow,* cord insertion.) See color version of figure.

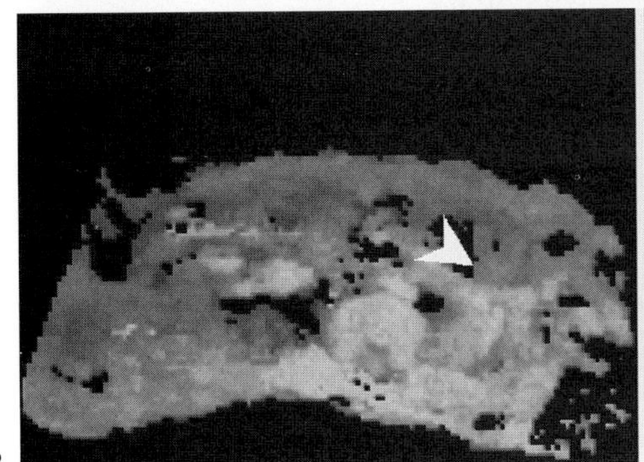

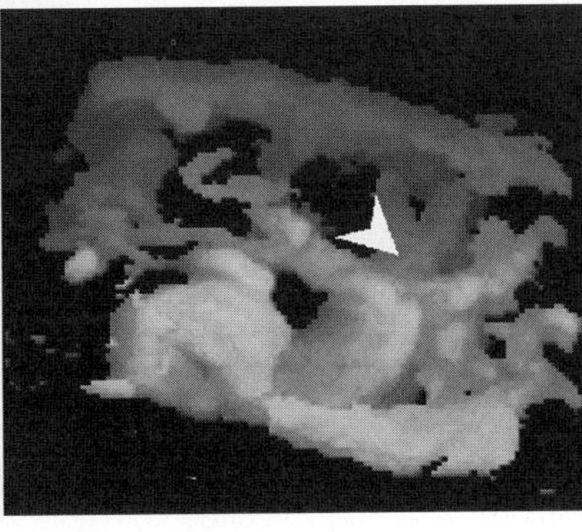

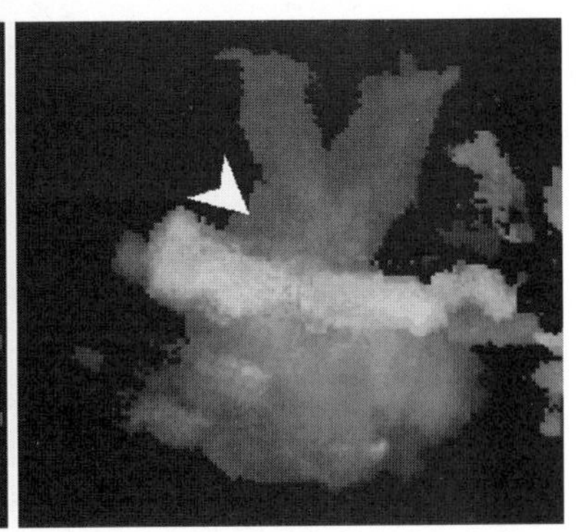

B

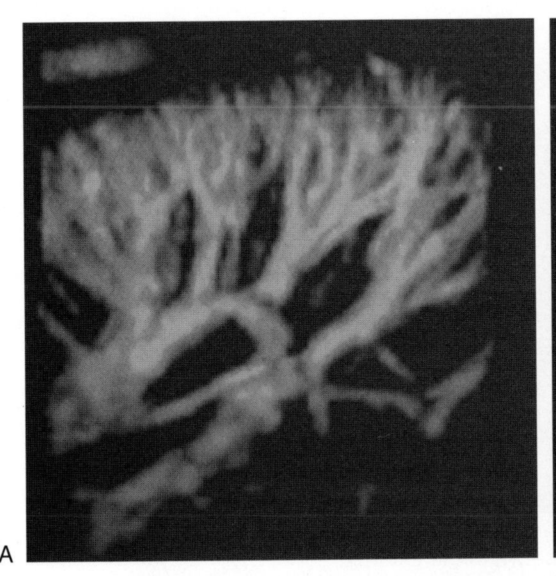

A

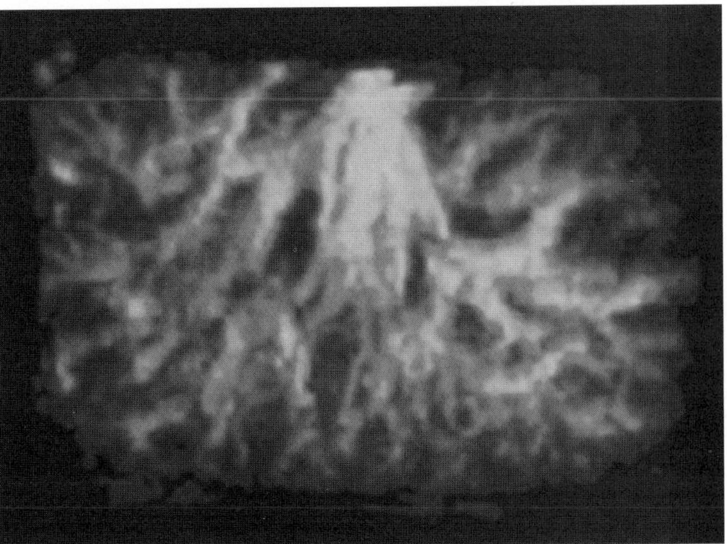

B

Figure 10-11. 3DUS power Doppler studies of the renal vasculature. **A:** A free-hand scan without position sensing of the arcuate and intralobar arteries in the kidney. (Image courtesy of ATL, Bothell, WA.) **B:** Free-hand scan using image-based position sensing. (Image courtesy of Siemens Ultrasound, Issaquah, WA.) See color version of figure.

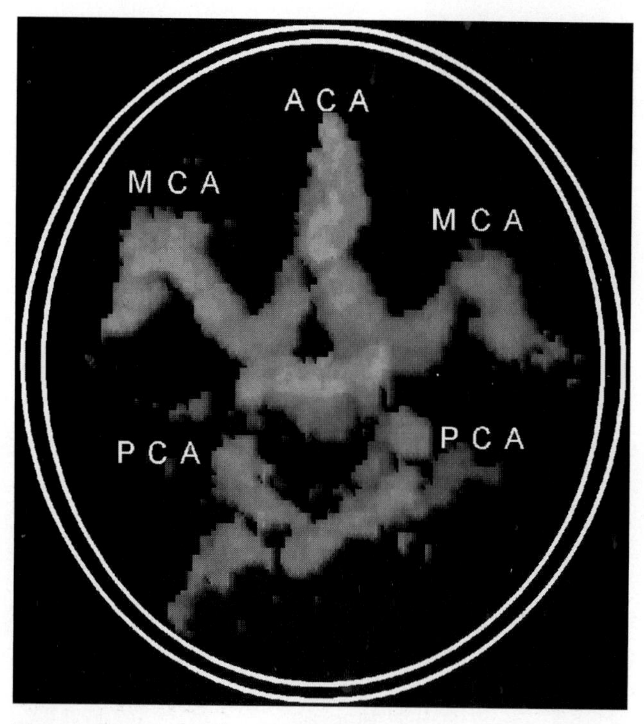

Figure 10-12. 3DUS power Doppler study of the cerebral circulation using free-hand scanning with position sensing. 2DUS image data were acquired from both sides of the head through the transtemporal window and combined using the position data. See color version of figure.

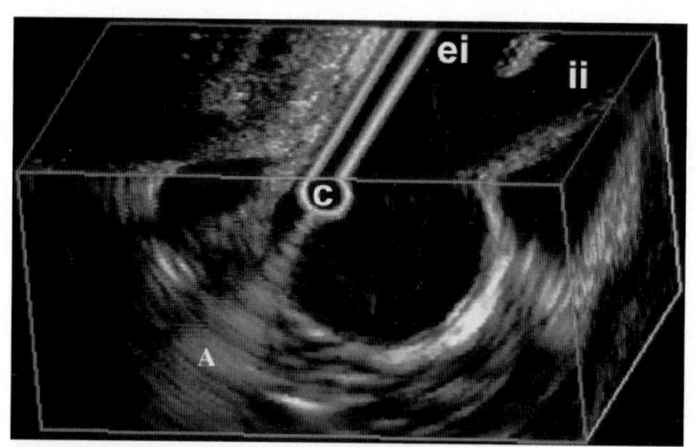

Figure 10-13. Images of the common external iliac artery from a 3D-IVUS study. The bifurcation between the external iliac (ei) and the internal iliac (ii) is clearly shown. C, catheter; A, artifact that widens as one goes further from the transducer. Note that the vessel wall detail is better assessed on these images than on a transcutaneous image. (Image courtesy of Dr. Ji-Bin Liu.)

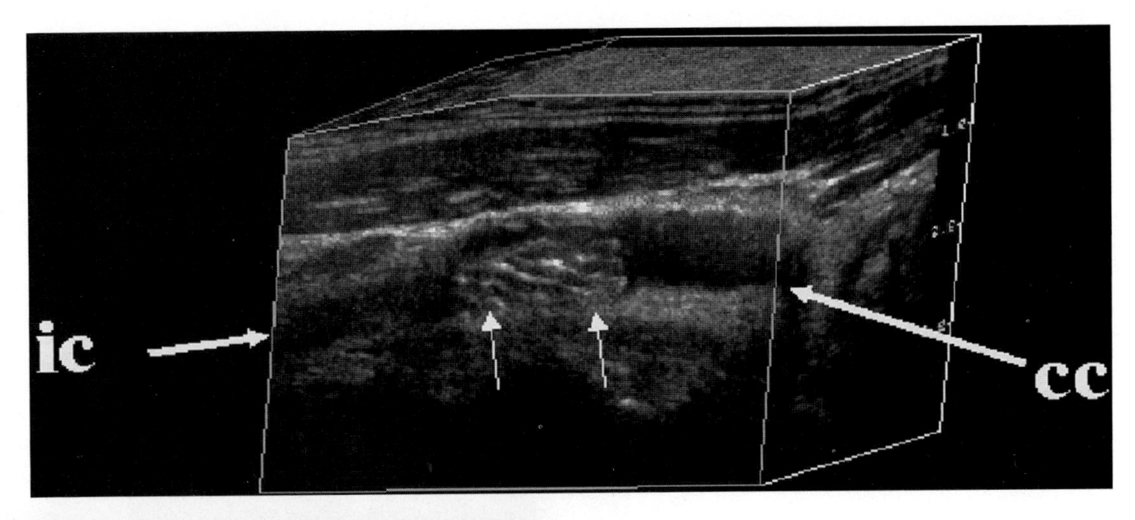

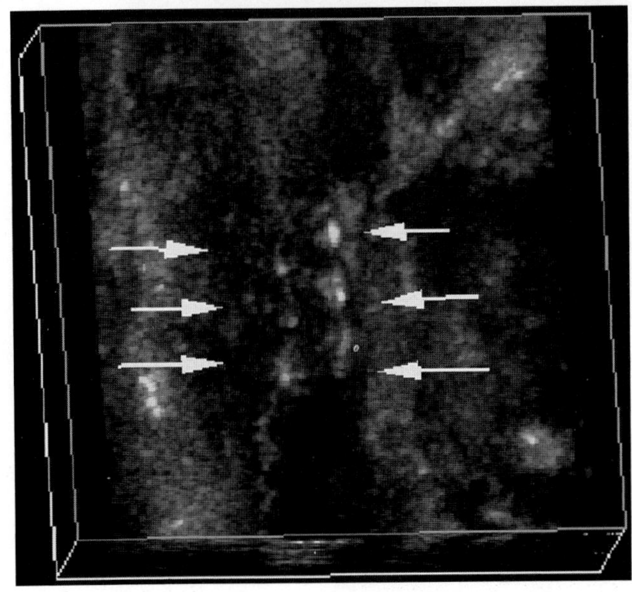

Figure 10-14. 3D-IVUS study showing two veins of a stent in a stenosed internal carotid artery. These data can be used to assess how well the stent expanded and its location longitudinally in the vessel. ic, internal carotid; cc, common carotid; *small arrows,* stent. (Image courtesy of Life Imaging Systems, Inc., London, Ontario, Canada)

and allow smaller vessels with slower flow to be visualized (7,8). Delcker et al. have shown promising results with contrast-enhanced transcranial Doppler studies (47).

3DUS color and power Doppler imaging has the potential to show physiologic information relating to flowing blood and can identify true flow volume (48,51). Laboratory results have shown that it is possible to accurately measure flow restrictions using power Doppler (48,52).

VASCULAR EVALUATION BY INTRAVASCULAR CATHETER IMAGING

3D intravascular ultrasound (3D-IVUS) imaging of blood flow and lumen structure has shown great promise for imaging vessels, including the coronary arteries, despite its invasive nature (12,16,19,27,28,31,53–58). Intravascular methods utilize higher frequencies that permit higher resolution, which can enhance visualization of flow dynamics (59), vessel (Figure 10-13) and graft patency (60), and stent deployment (Figure 10-14) (61,62). Thus far, technical issues regarding position localization of the catheter tip have provided volume data with limited geometric accuracy, because the steady pullback method assumes position data are co-linear, which often is not the case in tortuous vessels.

Luminal Stenosis and Plaque Assessment

3D-IVUS methods offer considerable potential to improve assessment of luminal stenoses (12–15,32,33,63). Multiple transverse images can be stacked together and displayed

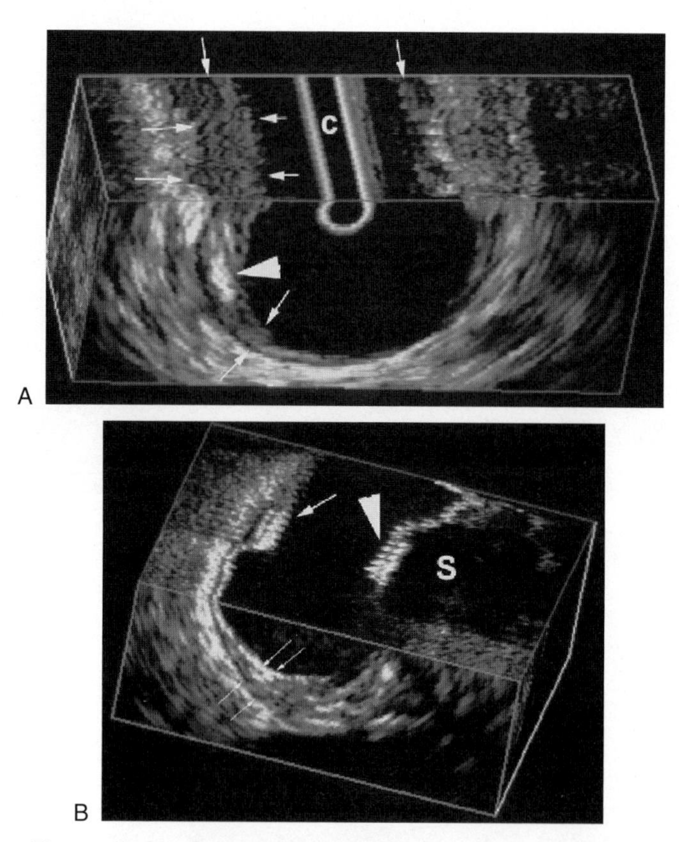

Figure 10-15. A: 3D-IVUS image showing soft plaque consisting predominantly of lipid materials (*arrows*). Images obtained from an intraluminal catheter (C). **B:** The larger echogenic, nonshadowing region in the vascular wall likely represents a more fibrotic plaque (*arrowhead*) with significant posterior shadowing (S). In the 10 o'clock position there is similar echogenic material but with minimal shadowing (*arrow*). This is probably mostly fibrous plaque. In the 7 o'clock position there is mural thickening with predominantly low-level echoes. This plaque is mostly lipid in content. Note the sawtooth-type artifact on the luminal surface superiorly caused by cardiac pulsations. (Images courtesy of Dr. JiBin Liu.)

in ways that allow both the narrowest region and the normal reference region to be seen simultaneously (32). 3D-IVUS B-mode imaging permits the length and diameter of the narrowing to be more precisely measured and compared with the diameter of the vessel at an arbitrary length away from the area of maximal luminal narrowing for reference purposes (Figure 10-15) (32). Accurate measurement of luminal narrowing is clinically important in the evaluation of patients being studied for plaque progression or regression and for evaluating restenosis (31–33). Dhawale showed that a cardiac gated pullback technique produced excellent area and volumetric assessment in phantoms and *in vivo* studies, with error rates of less than 10% for all parameters evaluated (54). Pasterkamp also described success in phantoms for a technique that includes 30-MHz imaging, automatic segmentation, and fast 3D visualization (57).

Coronary Artery Plaque Assessment

Although coronary artery percutaneous transluminal angioplasty (PTCA) procedures have proven successful, they still have limitations because of restenosis (30% to 40%), acute dissection, and thrombosis that may be related to inaccurate assessment of plaque formation and extent (29). To date, one of the major practical clinical benefits of 3D-IVUS techniques is in visualizing plaque severity (32,33). One advantage of 3D-IVUS is that it simplifies understanding of plaque size and internal complexity (1,32,54,58,64,65). High-frequency 3D-IVUS catheters have been shown to detect intraplaque calcification with a sensitivity of about 90% (1,66). Lipid plaques can be detected using 3D-IVUS with sensitivities between 51% and 78% and specificities between 75% and 97% (1). 3D-IVUS has been shown to provide better assessment of coronary artery plaque volume than 2D-IVUS (25,27,30,32,64,67). The architecture of complex plaque in an entire coronary artery segment can be visualized and measured more completely using 3D-IVUS. The interactive review afforded by 3D-IVUS provides more detailed insight and does not require visualizing the vessel geometry mentally (32). In addition, 3D-IVUS makes it possible to visualize and quantify the total extent of plaque calcification in arterial segments, including severely calcified plaque, something that is impossible with 2D-IVUS (32,33). Ge et al. have reported excellent interobserver and intraobserver precision in measuring plaque volumes (68).

There is considerable enthusiasm for the ability of 3D-IVUS to significantly improve the clinical utility of interventional IVUS techniques (30,32,33,64). 3D-IVUS provides information that helps characterize plaque composition and can assist in choosing the appropriate treatment strategy (e.g., whether fibrinolytic or mechanical recanalization is the most appropriate treatment) (66). Newer percutaneous techniques include directional coronary atherectomy, laser angioplasty, stent deployment, and rotablator treatment (66). 3D-IVUS can accurately measure the length of the lesion and accurately measures the luminal diameter of the reference segment of vessel, which helps determine the appropriate stent length to employ (32). 3D-IVUS is useful to assess if a stent has been appropriately expanded (29,32), to detect procedural complications, and to identify high-risk lesions such as plaques with extensive hemorrhage that are prone to rupture (66). 3D-IVUS also can help define an endpoint during coronary artery procedures by proving the vessel has been dilated a finite amount (33). Although this would have been possible with 2D-IVUS, it would have taken considerably longer (32).

Generally, it is extremely difficult to image with 2D-IVUS the same location within a blood vessel before and after an interventional procedure (32). 3D-IVUS solves this problem by monitoring the results of intervention immediately, during, and after long-term follow-up by permitting review of volume data to localize at the same position in the vessel (32). 3D-IVUS also may further our understanding of the therapeutic mechanisms of the various interventional devices (32) and identify subgroups of patients that may benefit from intensive lipid-lowering interventions prior to interventional treatment (66). Overall, assessment of vessel plaque morphology is best accomplished with IVUS catheters, although transcutaneous approaches have been evaluated (26).

Arterial Dissection

Ultrasound readily detects the intimal flap, provided the transducer can be closely positioned adjacent to the vessel. Using a pullback technique, 3D studies can be reconstructed using both types of imaging approaches. 3D reconstructions are desirable to see the full extent of the dissection and to appropriately classify the lesion and treat it. Although arterial dissection can be evaluated with an IVUS approach, it is most frequently done using a transesophageal (TEE) probe (69).

Buck reports that 3D-TEE provides the best-quality studies, compared with transcutaneous methods, because there is less of a problem with sound beam penetration (69). Unfortunately, the thoracic aorta must be imaged using at least two volume acquisitions—one for the ascending and one for the descending aorta. This means that some topographic information may be lost (69). Also, it may be difficult or impossible to image regions of the aortic arch and the distal descending aorta.

3D-IVUS can show the entire topographic extent of the dissection. However, marked echo dropout can be a problem with current IVUS transducers (69). Also, 3D-IVUS is more invasive than 3D-TEE, although few complications have been reported (70). Reports of the clinical utility of both techniques remain preliminary. The role of 3D Doppler exams in this pathology also remains to be evaluated.

3D-IVUS Disadvantages

At present 3D-IVUS examination time may be longer than what is required for a standard IVUS exam, which may pose a slightly increased risk of complications. As techniques improve, this issue is expected to diminish. Additionally, subtle changes in the wall can be appreciated only on the highest-quality B-mode images. Limitations in the data acquisition technique based on pullback methods or difficulty in precisely localizing the position of the transducer may introduce artifacts due to cardiac motion, misregistration, and other factors. As a result, image quality may be compromised sufficiently to obscure a small lesion, such as a pseudo-aneurysm. To date, most 3D-IVUS images of the coronary artery appear to withstand the reconstruction process and preserve their diagnostic information. Also, not all segments of the coronary arteries are currently amenable to the technique because the transducer is too large to be introduced into all segments of the coronary arterial tree.

HOW TO DO IT
Practical Clinical Tips for Vascular Evaluation

- 3DUS data are degraded by motion, leading to poor vessel definition:
  - Lumen/intimal interface.
  - Plaque/media interface.
- Highest-quality data sets are obtained using cardiac and respiratory gating:
  - May lengthen the exam time.
  - Increase chance of patient movement.
- An interventionalist and imager should be present during intravascular procedures:
  - Interventionalist—manipulates catheters.
  - Imager—operates 3DUS system.
- Internal vascular data acquisition methods:
  - Advance catheter beyond area of interest.
  - Acquire images by manually or mechanically ''pulling back'' transducer past area of interest.
  - Limit pullback rate to a few millimeters per second.
  - Distortions caused by nonuniform transducer rotation or noncoaxial position of catheter tip.
- External vascular data acquisition methods:
  - Mechanical positioners and volume transducers suitable for imaging short vessel segments.
  - Free-hand scanning is optimal for imaging longer vessel segments.
  - Best-quality images result with highest frequency transducer closest proximity to vessel.
- 3DUS data sets are best when degree of vessel narrowing small and plaque echogenicity high:
  - Plaque calcifications may shadow regions of interest, requiring multiple acquisitions.
- 3DUS power/color Doppler images require slow scans to minimize flash artifacts.
- On-line reconstruction provides essential feed-back during procedures.
- Multiplanar slices and rendering can show the lumen as either a cylinder or a column.
- Assessment of luminal diameter benefits from use of a semiautomated segmentation program.
- Artifacts simulating stenoses within vessel may be caused by:
  - Sudden shifts in catheter location within the vessel during pullback.
  - Patient motion.
  - Doppler signal dropout.
- Pulsatility can introduce sawtooth artifact variations in the diameter of the lumen.

REFERENCES

1. Ge J, Gorge G, Baumgart D, Liu F, Haude M, Erbel R. Intravascular ultrasound. In: Liu JB, Goldberg BB, eds. *Endoluminal ultrasound: vascular and nonvascular applications.* St. Louis: Mosby, 1997:37–80.
2. Beach KW. 1975–2000: a quarter century of ultrasound technology. *Ultrasound Med Biol* 1992;18:377–388.
3. Arbeille P, Desombre C, Aesh B, Philippot M, Lapierre F. Quantification and assessment of carotid artery lesions: degree of stenosis and plaque volume. *J Clin Ultrasound* 1995;23:113–124.
4. Horn M, Michelini M, Greisleer HP. Carotid endarterectomy without arteriography: the preeminent role of the vascular laboratory. *Ann Vasc Surg* 1994;8:221–224.
5. Polak JF. Role of Duplex US as a screening test for carotid atherosclerotic disease: benefit without cost? *Radiology* 1995;197:581–582.
6. Burns PN. Contrast agents for ultrasound imaging and Doppler. In: Rumack CM, Wilson SR, Charboneau JW, eds. *Diagnostic ultrasound.* St. Louis: Mosby, 1997:57–86.
7. Burns PN, Powers JE, Simpson DH, Uhlendorf V, Fritzsch T. Harmonic imaging—principles and preliminary results. *Angiology* 1997;47:S63–S73.
8. Burns PN, Powers JE, Simpson DH. Harmonic power mode Doppler using microbubble contrast agents: an improved method for small vessel flow imaging. IEEE ultrasonics symposium proceedings 1994;3: 1547–1550.
9. Forsberg F, Ji-Bin, Lui, Merton DA, Rawool NM, Goldberg BB. *In vivo* evaluation of a new ultrasound contrast agent. IEEE ultrasonics symposium proceedings 1994;3:1555–1558.
10. Goldberg BB, Liu J, Forsberg F. Ultrasound contrast agents: a review. *Ultrasound Med Biol* 1994;20: 319–333.
11. Blankenhorn DH, Chin HP, Strikwerda S, Bamberger J, Hestenes JD. Common carotid artery contours reconstructed in three dimensions from parallel ultrasonic images. *Radiology* 1983;148:533–537.
12. Delcker A, Diener HC. 3D ultrasound measurement of atherosclerotic plaque volume in carotid arteries. *Bildgebung* 1994;61:116–121.
13. Delcker A, Diener HC. Quantification of atherosclerotic plaques in carotid arteries by three-dimensional ultrasound. *Br J Radiol* 1994;67:672–678.
14. Delcker A. Influence of data acquisition on accuracy in carotid plaque volume measurements with 3D ultrasound. *Eur J Ultrasound* 1996;4:161–168.
15. Fenster A, Lee D, Sherebrin S, Rankin R, Spence D, Downey D. Three-dimensional ultrasound imaging of carotid occlusive disease. In: Klingelhofer J, Bartels E, Ringelstein EB, eds. *New trends in cerebral hemodynamics and neurosonology.* Amsterdam: Elsevier Science, 1997:17–24.
16. Franceschi D, Bondi JA, Rubin JR. A new approach for three-dimensional reconstruction of arterial ultrasonography. *J Vasc Surg* 1992;15:800–804.
17. Kitney RI, Moura L, Straughan K. 3D visualization of arterial structures using ultrasound and voxel modelling. *Int J Card Imaging* 1989;4:135–143.
18. Pretorius DH, Nelson TR, Jaffe JS. 3-Dimensional sonographic analysis based on color flow Doppler and gray scale image data: a preliminary report. *J Ultrasound Med* 1992;11:225–232.
19. Rosenfield K, Kaufman J, Pieczek A, Langevin RE Jr, Razvi S, Isner JM. Real-time three-dimensional reconstruction of intravascular ultrasound images of iliac arteries. *Am J Cardiol* 1992;70:412–415.
20. Selzer RH, Lee PL, Lai JY, et al. Computer-generated 3D ultrasound images of the carotid artery. Proceedings *Comput Cardiol* 1989:21–26.
21. Webber JD, Foster E, Heidenreich P, LaBerge J, Ring EJ, Schiller NB. Three-dimensional transabdominal ultrasound identification of aortic plaque. *Am J Card Imaging* 1995;9:245–249.
22. Spaulding KA, Kissner ME, Kim EK, Pretorius DH, Rose SC, Garrosi K, Nelson TR. 3-Dimensional gray scale ultrasound imaging of the celiac axis: a preliminary report. *J Ultrasound Med* 1998;17:239–248.
23. Freed KS, Brown LK, Carroll BA. The extracranial cerebral vessels. In: Rumack CM, Wilson SR, Charboneau JW, eds. *Diagnostic ultrasound.* St. Louis: Mosby, 1997:885–919.
24. Polak JF. The peripheral arteries. In: Rumack CM, Wilson SR, Charboneau JW, eds. *Diagnostic ultrasound.* St. Louis: Mosby, 1997;921–941.
25. Weissman NJ, Palacios IF, Nidorf SM, Dinsmore RE, Weyman AE. Three-dimensional intravascular ultrasound assessment of plaque volume after successful atherectomy. *Am Heart J* 1995;130:413–419.
26. Ge J, Erbel R. Characteristic plaque morphology. In: Erbel R, Roelandt JRTC, Ge J, Gorge G, eds. *Intravascular ultrasound.* St. Louis: Mosby, 1998:77–89.
27. Cavaye DM, White RA. Intravascular ultrasound imaging: development and clinical applications. *Int Angiol* 1993;12:245–255.
28. Chandrasekaran K, Sehgal CM, Hsu TL, et al. Three-dimensional volumetric ultrasound imaging of arterial pathology from two-dimensional intravascular ultrasound: an *in vitro* study. *Angiology* 1994;45:253–264.
29. Erbel R, Roelandt JRTC, Ge J, Gorge G. Introduction. In: Erbel R, Ge J, eds. *Intravascular ultrasound.* St. Louis: Mosby, 1998:1–9.
30. Prati F, DiMario C, Gil R, et al. Three-dimensional intravascular ultrasound for stenting. In: Erbel R, Ge J, eds. *Intravascular ultrasound.* St. Louis: Mosby, 1998;239–247.
31. Roelandt JRTC, di Mario C, Pandian NG, et al. Three-dimensional reconstruction of intracoronary ultrasound images: rationale, approaches, problems, and directions. *Circulation* 1994;90:1045–1053.
32. Von Birgelen C, Nicosia A, Serruys PW, Roelandt JRTC. Assessment of balloon angioplasty and directional atherectomy with three-dimensional intracoronary ultrasound. In: Erbel R, Roelandt JRTC, Ge J, Gorge G, eds. *Intravascular ultrasound.* St. Louis: Mosby, 1998:183–189.
33. Von Birgelen C, de Feyter PJ, Roelandt JRTC. Three-dimensional reconstruction of intracoronary ultrasound. In: Erbel R, Roelandt JRTC, Ge J, Gorge G, eds. *Intravascular ultrasound.* St. Louis: Mosby, 1998:51–60.
34. Bluth EI, Kay D, Merritt CRB. Sonographic characterization of carotid plaque: detection of hemorrhage. *AJR* 1986;146:1061–1065.
35. Leahy AL, McCollum PT, Feeley TM, et al. Duplex ultrasonography and selection of patients for carotid endarterectomy: plaque morphology or luminal narrowing? *J Vasc Surg* 1988;8:558–562.

36. Merritt CR, Bluth EI. The future of carotid sonography [comment]. *AJR* 1992;158:37–39.
37. Zwiebel WJ. Aortic and iliac aneurysm. *Semin Ultrasound CT MR* 1992;13:53–68.
38. Delcker A, Diener HC, Wilhelm H. Influence of vascular risk factors for atherosclerotic carotid artery plaque progression. *Stroke* 1995;26:2016–2022.
39. Downey DB. Power Doppler in prostate cancer. *Curr Opin Urol* 1997;7:93–99.
40. Downey DB, Fenster A. Vascular imaging with a three-dimensional power Doppler system. *AJR* 1995;165: 665–668.
41. Ritchie CJ, Edwards WS, Mack LA, Cyr DR, Kim Y. Three-dimensional ultrasonic angiography using power-mode Doppler. *Ultrasound Med Biol* 1996;22:277–286.
42. Pretorius DH, Nelson TR, Baergen RN, Pai E, Cantrell C. Imaging of placental vasculature using three-dimensional ultrasound and color power Doppler: a preliminary study. *Ultrasound Obstet Gynecol* 1998 (*in press*).
43. Delabays A, Sugeng L, Pandian NG, et al. Dynamic three-dimensional echocardiographic assessment of intracardiac blood flow jets. *Am J Cardiol* 1995;76:1053–1058.
44. Lyden PD, Nelson TR. Visualization of the cerebral circulation using three-dimensional transcranial power Doppler ultrasound imaging. *J Neuroimaging* 1997;7:35–39.
45. Burns PN, Powers JE, Fritzsch T. Harmonic imaging: a new imaging and Doppler method for contrast enhanced ultrasound. *Radiology* 1992;185(P):140–142.
46. Burns PN. Harmonic imaging with ultrasound contrast agents. *Clin Radiol* 1996;51:S50–S55.
47. Delcker A, Turowski B. Diagnostic value of three-dimensional transcranial contrast duplex sonography. *J Neuroimaging* 1997;7:139–144.
48. Guo Z, Moreau M, Rickey DW, Picot PA, Fenster A. Quantitative investigation of *in vitro* flow using three-dimensional colour Doppler ultrasound. *Ultrasound Med Biol* 1995;21:807–816.
49. Nelson TR, Mattrey RF, Steinbach GC, Lee Y, Lazenby J. 3D harmonic grey-scale ultrasound visualization of vascular anatomy. *JUM* 1998;17(3):S55.
50. Nelson TR, Mattrey RF, Steinbach GC, Lee Y, Lazenby J. Gated 3D ultrasound imaging of the coronary arteries using contrast and harmonic imaging. *JUM* 1998;17:S56.
51. Poulsen JK, Kim WY. Measurement of volumetric flow with no angle correction using multiplanar pulsed Doppler ultrasound. *IEEE Trans Biomed Eng* 1996;43:589–599.
52. Guo Z, Fenster A. Three-dimensional power Doppler imaging: a phantom study to quantify vessel stenosis. *Ultrasound Med Biol* 1996;22:1059–1069.
53. De Man F, De Scheerder I, Herregods MC, Piessesns J, De Geest H. Role of intravascular ultrasound in coronary artery disease: a new gold standard? An overview. *Acta Cardiol* 1994;49:223–231.
54. Dhawale PJ, Wilson DL, Hodgson JM. Volumetric intracoronary ultrasound: methods and validation. *Cathet Cardiovasc Diagn* 1994;33:296–307.
55. Koch L, Roth T. Technical aspects of intravascular ultrasound. In: Erbel R, Roelandt JRTC, Ge J, Gorge G, eds. *Intravascular ultrasound.* St. Louis: Mosby, 1998:17–30.
56. Ng KH, Evans JL, Vonesh MJ, et al. Arterial imaging with a new forward-viewing intravascular ultrasound catheter. II. Three-dimensional reconstruction and display of data. *Circulation* 1994;89:718–723.
57. Pasterkamp G, Borst C, Moulaert AF, et al. Intravascular ultrasound image subtraction: a contrast enhancing technique to facilitate automatic three-dimensional visualization of the arterial lumen. *Ultrasound Med Biol* 1995;21:913–918.
58. Von Birgelen C, di Mario C, Li W, et al. Mophometric analysis in three-dimensional intracoronary ultrasound: an *in vitro* and *in vivo* study performed with a novel system for the contour detection of lumen and plaque. *Am Heart J* 1996;132:516–527.
59. Chandran KB, Vonesh MJ, Roy A, et al. Computation of vascular flow dynamics from intravascular ultrasound images. *Med Eng Phys* 1996;18:295–304.
60. Ennis BM, Zientek DM, Ruggie NT, Billhardt RA, Klein LW. Characterization of a saphenous vein graft aneurysm by intravascular ultrasound and computerized three-dimensional reconstruction. *Cathet Cardiovasc Diagn* 1993;28:328–331.
61. Gil R, von Birgelen C, Prati F, et al. Usefulness of three-dimensional reconstruction for interpretation and quantitative analysis of intracoronary ultrasound during stent deployment. *Am J Cardiol* 1996;77:761–764.
62. Mintz GS, Pichard AD, Satler LF, et al. Three-dimensional intravascular ultrasonography: reconstruction of endovascular stents *in vitro* and *in vivo*. *J Clin Ultrasound* 1993;21:609–615.
63. Liu JB, Miller LS, Bagley DH, Bonn J, Forsberg F, Goldberg BB. Three-dimensional endoluminal ultrasound. In: Liu JB, Goldberg BB, eds. *Endoluminal ultrasound: vascular and nonvascular applications.* St. Louis: Mosby, 1997:325–346.
64. Von Birgelen C, di Mario C, Reimars B, et al. Three-dimensional intracoronary ultrasound imaging: methodology and clinical relevance for the assessment of coronary arteries and bypass grafts. *J Cardiovasc Surg (Torino)* 1996;37:129–139.
65. Von Birgelen C, van der Lugt A, Nicosia A, et al. Computerized assessment of coronary lumen and atherosclerotic plaque dimensions in three-dimensional intravascular ultrasound correlated with histomorphometry. *Am J Cardiol* 1996;78:1202–1209.
66. Ramo P, Spencer T. Tissue characterization. In: Erbel R, Roelandt JRTC, Ge J, Gorge G, eds. *Intravascular ultrasound.* St. Louis: Mosby, 1998:61–68.
67. Shah VT, Ge J, Ashry M, Erbel R. Distribution of atherosclerosis. In: Erbel R, Roelandt JRTC, Ge J, Gorge G, eds. *Intravascular ultrasound.* St. Louis: Mosby, 1998:81–90.
68. Ge J, Erbel R, Gerber T, et al. Intravascular ultrasound imaging of angiographically normal coronary arteries: a prospective study *in vivo*. *Br Heart J* 1994;71:572–578.
69. Buck T, Gorge G, Ge J, Erbel R. Three-dimensional reconstruction of transesophageal and intravascular imaging in aortic disease using IVUS catheters. In: Erbel R, Roelandt JRTC, Ge J, Gorge G, eds. *Intravascular ultrasound.* St. Louis: Mosby, 1998:253–260.
70. Weintraub AR, Erbel R, Gorge G, et al. Intravascular ultrasound imaging in acute aortic dissection. *J Am Coll Cardiol* 1994;24:495–503.

11

Cardiac System

OVERVIEW

Cardiac anatomy is complex and often difficult to visualize or comprehend. Three-dimensional echocardiography (3DE) integrates a series of 2D echo slices to develop a 3D impression of underlying cardiac anatomy or pathology. Potential advantages of 3DE over current echocardiography methods include improved visualization of normal and abnormal cardiac structures and evaluation of complex cardiac structures, reduced patient scanning times, and cost-effectiveness. This chapter will review some basic concepts behind 3DE and describe clinical applications.

KEY CONCEPTS

- Cardiac anatomy is complex and often difficult to visualize or comprehend.
- Evaluation of complex or abnormal structures can be time-consuming and tedious.
- Repeated scanning often is required to visualize and comprehend cardiac anatomy.
- Scanning methods:
  - Transthoracic:
    - Integrated scan heads with position sensor.
    - External position sensor attached to transducer.
  - Transesophageal:
    - Integrated transesophageal (TEE) rotational probes.
  - Intravascular.
- Viewing methods:
  - Planar slices:
    - Reoriented to standard orientation.
    - Intersecting orthogonal slices help localize position.
    - Animation of cardiac motion assists evaluation of cardiac structures.
  - Rendered images:
    - Surface rendering of valves and chambers shows anatomic continuity.

KEY CONCEPTS, *continued*

- Volume rendering of chambers and vessels shows anatomic continuity.
- Animation of cardiac motion assists evaluation of cardiac structures.
- Clinical applications in adults, children. and fetuses:
  - Improved visualization of chambers, valves, vessels, and myocardium.
  - Better understanding of overall cardiac anatomy throughout cardiac cycle with gating.
  - Evaluation of congenital heart disease (CHD).
  - Improved assessment of dynamic cardiovascular anatomy.
  - Evaluation of valve function and anatomy.
  - Visualization of regurgitant jets using color/power Doppler.
  - Accurate measurements of length, area, and volume throughout cardiac cycle.
  - Surgical planning.
  - Simulation of surgical approaches prior to operating.
  - Evaluation of pediatric patients with CHD.

Cardiac anatomy is complex and often difficult to visualize or comprehend. Echocardiography is a valuable diagnostic tool for evaluation of cardiac disease that can provide important information about cardiac function and anatomy at minimal risk to the patient. A typical approach to patient imaging is to scan repeatedly through the heart to clarify the exact spatial relationships, which can be time-consuming and tedious, particularly with complex or abnormal structures.

Three-dimensional echocardiography (3DE) represents a natural extension of conventional echocardiography methods, which requires integrating a series of 2D image slices to develop a 3D impression of underlying anatomy or pathology (Figure 11-1). Although 3DE was among the earliest areas of three-dimensional ultrasound (3DUS) to be explored (1–9), the technical challenges in obtaining good volumetric data from the moving heart have limited clinical utilization. 3DE acquisition, visualization, and measurement methods are similar to those described in Chapters 2–4. This chapter will review some of the special concepts behind 3DE and describe clinical applications.

The recent availability of affordable high-performance computing systems has greatly

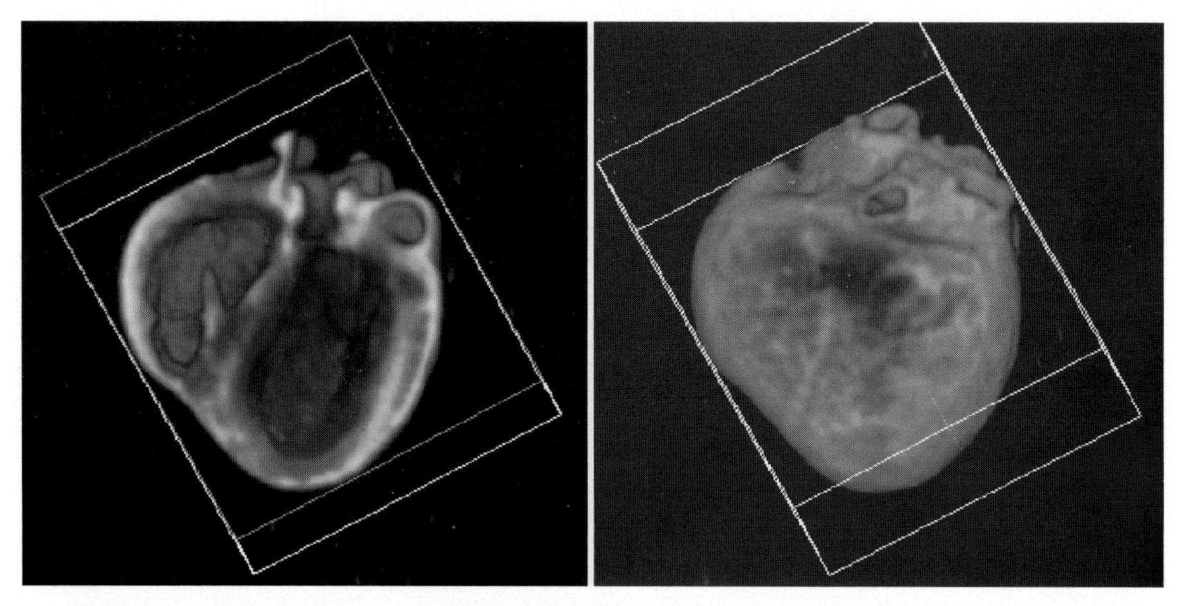

Figure 11-1. Two rendered images of an *in vivo* 3DE study of a canine heart. The intracardiac and external spatial relationships are clearly seen. (Image courtesy of Dr. J. Greenleaf.)

expanded the potential of 3DE to provide multidimensional diagnostic information interactively in the clinical setting. Interactive visualization and manipulation of cardiac volume data by rotation and zooming in on localized features or viewing optimized cross-sectional slices greatly assist comprehension of patient anatomy and extraction of vital anatomic information and quantitative data, providing better identification of complex anatomy and pathology. 3DE imaging also provides the physician with the ability to evaluate patient data after he or she has left the scanning suite and to review the diagnosis with experts at remote locations via network communications.

3DE poses some unique challenges because of the desirability of obtaining information about cardiac anatomy and functional dynamics throughout the cardiac cycle. Currently, most 3DE imaging of cardiac dynamics or blood flow requires obtaining a series of volumes throughout the cardiac cycle by synchronizing data to the appropriate time in the cardiac cycle using ECG-gating or motion analysis methods (Figure 11-2). Accurate visualization of cardiac anatomy requires synchronizing 3DE data collection to the cardiac cycle. Adult and pediatric 3DE imaging uses the electrocardiogram (ECG) signal from patient electrodes to provide a trigger signal to synchronize with the data acquisition.

Obtaining an ECG from the fetus is much more difficult. One approach is to perform data analysis on the acquired data to extract information about the motion in the image field (e.g., the periodic contraction of the heart). Temporal Fourier analysis of the cardiac motion has been used to identify the fundamental frequency of the heart motion (10,11), and phase information is used to identify the location of each acquired 2D image within the cardiac cycle (Figure 11-3). Similar approaches use M-mode (12) and Doppler

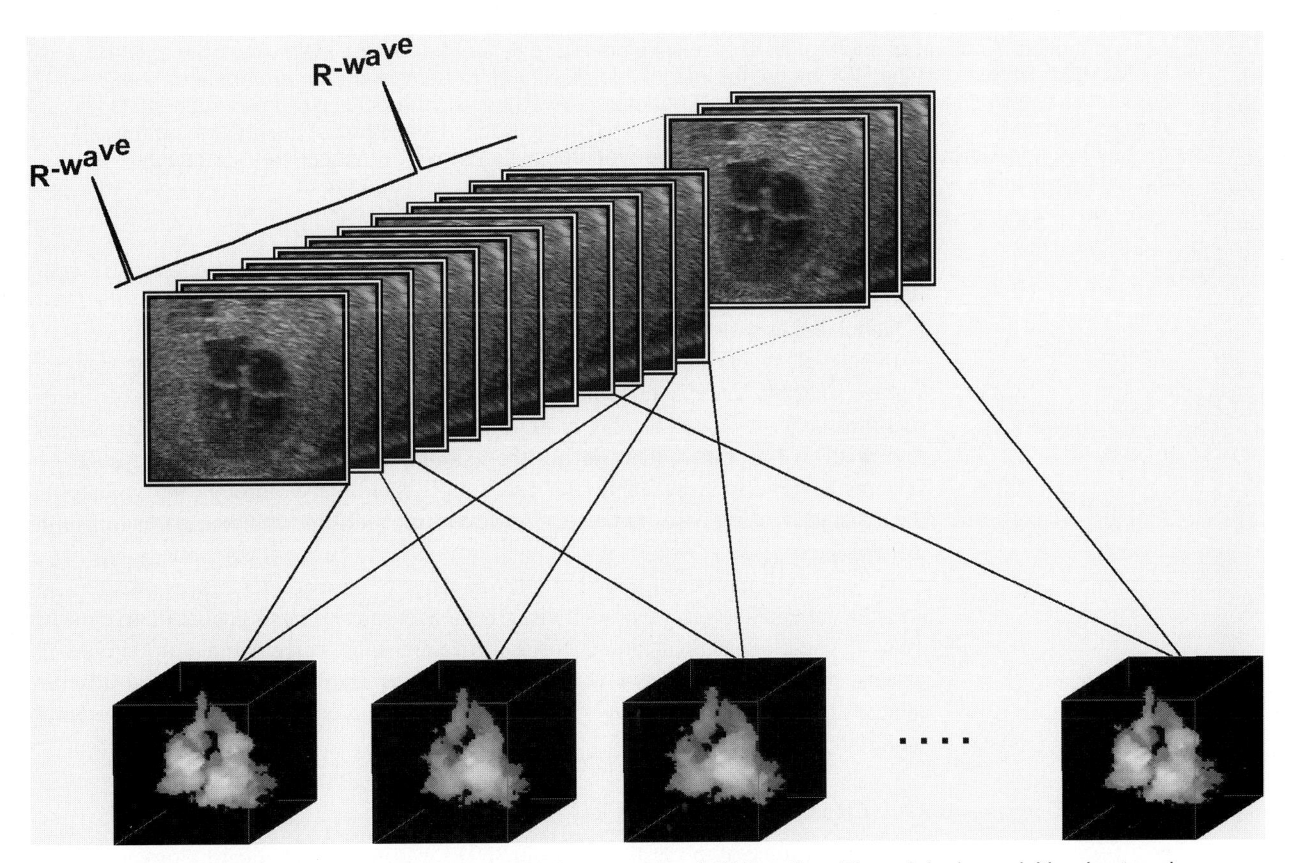

Figure 11-2. A diagram showing the basic concepts of 3DE gating. The original acquisition is stored in the order acquired along with the ECG. Subsequently, each image is assigned to a volume corresponding to its location in the cardiac cycle based on the ECG R-wave. The volumes represent a specific time in the cardiac cycle and can be displayed individually or as part of an animated "cine-loop" display to show the dynamics of valve motion and myocardial contraction.

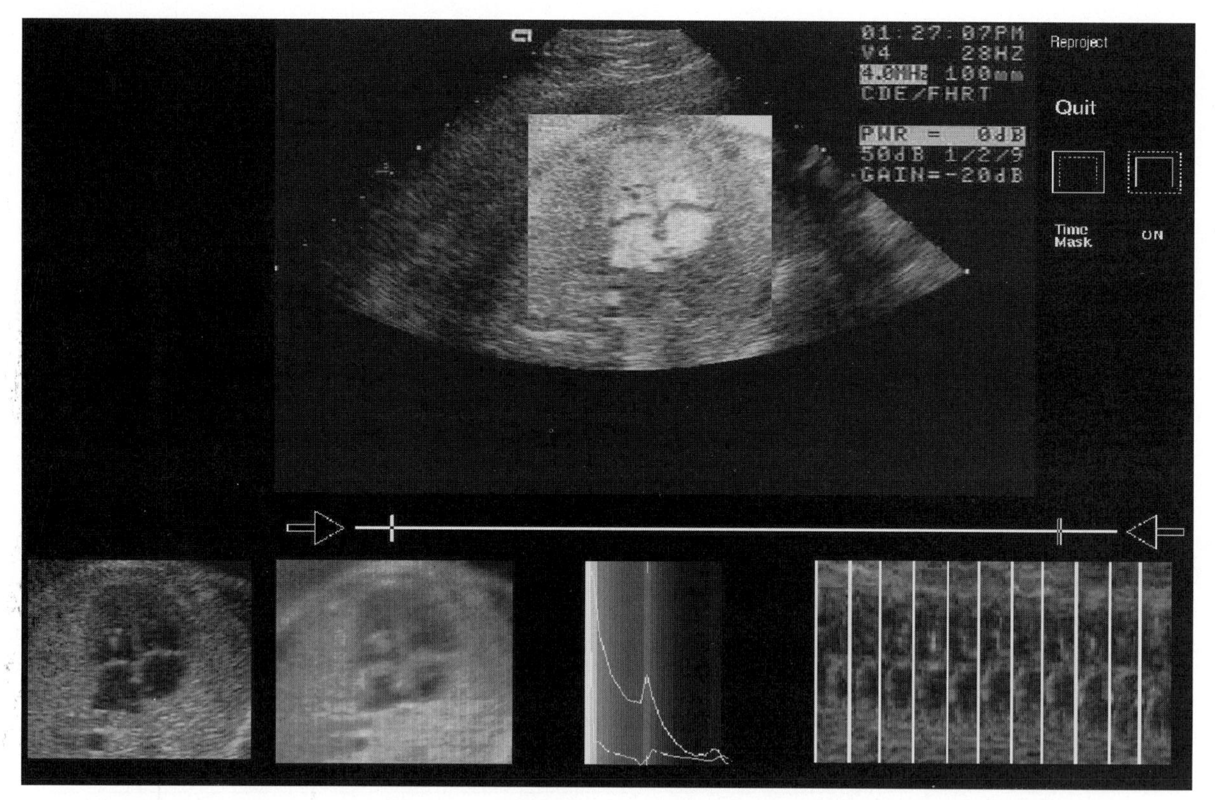

Figure 11-3. Determination of the cardiac cycle timing using temporal Fourier analysis of the periodic cardiac motion. A region of interest (ROI) is positioned over the fetal heart. The lower row of images shows (from left to right) the ROI image, the magnitude display of the temporal Fourier transform, the cumulative amplitude spectrum of the temporal Fourier transform showing the location of the fundamental cardiac beat frequency, and a time slice through the 2DUS acquisition in the region of the heart showing the heart motion with the Fourier-based cardiac cycle synchronization shown as vertical lines corresponding to end-diastole.

(13) methods to determine the heart rate. These methods do not require electrical connection to the patient and are retrospective, utilizing acquired data to determine the periodic behavior rather than being real-time during the acquisition.

In general, synchronization of images to physiologic triggers is demanding from a data storage, analysis, and display point of view. Because the quantity of data can be immense, real-time acquisition and display of 3DE has been difficult to obtain and most studies are reviewed off-line rather than during the examination. Recently, prototype commercial systems have become available that can sample three planes through the volume under the transducer in real time. At present, however, true real-time volumetric imaging remains a future goal.

To date, 3DE has been applied to a broad range of patients (fetal, pediatric, and adult) and clinical applications (anatomic visualization, surgical planning, quantitative measurements). The clinical applications for the different patient populations are similar in all groups, particularly data review and postprocessing analysis. The principal differences relate to the unique requirements to image the fetal and pediatric patients, who represent important constituencies.

GENERAL APPROACHES TO 3DE IMAGING

Although some of the earliest work in 3DUS was in the area of cardiology, as improved equipment performance and advanced catheters have become available, three- and four-dimensional imaging of the heart remains an active area of research and clinical investigation (14–30). In addition to imaging, functional measurements of blood flow and ejection fraction provide quantitative information essential to understand heart function (31,32).

Table 11-1. *Clinical benefits of 3D echocardiography—adult*

Valvular disease	Clinical contributions	Management benefits
Stenosis Prolapse Regurgitation	Multiple views of valve Assessment of valve geometry Visualization of jet morphology Assess regurgitation severity Surgeon's view of anomaly Preoperative assessment	Surgical planning Post-op review
Intracardiac masses	**Clinical contributions**	**Management benefits**
Myxomas/tumors	Clarify anatomic relationships Precise localization Volume measurement Preoperative assessment	Surgical planning Post-op follow-up
Aortic diseases	**Clinical contributions**	**Management benefits**
Dissection	Visualize vessel lumen Optimize view of anatomy Preoperative assessment	Surgical planning

Adult Cardiac Imaging

3DE is a valuable diagnostic tool for evaluation of adult cardiac disease (Table 11-1). 3DE adult cardiac imaging methods have shown improved visualization of the heart atria and septum (15), overall cardiac anatomy (24,33,34), and congenital heart disease (25,26). 3DE of cardiac structures incorporating color/power Doppler techniques offers the possibility of visualizing valvular insufficiency and regurgitant jets (35–38). Valve function and morphology also are readily evaluated with 3DE methods (Figures 11-4, 11-5, 11-6) (39–41). 3DE evaluation of congenital heart defects such as atrial septal defects (42) and ventricular septal defects assists in identifying the location and extent of the defects and may be of assistance in adults (43). 3DE also provides valuable information that can assist in assessing tumors and masses (Figure 11-7), plus surgical planning, including visualizing surgical approaches prior to operating (30).

3DE cardiac methods utilize both transthoracic (25,44) and transesophageal (41,45,46)

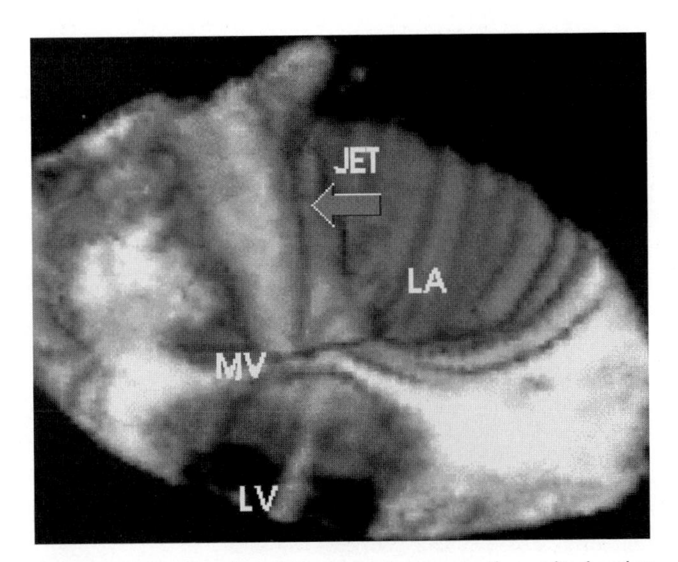

Figure 11-4. Rendered images demonstrating mitral valve prolapse and regurgitation. Rendered views assist in evaluating the extent of the prolapse and regurgitant jet and planning corrective surgical procedures. (Image courtesy of TomTec, Inc., Lafayette, Colorado.)

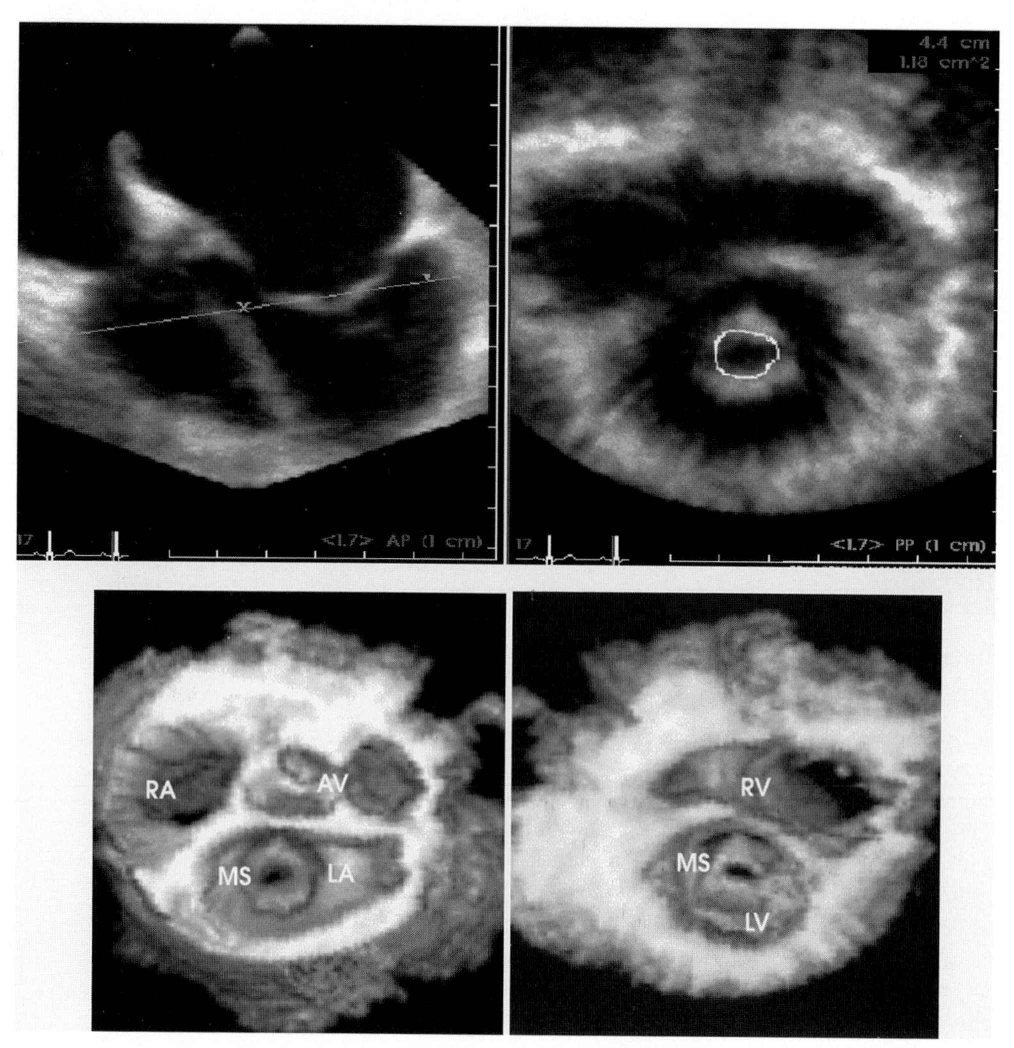

Figure 11-5. Rendered images demonstrating mitral valve stenosis. Rendered views assist in evaluating the extent of the stenosis and planning corrective surgical procedures. The upper left image shows a cross-section of the valve, with the *line* showing the location of the plane in the upper right image. The two lower images show rendered views of the mitral valve. The lower left image shows the mitral stenosis from above looking down, and the lower right image shows the mitral stenosis from below looking up. The rendered images give an overall better impression of the valvular anatomy than the single planar slice and approximate what the surgeon will see during a procedure. (Image courtesy of TomTec, Inc., Lafayette, Colorado)

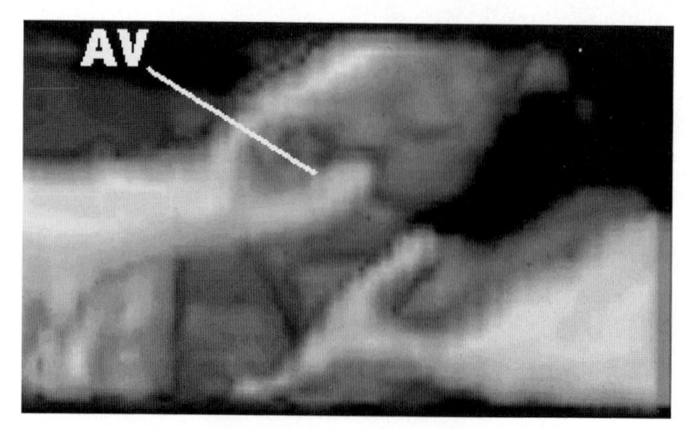

Figure 11-6. Rendered images of a long axis view, including the aortic valve leaflets. (Image courtesy of TomTec, Inc., Lafayette, Colorado.)

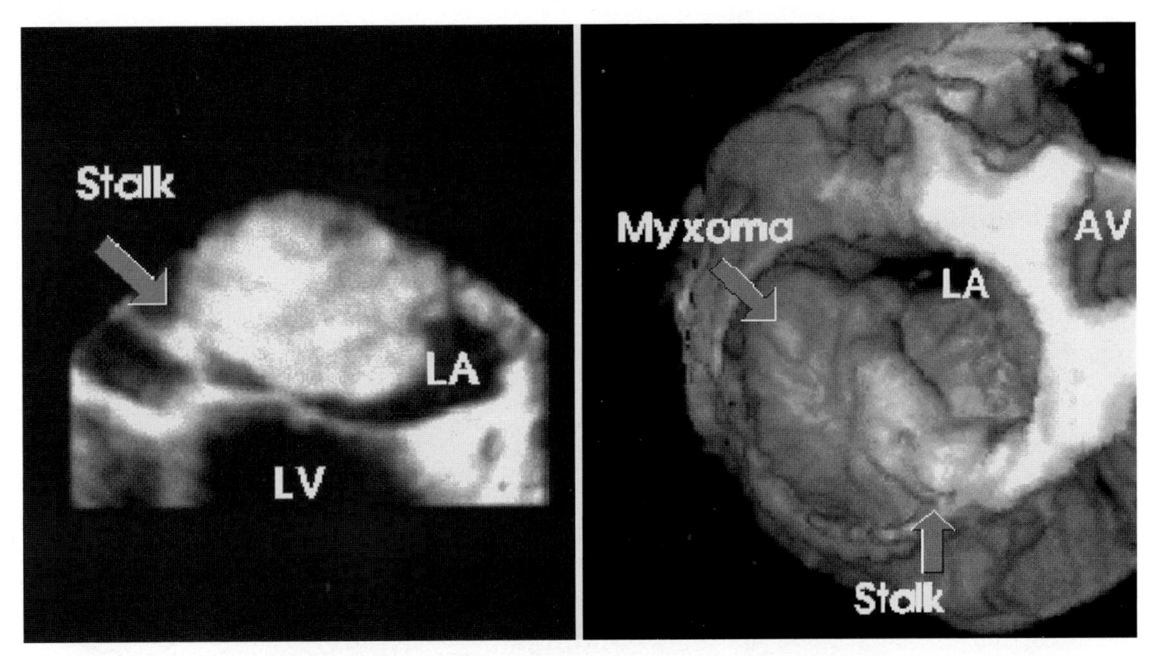

Figure 11-7. Rendered images demonstrating an intracardiac myxoma. Rendered views assist in evaluating the extent of the mass and planning corrective surgical procedures. (Image courtesy of TomTec, Inc., Lafayette, Colorado.)

methods to obtain their information. 3DUS methods also can provide accurate measurements of cardiac chamber (28,29,47–49) and lumen (50) volumes. Volume-rendering methods can improve display and comprehension of cardiac anatomy and, when coupled with cine animation sequences, can provide important information about myocardial and valvular function (Figure 11-8) (11,51).

Pediatric Cardiac Imaging

Pediatric patients have shown good results with both transthoracic and transesophageal 3DE to evaluate congenital heart disease (Table 11-2) (17,28). Acquisition of high-quality 3DE data in infants and children may require a different approach from that with adults, depending on the age of the child. ECG signals provide gating information for volume synchronization similar to that for adults. However, in younger children some transesophageal probes may be too large to work reasonably with children or may require anesthesia, which significantly complicates the study. In general, transthoracic and subcostal imaging approaches can be used to obtain excellent images (28,42,52,53). Pediatric patients often are imaged as part of follow-up to prenatal detection of congenital heart disease or used to plan management options, including surgery. Postsurgical imaging evaluation of patient status can be rapid and noninvasive, facilitating close monitoring of recovery and function. Imaging can be performed using integrated probes in addition to free-hand position tracking techniques.

Fetal Cardiac Imaging

Although early efforts primarily focused on 3DE in the adult and pediatric patient, 3DE in the fetal patient also is an important and challenging area, with most CHD undiagnosed during prenatal diagnostic evaluation (Table 11-3) (54). Most conventional 3DUS imaging equipment can produce static volume images of the fetal heart (Figure 11-9).

The fetal heart is difficult to evaluate with 2DUS for several reasons: (a) The fetus frequently moves during the examination, often requiring brief rapid examination of specific structures; (b) the fetal heart is rarely positioned optimally for intuitive clear understanding of anatomy; (c) the fetal heart contracts approximately 140 times per minute,

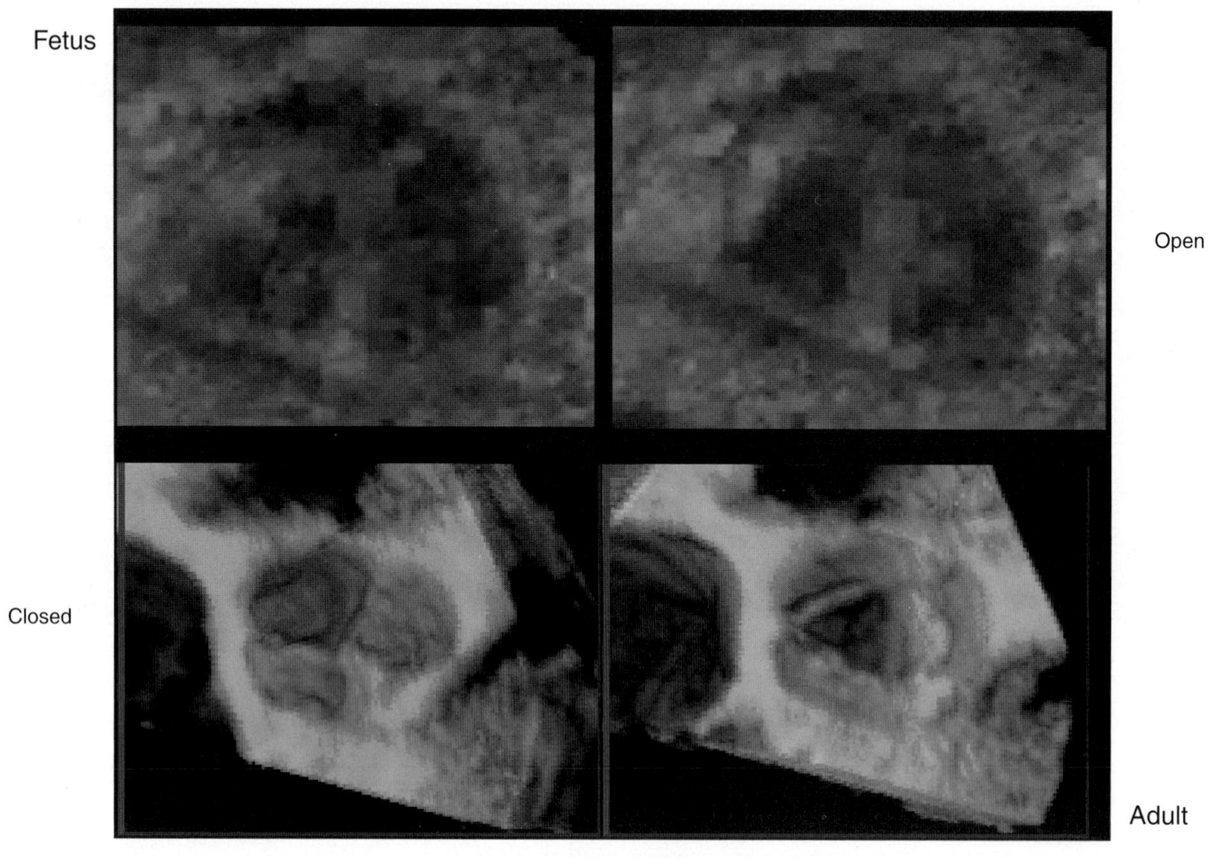

Figure 11-8. Images of the aortic valve in the open and closed position for fetal and adult hearts. The difference in size of the structure relative to the spatial resolution of the scanner is an important determinant of the resultant image quality. Gated data permit evaluation of the dynamics in addition to the anatomic relationships. (Image of fetal aortic valve courtesy of Dr. T. Nelson. Image of adult aortic valve courtesy of TomTec, Inc., Lafayette, Colorado.)

Table 11-2. *Clinical benefits of 3D echocardiography—pediatric*

Congenital heart disease	Clinical contributions	Management benefits
ASD VSD Aortic stenosis Transposition of great vessels	Clarify anatomic relationships Precise localization Assessment of size and geometry Surgeon's view of anomaly Preoperative assessment	More accurate Diagnosis Surgical planning

Table 11-3. *Clinical benefits of 3D echocardiography—fetal*

Screening and diagnosis	Clinical contributions	Management benefits
Normal fetal heart and great vessels anatomy Congenital heart disease detection	Visualize chambers and vessels Optimize view of anatomy Measure fetal heart volume accurately Less dependence on acquisition orientation Gated 3DUS more beneficial than nongated studies	Demonstrate fetal cardiac anatomy more consistently than with static 3DE or 2DE methods Confirm normalcy Optimize delivery and postnatal management

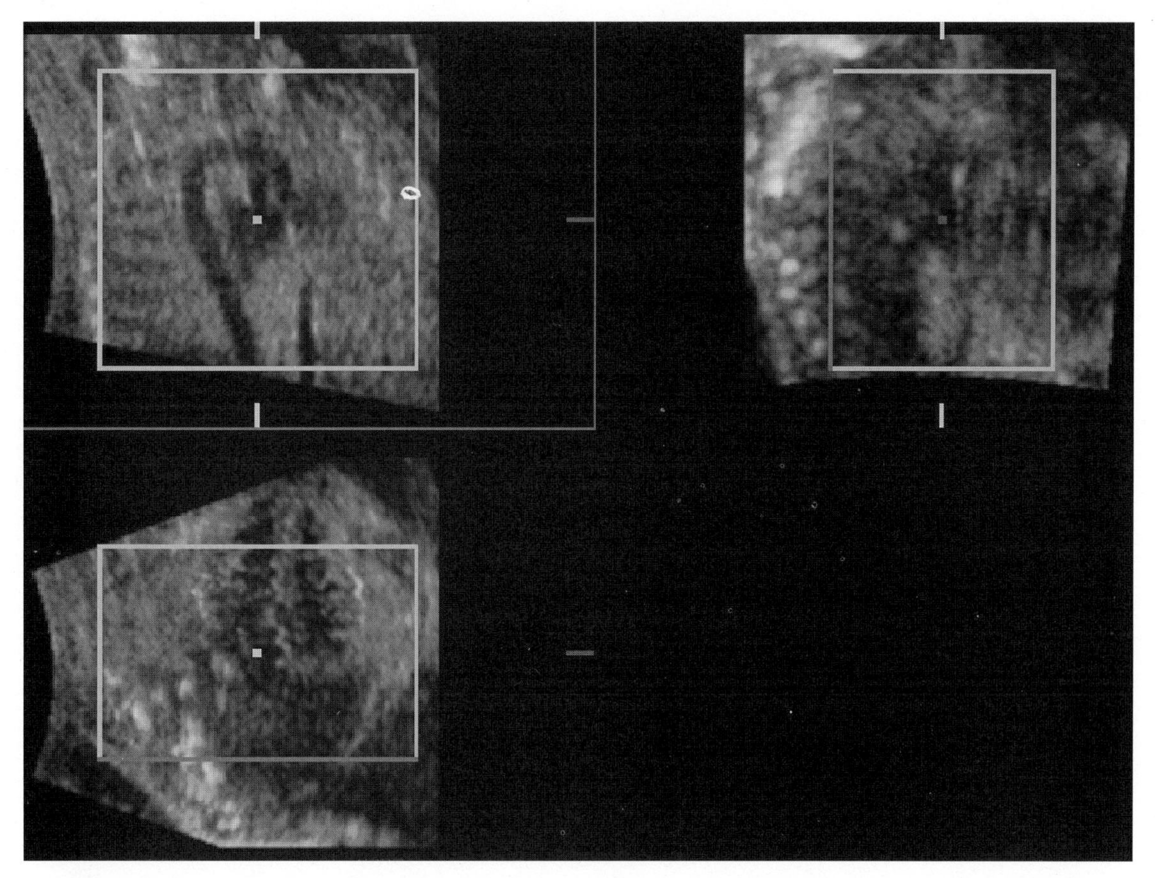

Figure 11-9. Images for a nongated 3DE study of the fetal heart using the Voluson 530D (Medison, Korea). The intracardiac structures are not well visualized because of motion occurring during the imaging process, as can be seen in the lower left image. The great vessels and surrounding landmarks are well seen.

requiring constant assessment of moving structures; (d) complex congenital heart disease patterns are not familiar to the clinical obstetricians and radiologists who perform most obstetric sonography.

3DE can assist in improving the evaluation of the fetal heart by acquiring volume data rapidly that can be rotated into a standard orientation for viewing, providing the ability to slow down the heart and evaluate it systematically throughout the cardiac cycle and permitting examiners to compare abnormal heart anatomy with familiar normal heart anatomy more rapidly.

There have been some studies that have shown promise in imaging the great vessels, and relatively nonmoving areas using nongated 3DUS methods (55). In static studies critical areas of cardiac anatomy are generally poorly visualized because of motion (56) and the small size of the fetal heart, which limits spatial resolution (57). Several cardiac lesions have not been evaluated adequately with nongated 3DE because of problems with cardiac motion (56,58). In general, motion-related artifacts and image blurring in nongated studies can degrade image quality in the interior regions of the heart (59).

Some type of gating is generally required to produce detailed information about function and intracardiac anatomy. The difficulty in obtaining fetal ECG data necessitates using other approaches to synchronize volume data to the cardiac cycle. Fourier (11), M-mode (12), and Doppler (13) methods have been used to determine the fetal heart rate and produce gated fetal cardiac images at each time point throughout the cardiac cycle. Gated 3DE data for the fetal heart have demonstrated fetal cardiac anatomy more consistently than 2DE methods and have shown less dependence on the orientation of the acquisition (Figure 11-10) (60).

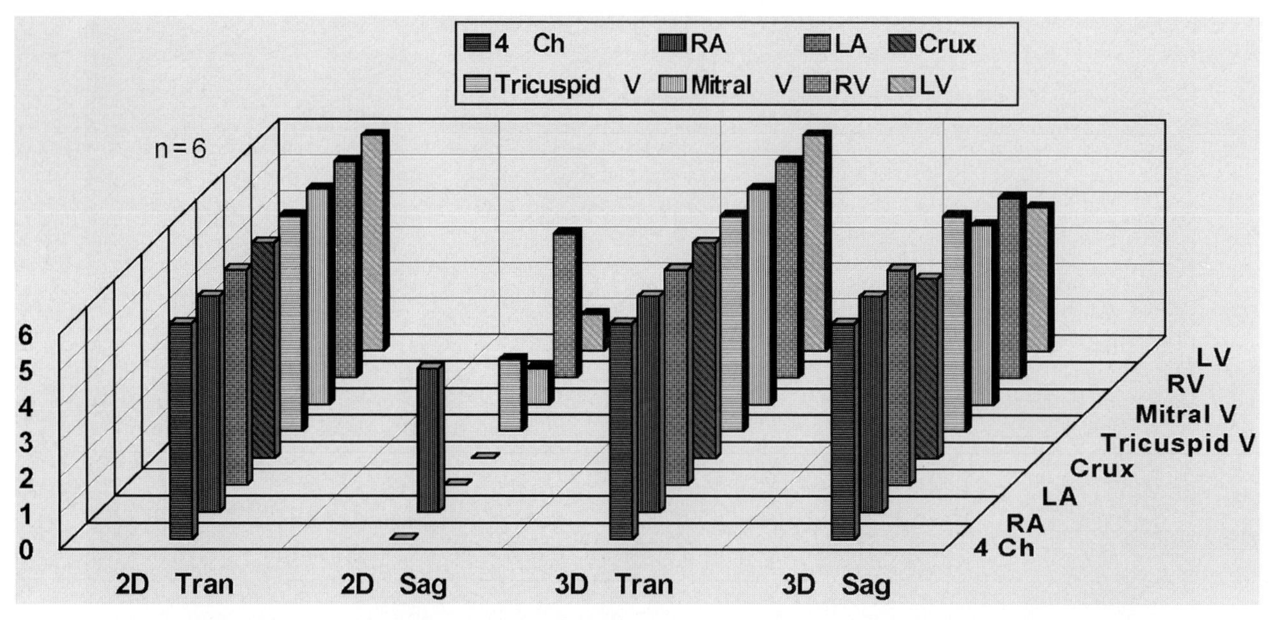

Figure 11-10. Comparison of 2DE and 3DE for the relative visualization of cardiac structures for different acquisition orientations in the fetal heart. 3DE provides a more consistent ability to visualize important structures regardless of the orientation of the original acquisition. (Adapted from Ref. 60.)

QUANTITATIVE MEASUREMENTS

Although visual assessment of cardiac anatomy and function is valuable, an important benefit of 3DE imaging is the availability of data of sufficient quality to accurately measure the length, area, or volume of chambers and vessels and follow temporal changes. Volume measurement is accomplished by some type of segmentation, including masking using either individual planes or an interactive volume-editing tool to limit the volume region of interest to only the object of interest. After the object is segmented, the voxels are summed and the matrix voxel scaling factors are applied to determine the volume. The improved measurement accuracy afforded by 3DE methods makes possible accurate quantitative measurement of heart chambers, vessel dimensions, and organ volumes.

3DUS measurement of volume has shown an accuracy of better than 5% for regular, irregular, and disconnected objects and *in vivo* organs relatively independent of the object size over several orders of magnitude (61–63). Gated 3DE data also have shown more accurate measurements of cardiac chamber volume (28,29,47–49) and lumen volumes (50) than either 2D or ellipsoidal models. Quantitative 3DE data provide a more accurate basis for decision making and comparison against previous studies or reference data bases.

Measurement of nongated fetal heart volumes has been shown to be accurate and able to detect changes throughout gestation (64). Individual fetal chamber volumes that are quite small (<1 ml) also can be accurately measured throughout the cardiac cycle with gated 3DE techniques (Figure 11-11) (65).

CORONARY ARTERY IMAGING

Imaging of the coronary arteries is an essential part of cardiac evaluation. The advent of intravascular ultrasound catheters is providing good-quality images of the interior of coronary vessels that lend themselves to 3D techniques (66). Although 3D intravascular (3D-IVUS) methods are invasive, the ability to image the interior of vessels provides important information about plaque formation and other occlusive processes and has an important role in diagnosis and management of coronary artery disease (Figure 11-12).

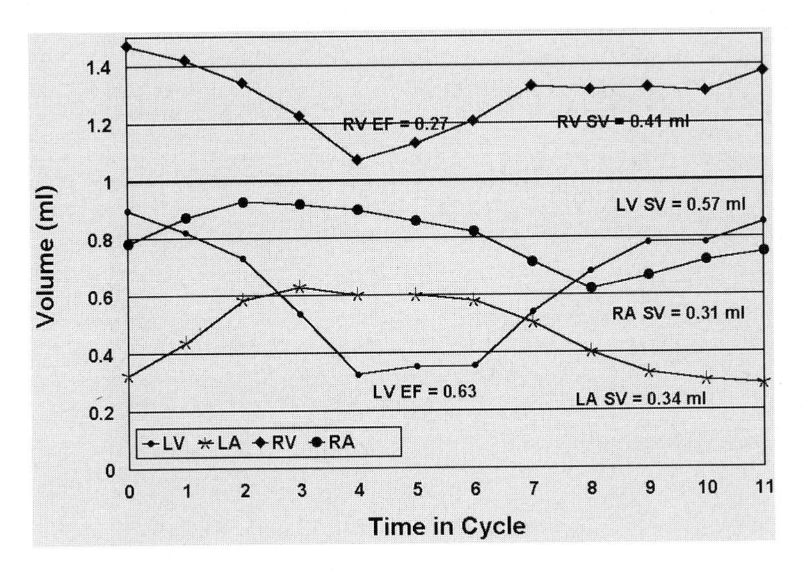

Figure 11-11. Fetal cardiac chamber volume measurement for each chamber throughout the cardiac cycle. LV, left ventricle, RV, right ventricle; LA, left atrium; RA, right atrium; SV, stroke volume; EF, ejection fraction. (Image courtesy of ref. 65.)

3D-IVUS can assist in visualizing plaque severity, which simplifies understanding of plaque size and internal complexity (66–69). 3D-IVUS can assess the architecture of complex plaques in an entire coronary artery segment more completely and provide a better assessment of coronary artery plaque volume than 2D methods (69–74). 3D-IVUS can detect intraplaque calcification with a sensitivity of about 90% (67,75).

3D-IVUS also can provide information that helps characterize plaque composition and can assist in choosing the appropriate treatment strategy (e.g., whether fibrinolytic or mechanical recanalization is the most appropriate treatment) (75). Newer percutaneous techniques include directional coronary atherectomy, laser angioplasty, stent deployment, and rotablator treatment (75). 3D-IVUS also can assist in determining appropriate stent length to employ (69) and assess if the stent has been appropriately expanded (69,76) and if there are procedural complications, and it can identify high-risk lesions such as plaques with extensive hemorrhage that are prone to rupture (75). 3D-IVUS also is useful to help define procedure end points by proving the vessel has been dilated a finite amount more readily than 2D methods (77).

In addition to 3D-IVUS techniques, recent advances in contrast materials and harmonic

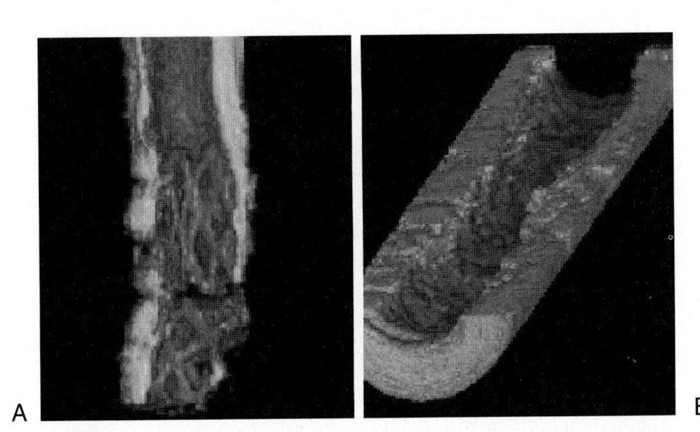

Figure 11-12. A: Stent in coronary artery, long axis. (Images courtesy of University Clinic Bonn, W. Fehske, and TomTec, GmBh, FRG.) **B:** Plaque in coronary artery, long axis. (Images courtesy of RWTH Aachen, P. Hanrath, Klus, and TomTec, GmBh, FRG.)

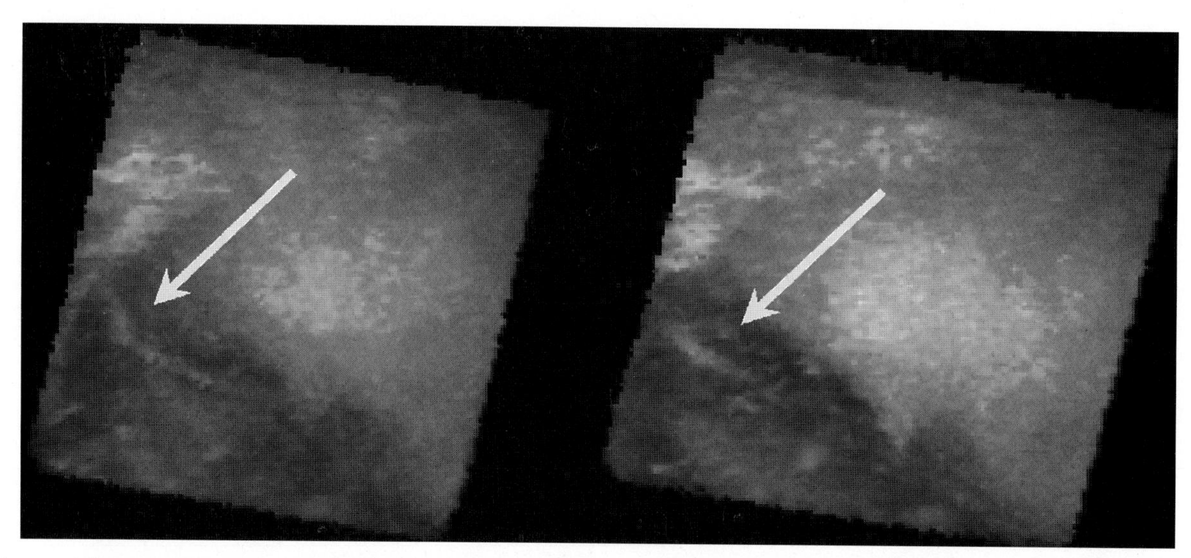

Figure 11-13. Volume-rendered images of dog coronary arteries imaged at 7 MHz from a gated 3DUS acquisition using harmonic imaging on an Elegra (Siemens Ultrasound, Issaquah, WA) with a microbubble contrast agent (Imagent, Alliance Pharmaceuticals, San Diego, CA). The morphology of the coronary vessel is clearly appreciated as it progresses into the myocardium. (Images courtesy of Dr. T. Nelson.)

imaging suggest noninvasive methods of visualizing coronary artery blood flow may be possible in the future (Figure 11-13) (78).

CARDIAC VISUALIZATION

The challenge in 3DE is to permit the physician to completely understand cardiac anatomy and function. There are basically two approaches for 3DE visualization (79): slice projection and volume rendering that includes ray casting. In addition, animation greatly assists visualizing the dynamics of cardiac motion. The general features of these techniques are discussed in more detail in Chapter 3.

Identification of cardiac anatomy can be facilitated by reorienting volume data to a standard anatomic orientation to provide a consistent reference frame between patients and analysis methods. Reorientation of the volume may be used to provide a standardized presentation (e.g., apex down, base up, right and left standard anatomy presentation) or a viewing perspective similar to that commonly found during surgical or interventional procedures (e.g., a surgeon's eye view) (see Figure 11-5). Utilization of information display that most closely replicates the information required for management, whether surgical planning or functional evaluation, can greatly assist comprehension of the patient's condition.

For certain types of evaluation, such as segmentation of cardiac surfaces, the image quality often can be improved by filtering the volume data using either 3D (or 4D) median and Gaussian filters prior to processing or display (80,81). Because ultrasound images often have significant noise or speckle, filtering assists segmentation algorithms used to measure volume or extract cardiac features. For example, images of the chambers and blood pool can be extracted using algorithms that identify the lower signal intensity of blood compared with myocardial tissue, similar to thresholding. Subsequently, the segmented data can be volume rendered to show chambers and great vessels in a single image (Figures 11-14, 11-15). This approach has been used to analyze cardiac function and measure chamber volume, stroke volume, and ejection fraction.

Review of 3DE data often begins with display of image planes viewed at arbitrary orientations in the 3D data set. Interactive viewing of planes in orientations other than those available during data acquisition makes it possible to optimize the viewing plane

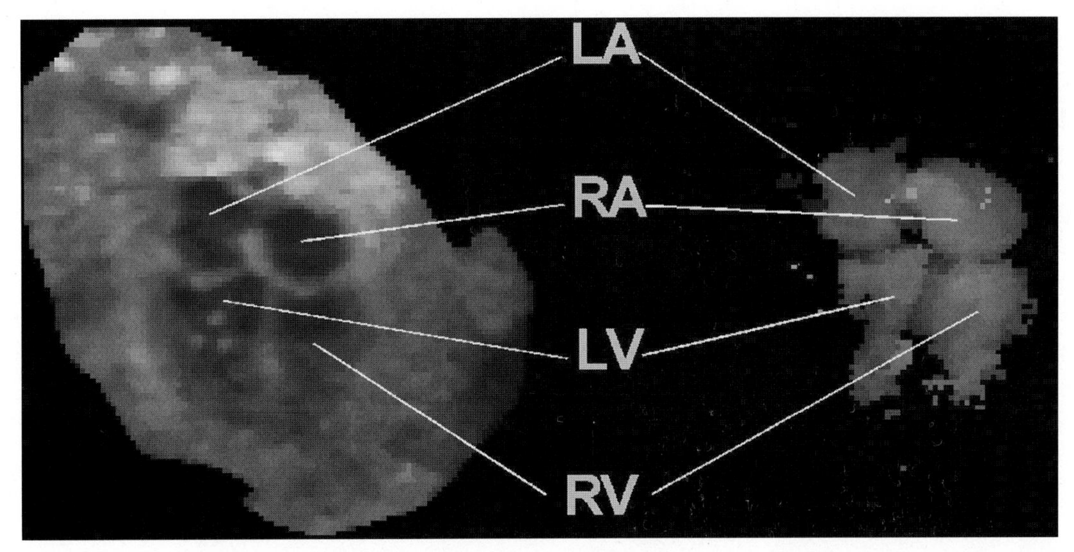

Figure 11-14. Imaging of the fetal heart and chambers showing volume-rendered images of the cardiac chambers and vessels. The signal void in the original acquisition has been used to show only the blood signal. Chamber and vessel rendering display uses a modified maximum-intensity method with depth-coding. LV, left ventricle; RV, right ventricle; LA, left atrium; RA, right atrium. (Image courtesy of ref. 65.)

to the proper position in the heart. Interactive slicing methods replicate the scanner operational ''feel,'' permitting rapid localization to the region of interest; require minimal processing; and provide retrospective evaluation of cardiac anatomy. Review of multiple slices displayed simultaneously can be particularly valuable in assisting understanding of patient anatomy (Figure 11-16). Multiplane displays can be combined with rendered images of the entire heart to further assist in localization of slice plane location and identification of key landmarks.

Volume-rendering methods are useful to display global features of the heart, including chamber geometry, valve configuration, and great vessel anatomy. Animation sequences such as rotation and gated ''cine-loop'' displays throughout the cardiac cycle greatly help in understanding cardiac anatomy and function. Analysis of dynamic function can increase the diagnostic value of many studies (59,82). Without animation it often is difficult to appreciate 3D and 4D information from 2D images. An additional advantage is that motion can be viewed at normal, accelerated, or reduced speed to enhance comprehension.

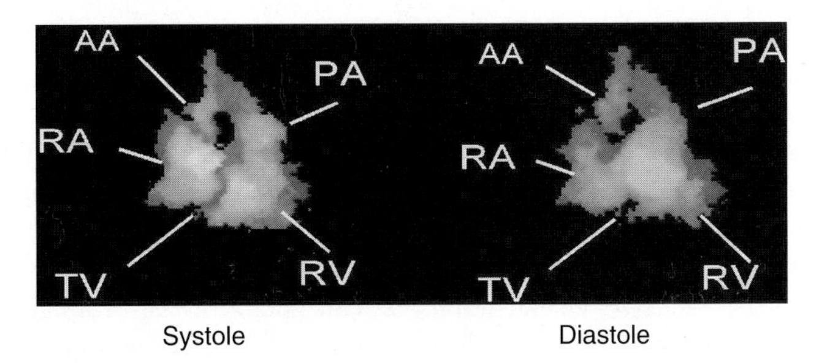

Figure 11-15. A series of volume-rendered images at end-diastole and end-systole in the cardiac cycle. Note the tricuspid valve (TV), which is clearly seen in each image. The TV is closed in ventricular systole and open during ventricular diastole. Interactive display of cardiac dynamics greatly assists comprehension of cardiac anatomy and function. LV, left ventricle; RV, right ventricle; LA, left atrium; RA, right atrium. (Image courtesy of ref. 11.)

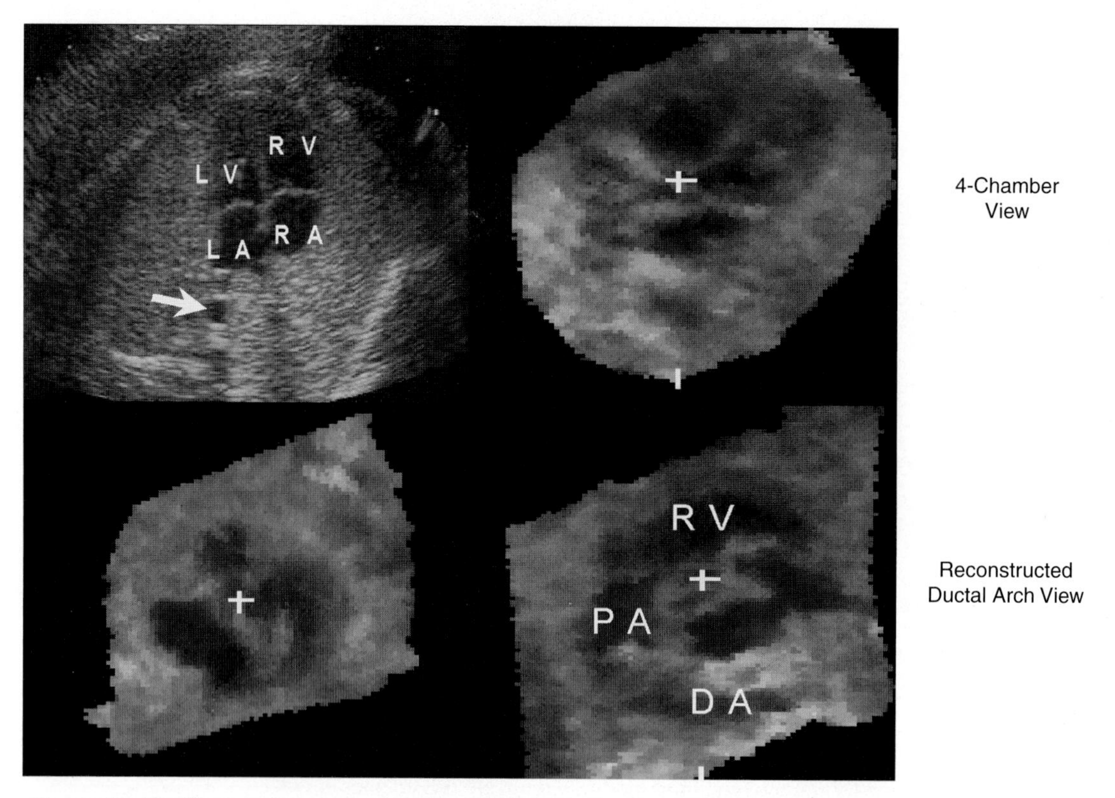

Figure 11-16. Three orthogonal slices through the volume for a gated transverse acquisition. The upper left view shows one frame of the 2DUS acquisition. The other three planes are from one volume of the gated reconstruction showing orthogonal intersecting slices that demonstrate standard cardiac views. The *cross* indicates the point of common intersection. (With permission from Ref. 65.)

OPTIMIZATION OF 3D ECHOCARDIOGRAPHIC VISUALIZATION

Identifying the optimal method for display of 3DE data is challenging and an area that continues to develop. Each choice of visualization method offers trade-offs. Viewing planar slices from optimized orientations is a straightforward method of display that provides interactive review that closely resembles clinical scanning procedures, giving a virtual rescan of the patient after the examination is complete. In addition, planar slices extracted from volume data offer projections that may not be available during patient scanning. Slice displays do not necessarily require intermediate processing or filtering of ultrasound image data, which further minimizes delay time to viewing.

Volume-rendering methods, on the other hand, produce high-quality images that are relatively tolerant of noise in the ultrasound data but may obscure fine details. Filtering can improve results as long as it does not obscure fine detail. Careful selection of opacity values helps provide an accurate rendition of the structure being studied and can be particularly valuable for evaluating complicated anatomic structures such as valve morphology or congenital anomalies. Stereoscopic viewing can further enhance identification of small structures with greater confidence in less time.

At present, physician involvement in interactively optimizing display parameters is an essential part of cardiac 3DE visualization that can greatly assist evaluation of complex cardiac structures for which it is difficult to develop a 3D understanding. Review of 3DE data using high-performance workstations can reduce patient scanning times compared with current 2D techniques, thus increasing the number of patients scanned, improving operational efficiency, and making more cost-effective use of sonographers and equipment. Standardization of echocardiographic examination protocols also could lead to more uniform examinations. Regardless of which viewing technique is used, a key benefit of 3DUS is that once the patient has been scanned, the original data may be analyzed for the entire region or magnified subregions without the need to rescan the patient.

HOW TO DO IT
Practical Clinical Tips for Cardiac Imaging

- Volume acquisition:
  - Connect ECG leads if available (adult and pediatric patients).
  - Select appropriate probe (as high a frequency as possible).
  - Select acoustic window to best visualize cardiac anatomy.
  - Optimize scanner settings.
  - Acquire scan data:
    - Steady sweep (fan, rotation, etc.) in constant direction.
    - Slow enough to acquire entire volume throughout cardiac cycle.
    - Minimize patient motion:
      - Breath-holding (if possible).
- Volume reconstruction:
  - Select region of interest.
  - Determine heart rate and gating strategy.
- Volume review:
  - View planar slices:
    - Individual planes or multiple orthogonal planes.
    - Reorientation to standard anatomic orientation.
    - Identification of area of interest.
    - Viewing of rendered image.
    - Selection of type of rendering appropriate for study.
    - Ray casting.
    - Maximum/minimum projection.
    - Surface fitting.
    - Threshold/segment data.
    - Selection of rendering parameters.
    - Interactive adjustment of viewing parameters and orientation.
    - Use of animation whenever possible to help identify cardiac structures.

REFERENCES

1. Dekker DL, Piziali RL, Dong E Jr. A system for ultrasonically imaging the human heart in three dimensions. *Comput Biomed Res* 1974;7:544–553.
2. Fine DG, Sapoznikov D, Mosseri M, Gotsman MS. Three-dimensional echocardiographic reconstruction: qualitative and quantitative evaluation of ventricular function. *Comput Methods Prog Biomed* 1988;26: 33–44.
3. Geiser EA, Christie LG, Conetta DA, Conti CR, Gossman GS. A mechanical arm for spatial registration of two-dimensional echocardiographic sections. *Cathet Cardiovasc Diagn* 1982;8:89–101.
4. Greenleaf JF. Three-dimensional imaging in ultrasound. *J Med Systems* 1982;6:579–589.
5. Matsumoto M, Inoue M, Tamura S, et al. Three-dimensional echocardiography for spatial visualization and volume calculation of cardiac structures. *J Clin Ultrasound* 1981;9:157–165.
6. Meritz WE, Pearlman AS, McCabe DH, et al. An ultrasonic technique for imaging the ventricle in three dimensions and calculating its volume. *IEEE Trans Biomed Eng* 1983;30:482–491.
7. Nixon JV, Saffer SI, Lipscomb K, Blomqvist CG. Three-dimensional echoventriculography. *Am Heart J* 1983;106:435–443.
8. Sawada H, Fujii J, Kato K, Onoe M, Kuno Y. Three dimensional reconstruction of the left ventricle from multiple cross sectional echocardiograms: value for measuring left ventricular volume. *Br Heart J* 1983; 50:438–442.
9. Stickels KR, Wann LS. An analysis of three-dimensional reconstructive echocardiography. *Ultrasound Med Biol* 1984;10:575–580.
10. Nelson TR. Synchronization of time-varying physiological data in 3DUS studies. *Med Phys* 1995;22(6): 973.
11. Nelson TR, Pretorius DH, Sklansky M, Hagen-Ansert S. Three-dimensional echocardiographic evaluation of fetal heart anatomy and function: acquisition, analysis, and display. *J Ultrasound Med* 1996;15:1–9.
12. Deng J, Gardener JE, Rodeck CH, Lees WR. Fetal echocardiography in three and four dimensions. *Ultrasound Med Biol* 1996;22(8):979–986.
13. Kwon J, Shaffer E, Shandas R, et al. Acquisition of three dimensional fetal echocardiograms using an external trigger source. *J Am Soc Echocardiog* 1996;9(3):389.

14. Belohlavek M, Dutt V, Greenleaf JF, Foley DA, Gerber TC, Seward JB. Multidimensional ultrasonic visualization in cardiology. Ultrasonics Symposium (Cat. No. 92CH3118–7) *IEEE* 1992:1137–1145.
15. Belohlavek M, Foley DA, Gerber TC, Greenleaf JF, Seward JB. Three-dimensional ultrasound imaging of the atrial septum: normal and pathologic anatomy. *J Am Coll Cardiol* 1993;22:1673–1678.
16. Belohlavek M, Foley DA, Gerber TC, et al. Three- and four-dimensional cardiovascular ultrasound imaging: A new era for echocardiography. *Mayo Clin Proc* 1993;68:221–240.
17. Fulton DR, Marx GR, Pandian NG, et al. Dynamic three-dimensional echocardiographic imaging of congenital heart defects in infants and children by computer-controlled tomographic parallel slicing using a single integrated ultrasound instrument. *Echocardiography* 1994;11:155–164.
18. Greenleaf JF, Belohlavek M, Gerber TC, Foley DA, Seward JB. Multidimensional visualization in echocardiography: an introduction. *Mayo Clin Proc* 1993;68:213–219.
19. Levine, RA, Weyman AE, Handschumacher MD. Three-dimensional echocardiography: techniques and applications. *Am J Cardiol* 1992;69:121H–130H.
20. McCann HA, Chandrasekaran K, Hoffman EA, et al. A method for three-dimensional ultrasonic imaging of the heart *in vivo*. *Dynamic Cardiovasc Imaging* 1987;1:97–109.
21. McCann HA, Sharp JC, Kinter TM, et al. Multidimensional ultrasonic imaging for cardiology. *Proceedings IEEE* 1988;76:1063–1073.
22. Ofili OE, Nanda NC. Three-dimensional and four-dimensional echocardiography. *Ultrasound Med Biol* 1994;20(8):669–675.
23. Pandian NG, Nanda NC, Schwartz SL, Fan P, Cao Q. Three-dimensional and four-dimensional transesophageal echocardiographic imaging of the heart and aorta in humans using a computed tomographic imaging probe. *Echocardiography* 1992;9:677–687.
24. Salustri A, Roelandt JRTC. Ultrasonic three-dimensional reconstruction of the heart. *Ultrasound Med Biol* 1995;21:281–293.
25. Salustri A, Spitaels S, McGhie J, et al. Transthoracic three-dimensional echocardiography in adult patients with congenital heart disease. *J Am Coll Cardiol* 1995;26:759–767.
26. Seward JB, Belohlavek M, O'Leary PW, Foley DA, Greenleaf JF. Congenital heart disease: wide-field, three-dimensional, and four-dimensional ultrasound imaging. *Am J Cardiac Imaging* 1995;9(1):38–43.
27. Tardif JC, Vannan MA, Pandian NG. Biplane and multiplane transesophageal echocardiography: methodology and echo-anatomic correlations. *Am J Cardiac Imaging* 1995;9:87–99.
28. Vogel M, Losch S. Dynamic three-dimensional echocardiography with a computed tomography imaging probe: initial clinical experience with transthoracic application in infants and children with congenital heart defects. *Br Heart J* 1994;71:462–467.
29. Vogel M, White PA, Redington AN. *In vitro* validation of right ventricular volume measurement by three dimensional echocardiography. *Br Heart J* 1995;74:460–463.
30. Vogel M, Ho SY, Buhlmeyer K, Anderson RH. Assessment of congenital heart defects by dynamic three-dimensional echocardiography: methods of data acquisition and clinical potential. *Acta Paediatr* 1995; 410[Suppl]:34–39.
31. Ariet M, Geiser EA, Lupkiewicz SM, Conetta DA, Conti CR. Evaluation of a three-dimensional reconstruction to compute left ventricular volume and mass. *Am J Cardiol* 1984;54:415–420.
32. Nikravesh PE, Skorton DJ, Chandran KB, et al. Computerized three-dimensional finite element reconstruction of the left ventricle from cross-sectional echocardiograms. *Ultrason Imaging* 1994;6:48–59.
33. Binder T, Globits S, Zangeneh M, Gabriel H, Rothy W, Koller J, Glogar D. Three-dimensional echocardiography using a transoesophageal imaging probe: potentials and technical considerations. *Eur Heart J* 1996;17: 619–628.
34. Flachskampf F. Recent progress in quantitative echocardiography. *Curr Opin Cardiol* 1995;10:634–639.
35. Belohlavek M, Foley DA, Gerher TC, Greenleaf JF, Seward JB. Three-dimensional reconstruction of color Doppler jets in the human heart. *J Am Soc Echocardiogr* 1994;7:553–560.
36. Delabays A, Sugeng L, Pandian NG, et al. Dynamic three-dimensional echocardiographic assessment of intracardiac blood flow jets. *Am J Cardiol* 1995;76:1053–1058.
37. Fox MD, Gardiner WM. Three-dimensional Doppler velocimetry of flow jets. *IEEE Trans Biomed Eng* 1988;35:834–841.
38. Shiota T, Sinclair B, Ishii M, et al. Three-dimensional reconstruction of color Doppler flow convergence regions and regurgitant jets: an *in vitro* quantitative study. *J Am Coll Cardiol* 1996;27:1511–1518.
39. Kupferwasser I, Mohr-Kahaly S, Menzel T, et al. Quantification of mitral valve stenosis by three-dimensional transesophageal echocardiography. *Int J Card Imaging* 1996;12:241–247.
40. Levine RA, Handschumacher MD, Sanfilippo AJ, et al. Three-dimensional echo-cardiographic reconstruction of the mitral valve, with implications for the diagnosis of mitral valve prolapse. *Circulation* 1989;80:589–598.
41. Nanda NC, Roychoudhury D, Chung SM, et al. Quantitative assessment of normal and stenotic aortic valve using transesophageal three-dimensional echocardiography. *Echocardiography* 1994;11:617–625.
42. Marx GR, Fulton DR, Pandian NG, et al. Delineation of site, relative size and dynamic geometry of atrial septal defects by real-time three-dimensional echocardiography. *J Am Coll Cardiol* 1995;25:482–490.
43. Rivera JM, Siu SC, Handschumacher MD, Lethor JP, Guerrero JL. Three-dimensional reconstruction of ventricular septal defects: validation studies and *in vivo* feasibility. *J Am Coll Cardiol* 1994;23:201–208.
44. Maehle J, Bjoernstad K, Aakhus S, Torp HG, Angelsen BA. Three-dimensional echocardiography for quantitative left ventricular wall motion analysis: a method for reconstruction of endocardial surface and evaluation of regional dysfunction. *Echocardiography* 1994;11:397–408.
45. Roelandt JR, Di Mario C, Pandian NG, et al. Three-dimensional reconstruction of intracoronary ultrasound images: rationale, approaches, problems, and directions. *Circulation* 1994;90:1044–1055.
46. Ross JJ Jr, D'Adamo AJ, Karalis DG, Chandrasekaran K. Three-dimensional transesophageal echo imaging of the descending thoracic aorta. *Am J Cardiol* 1993;71:1000–1002.
47. Buck T, Erbel R. [Diagnosis of coronary heart disease using echocardiography 3D reconstructions. Diagnosis of global and regional left ventricular function] Diagnostik der koronaren Herzerkrankung mittels echokardiographischer 3D-Rekonstruktionen. Globale und regionale linksventrikulare Funktionsdiagnostik. *Herz* 1995; 20:252–262.

48. Martin RW, Bashein G. Measurement of stroke volume with three-dimensional transesophageal ultrasonic scanning: comparison with thermodilution measurement. *Anesthesiology* 1989;70:470–476.
49. Martin RW, Bashein G, Detmer PR, Moritz WE. Ventricular volume measurement from a multiplanar transesophageal ultrasonic imaging system: *in vitro* study. *IEEE Trans Biomed Eng* 1990;37:442–449.
50. Matar FA, Mintz GS, Douek P, et al. Coronary artery lumen volume measurement using three-dimensional intravascular ultrasound: validation of a new technique. *Cathet Cardiovasc Diagn* 1994;33:214–220.
51. Cao QL, Pandian NG, Azevedo J, et al. Enhanced comprehension of dynamic cardiovascular anatomy by three-dimensional echocardiography with the use of mixed shading techniques. *Echocardiography* 1994; 11:627–633.
52. Bates J, Tantengco M, Ryan T, Fiegenbaum H, Ensing G. A systematic approach to echocardiographic image acquisition and three-dimensional reconstruction with a subxyphoid rotational scan. *J Am Soc Echocardiogr* 1996:9:257–265.
53. Vogel M, Ho S, Lincoln C, Yacoub M, Anderson R. Three-dimensional echocardiography can simulate intraoperative visualization of congenitally malformed hearts. *Ann Thorac Surg* 1995;60:1282–1288.
54. Crane JP, LeFevre ML, Winborn RC, et al. Randomized trial of prenatal ultrasound screening: impact on detection, management and outcome of anomalous fetuses. *Am J Obstet Gynecol* 1994;171:392–399.
55. Zosmer N, Jurkovic D, Jauniaux E, et al. Selection and identification of standard cardiac views from three-dimensional volume scans of the fetal thorax. *J Ultrasound Med* 1996;15:25–32.
56. Merz E, Bahlmann F, Weber G. Volume scanning in the evaluation of fetal malformations: a new dimension in prenatal diagnosis. *Ultrasound Obstet Gynecol* 1995;5:222–227.
57. Meyer-Wittkopf M, Cook A, McLennan A, et al. Evaluation of three-dimensional ultrasonography and magnetic resonance imaging in assessment of congenital heart anomalies in fetal cardiac specimens. *Ultrasound Obstet Gynecol* 1996;8:303–308.
58. Leventhal M, Pretorius DH, Sklansky MS, Budorick NE, Nelson TR, Lou K. Three-dimensional ultrasonography of normal fetal heart: comparison with two-dimensional imaging. *J Ultrasound Med* 1998;17(6): 341–348.
59. Sklansky MS, Nelson TR, Pretorius DH. Three-dimensional fetal echocardiography: gated versus nongated techniques. *J Ultrasound Med* 1998;17(7):451–457.
60. Sklansky MS, Nelson TR, Pretorius DH. Usefulness of gated three-dimensional fetal echocardiography to reconstruct and display structures not visualized with two-dimensional imaging. *Am J Cardiol* 1997;80(5): 665–668.
61. Nelson TR, Pretorius DH. 3-dimensional ultrasound volume measurement. *Med Phys* 1993;201(3):927.
62. Riccabona M, Nelson TR, Pretorius DH, Davidson TE. Distance and volume measurement using three-dimensional ultrasonography. *J Ultrasound Med* 1995;14:881–886.
63. Riccabona M, Nelson TR, Pretorius DH, Davidson TE. Three-dimensional sonographic measurement of bladder volume. *J Ultrasound Med* 1996;15(9):627–632.
64. Chang FM, Hsu KF, Ko HC, et al. Fetal heart volume assessment by three-dimensional ultrasound. *Ultrasound Obstet Gynecol* 1997:942–948.
65. Nelson TR. Three-dimensional echocardiography. *Prog in Biophysics and Molecular Biology* 1998;69: 257–272.
66. Von Birgelen C, di Mario C, Li W, et al. Morphometric analysis in three-dimensional intracoronary ultrasound: an *in vitro* and *in vivo* study performed with a novel system for the contour detection of lumen and plaque. *Am Heart J* 1996;132:516–527.
67. Ge J, Gorge G, Baumgart D, Liu F, Haude M, Erbel R. Intravascular ultrasound. In: Liu JB, Goldberg BB, eds. *Endoluminal ultrasound: vascular and nonvascular applications.* St. Louis: Mosby, 1997:37–80.
68. Dhawale PJ, Wilson DL, Hodgson JM. Volumetric intracoronary ultrasound: methods and validation. *Cathet Cardiovasc Diagn* 1994;33:296–307.
69. von Birgelen C, Nicosia A, Serruys PW, Roelandt JRTC. Assessment of balloon angioplasty and directional atherectomy with three-dimensional intracoronary ultrasound. In: Erbel R, Roelandt JRTC, Ge J, Gorge G, eds. *Intravascular ultrasound.* St. Louis: Mosby, 1998:183–189.
70. Cavaye DM, White RA. Intravascular ultrasound imaging: development and clinical applications. *Int Angiol* 1993;12:245–255.
71. Prati F, DiMario C, Gil R, et al. Three-dimensional intravascular ultrasound for stenting. In: Erbel R, Roelandt JRTC, Ge J, Gorge G, eds. *Intravascular ultrasound.* St. Louis: Mosby, 1998:239–247.
72. Shah VT, Ge J, Ashry M, Erbel R. Distribution of atherosclerosis. In: Erbel R, Roelandt JRTC, Ge J, Gorge G, eds. *Intravascular ultrasound.* St. Louis: Mosby, 1998:81–90.
73. Von Birgelen C, di Mario C, Reimers B, et al. Three-dimensional intracoronary ultrasound imaging: methodology and clinical relevance for the assessment of coronary arteries and bypass grafts. *J Cardiovasc Surg (Torino)* 1996;37:129–139.
74. Weissman NJ, Palacios IF, Nidorf SM, Dinsmore RE, Weyman AE. Three-dimensional intravascular ultrasound assessment of plaque volume after successful atherectomy. *Am Heart J* 1995;130:413–419.
75. Ramo P, Spencer T. Tissue characterization. In: Erbel R, Roelandt JRTC, Ge J, Gorge G, eds. *Intravascular ultrasound.* St. Louis: Mosby, 1998:61–68.
76. Erbel R, Roelandt JRTC, Ge J, Gorge G. Introduction. In: Erbel R, Roelandt JRTC, Ge J, Gorge G, eds. *Intravascular ultrasound.* St. Louis: Mosby, 1998:1–9.
77. Von Birgelen C, de Feyter PJ, Roelandt JRTC. Three-dimensional reconstruction of intracoronary ultrasound. In: Erbel R, Roelandt JRTC, Ge J, Gorge G, eds. *Intravascular ultrasound.* St. Louis: Mosby, 1998:51–60.
78. Nelson TR, Mattrey RF, Steinbach GC, Lee Y, Lazenby J. Gated 3D ultrasound imaging of the coronary arteries using contrast and harmonic imaging. *JUM* 1998;17(3):S56.
79. Nelson TR, Elvins TT. Visualization of 3D ultrasound data. *IEEE Comput Graph Applic* 1993;13:50–57.
80. Pratt WK. *Digital image processing.* New York: Wiley, 1991.
81. Russ JC. *The image processing handbook.* Boca Raton, FL: CRC Press, 1992.
82. Schwartz SL, Cao Q-L, Azevedo J, Pandian NG. Simulation of intraoperative visualization of cardiac structures and study of dynamic surgical anatomy with real-time three-dimensional echocardiography. *Am J Cardiol* 1994;73:501–507.

12

Breast

OVERVIEW

This chapter will provide an overview of the applications of three-dimensional ultrasound (3DUS) imaging methods in the evaluation of the breast. Two-dimensional ultrasound (2DUS) imaging of the female breast is useful in determining if a mass is cystic or solid but is less useful for cancer screening. Recent work suggests that 3DUS, alone or in combination with color/power Doppler and contrast agents, may be helpful in differentiating benign from malignant lesions, including both surface features and internal architecture. Arbitrary plane evaluation of a breast volume permits planes parallel to the chest wall to be evaluated, providing a more complete evaluation than is possible with 2DUS.

KEY CONCEPTS

- Differentiate benign from malignant masses.
- Arbitrary planar review permits thorough evaluation of masses:
  - Planes parallel to chest wall can be evaluated.
  - Surface features of masses can be evaluated.
- Volume rendering of masses provides comprehensive images.
  - Surface features can be evaluated.
  - Vascularity of masses can be evaluated in 3D.

Breast imaging is an important area of clinical application and research. Although 2DUS has been shown to be valuable for differentiating some aspects of benign/malignant disease (1,2), 3DUS imaging of the breast may assist this area significantly. Initial work by Rotten et al. (3) showed the feasibility of the technique and reported that surface evaluation of breast masses with 3DUS is more complete than that with 2DUS because the entire surface of masses can be evaluated. Carson et al. (4) demonstrated differences in the vascularity of breast tissues and masses. Additional efforts at correcting refraction and motion artifacts

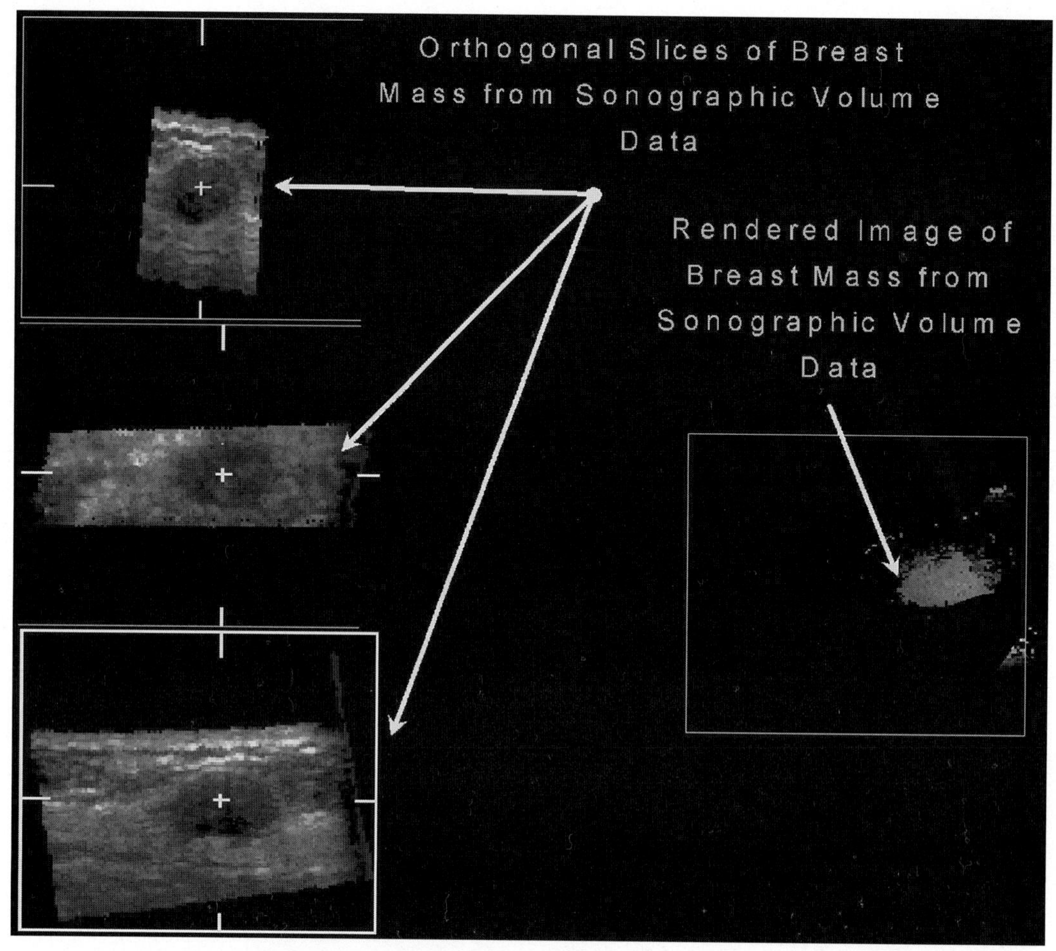

Figure 12-1. Breast mass displayed as a simple cyst in three orthogonal planes. The plane parallel to the chest wall is the middle image. The mass is shown in a rendered image on the right.

have the potential to improve resolution image quality (5), but additional work is needed to fully realize the potential of 3DUS. 3DUS provides improved evaluation of focal infiltrates, intracystic structures (Figure 12-1), and multifocal disease plus reevaluation of data after the patient has left the clinic (6). 3DUS also can be useful for assisting needle localization and guidance for biopsy purposes.

3DUS imaging of the breast should allow for direct visualization of sonographically detectable mass lesions, resulting in better analysis of those lesions both with gray-scale imaging and with color Doppler flow imaging using both arbitrary planar review and volume-rendering techniques (Figure 12-2). It is important to first identify the parameters used to evaluate breast masses with 2DUS: internal architecture, shape, boundary echo margins, and direction of the long axis of the mass relative to the skin surface of the breast (1). 3DUS can be used to evaluate these same features more completely.

Rotten et al. (1) suggested a framework for the value of 3DUS in evaluating breast masses in 1991 based on a report of four breast tumors that were scanned with 3DUS. First, three-dimensional (3D) reformatted planar sections provided new information unavailable with 2DUS: planes were evaluated that were positioned along the equatorial plane of the tumor, which was often parallel to the fibroglandular layer of the breast. Second, the lesion surface could be evaluated to determine whether it was continuous, discontinuous, or jagged/irregular. Third, the rim around the mass could be evaluated. Rotten et al. suggested that benign tumors are often surrounded by a continuous hyperechoic rim, whereas medullary carcinomas may have a discontinuous hyperechoic rim (1).

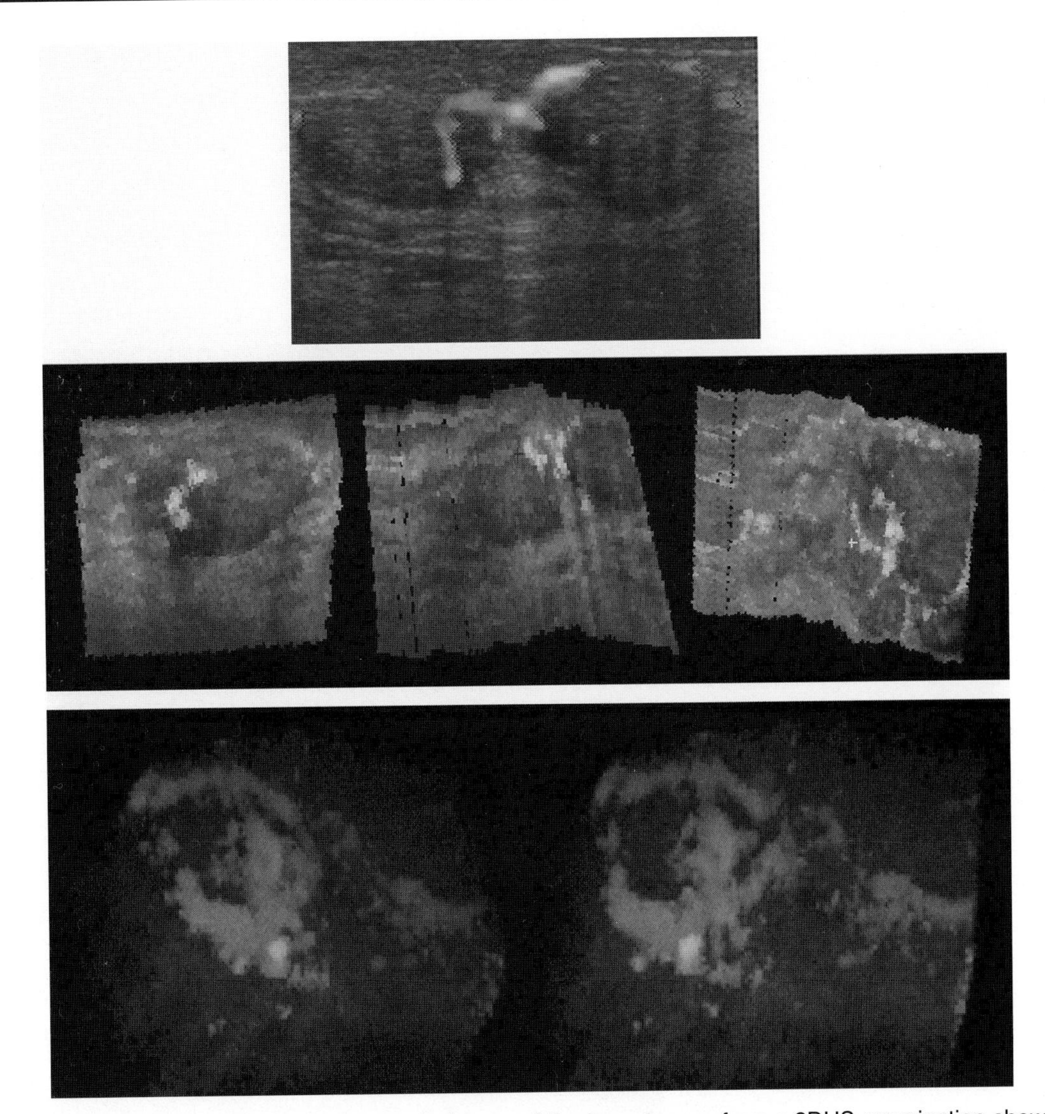

Figure 12-2. Fibroadenoma in breast. An image of the breast mass from a 2DUS examination shows a bilobed mass with vascularity seen anteriorly as well as within the mass. The middle row of images shows three orthogonal planar images from the 3DUS scan. The borders are slightly irregular. Volume-rendered images from two orientations are seen in the lower row of images, showing continuity of the vascularity of the lesion. See color version of figure.

Blohmer et al. (6) studied 50 patients with breast lesions (19 with breast cancer and 31 with benign disease) using a 10-MHz volume transducer (Voluson, Kretz Technik, Austria). In two cases with malignant lesions, 2DUS was incorrect, resulting in a false-negative interpretation. Conversely, in four cases of benign disease, 3DUS falsely suggested carcinoma. Advantages of 3DUS identified by Blohmer et al. included better judgment on the evidence of focal infiltration of tumor, existence and form of intracystic structures and of multifocal disease, short duration of the examination, and ability to review stored data (6). Sohn et al. performed a preliminary study using 3DUS of 35 patients with benign and malignant breast masses and described similar results to Rotten et al. In their study, 95% of malignant tumors and 80% of benign tumors were diagnosed correctly (7).

Volume rendering allows for visualization of more than a plane through a mass. Curvature and boundaries of the surface of the mass may be displayed with shading techniques using gray scale data. Ductal carcinomas may show spiked or jagged margins with these volume-rendering methods (3).

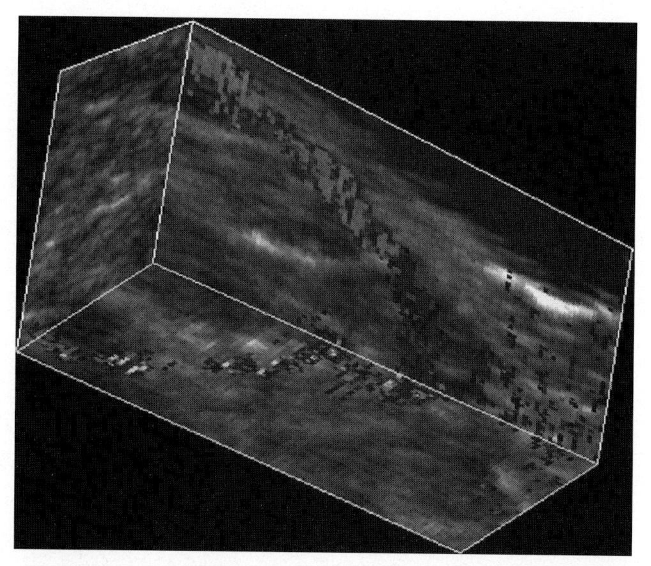

Figure 12-3. 3DUS breast combined B-mode and color Doppler images of a normal volunteer. The course of the vessel can be followed by slicing obliquely into the volume. The course of the vessel could not be seen in a single image with conventional 2DUS. (Image courtesy of Life Imaging Systems, Inc., London, Ontario, Canada.) See color version of figure.

COLOR FLOW IMAGING

Color flow imaging of the breast offers potential in differentiating benign (Figure 12-3) from malignant disease (8). 3DUS techniques applied to color flow imaging also offer promise in further clarifying the vascularity of both benign and malignant lesions (Figure 12-4). Carson et al. have studied this area in depth (4,5,9). Initially, they studied breast lesions with both continuous-wave and pulsed color Doppler employing a hand-held transmitter/receiver pair with crossed-beam patterns. The 3D display of color flow data aided

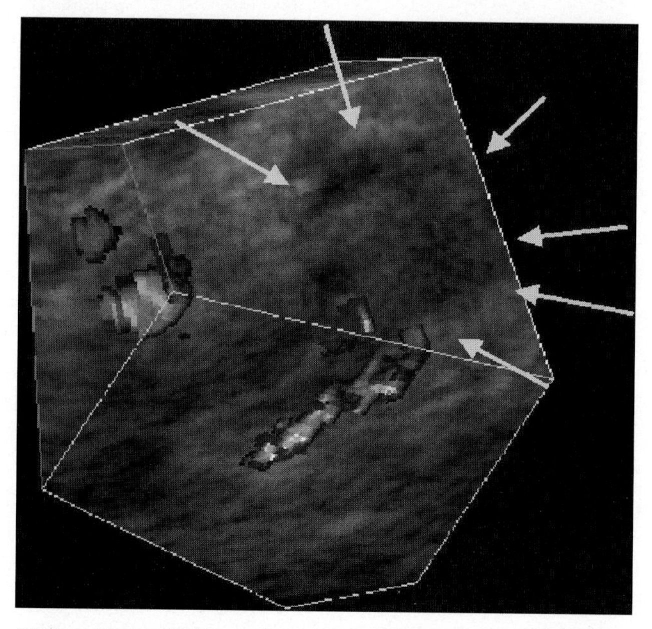

Figure 12-4. 3DUS study of a ductal carcinoma using a cube view showing various planes of the ductal carcinoma. Note the vessels approaching the irregular, jagged border of the mass. (Courtesy of Dr. D.B. Downey.) See color version of figure.

Table 12-1. *Vascular criteria used to rate malignancy characteristics of breast lesions*

Criterion description	Benign extreme (1)	Malignant extreme (5)
Vascularity in mass (outer cm)	None	Extensive (>25% of mass)
Vascularity outside mass (to 1 cm away)	None	Extensive (>25% of area)
Visible shunt vessels	None	Very clear or more than 2
Vessels wrapping around mass	None	Total involvement >240°
Tortuosity of vessels	Gently curving	Very tortuous
Apparently related vessels, beyond 1 cm	None	Unusually large, fast flow
Unusual vascularity, no associated mass	None	Multiple, large, fast flow
General enlargement of breast vessels	None	Extensive

A scale of 1 to 5 was used, with 5 being the most malignant
Adapted from Table 2 in ref. 9.

in visual detection of abnormal vascular morphology. It was apparent that only small traces of color dots seen on 2DUS images were observed curving around a lesion in the 3D display (4). They later reported that compound display of breast lesions using 3D linear or affine transforms and 3D nonlinear warped transforms provided adequate volume registration for evaluation of masses; advantages of this technique were related to significant increases in lesion and structural conspicuity due to a reduction in speckle noise (5). In a follow-up clinical study, 24 women with breast masses undergoing surgical biopsy were studied with both 2DUS and 3DUS. The studies were examined by radiologists and classified on a scale from 1 to 5 (5 = most suspicious for malignancy) for each of six conventional gray-scale and eight new vascular criteria. The vascular criteria are given in Table 12-1. Vascular structures were displayed as rotating color volumes in 3D, as they were superimposed on gray-scale images. The authors found that 3D images provided a stronger subjective appreciation of vascular morphology and allowed somewhat better ultrasound discrimination of malignant masses than did 2D images or videotapes (specificities of 85%, 79%, and 71%, respectively, at a sensitivity of 90%). A trend toward significance in lesion differentiation was identified only with regard to 3D display of vascularity (10).

Wide-band holographic techniques are also being combined with 3D ultrasound imaging to evaluate breast lesions. Preliminary work using breast and liver phantoms is encouraging. High resolution is obtained simultaneously in all three dimensions. Limitations of this technique are related to relatively large data sets, which require significant time for reconstruction and viewing, and frame rate limitations due to the velocity of sound (10). Davies et al. have evaluated breast lesions with computerized volume data, combining 50-MHz acoustic or near-infrared images with four-view tetrahedral radiography and x-ray projection microscopy to produce 3D images of tissue ranging from 500 μm to approximately 4 mm in thickness (12). Although these techniques do not directly impact diagnostic 3DUS imaging of the breast as we know it today, they do suggest that this technology can be used in many ways to improve the diagnostic work-up of disease.

HOW TO DO IT
Practical Clinical Tips for Breast Imaging

- Tissues must remain stationary during scanning for accurate registration of image slices:
  - Compressibility of breast tissue may complicate slice registration.

HOW TO DO IT, continued

- Masses must be stabilized if they are mobile, either using a hand or a device
  - Easily stabilized by positioning between thumb and finger of one hand.
  - Scan with other hand by sweeping across mass.
- Vascular evaluation using power/color Doppler requires a slow sweep across the mass.
  - Reduces flash and motion artifacts.
  - Harmonic/contrast imaging may improve situation.
- Multiplanar and volume rendering are valuable in evaluating breast masses.
- Evaluate vascularity with a gray-scale image background.
- Rotation of data provides visual clues to assess relationship of vascularity to breast mass.

REFERENCES

1. Stavros AT, Thickman D, Rapp CL, Dennis MA, Parker SH, Sisney GA. Solid breast nodules: use of sonography to distinguish between benign and malignant lesions. *Radiology* 1995;196:123–134.
2. Tohno E, Cosgrove D, Sloane J. *Ultrasound diagnosis of breast diseases.* New York: Churchill-Livingstone, 1994.
3. Rotten D, Levaillant JM, Constancis E, Billon AC, Le Guerinel Y, Rua P. Three-dimensional imaging of solid breast tumors with ultrasound: preliminary data and analysis of its possible contribution to the understanding of the standard two-dimensional sonographic images. *Ultrasound Obstet Gynecol* 1991;1:384–390.
4. Carson PL, Adler DD, Fowlkes JB, Harnist K, Rubin J. Enhanced color flow imaging of breast cancer vasculature: continuous wave Doppler and three-dimensional display. *J Ultrasound Med* 1992;11:377–385.
5. Moskalik A, Carson PL, Meyer CR, Fowlkes JB, Rubin JM, Roubidoux MA. Registration of three-dimensional compound ultrasound scans of the breast for refraction and motion correction. *Ultrasound Med Biol* 1995;21(6):769–778.
6. Blohmer JU, Bollmann R, Heinrich G, Paepke S, Lichtenegger W. Three-dimensional ultrasound study (3-D sonography) of the female breast. *Geburtshilfe Frauenheilkunde* 1996;56(4):161–165.
7. Sohn CH, Stolz W, Kaufmann M, Bastert G. Three-dimensional ultrasound imaging of benign and malignant breast tumors—initial clinical experiences. *Geburtshilfe Frauenheilkunde* 1992;52(9):520–525.
8. Cosgrove DO, Kedar RP, Bamber JC, et al. Breast diseases: color Doppler ultrasound in differential diagnosis. *Radiology* 1993;189(1):99–104.
9. Ferrara KW, Zgar B, Sokic-Melgar J, Algazi UR. High resolution 3D color flow mapping: applied to the assessment of breast vasculature. *Ultrasound Med Biol* 1996;22:293–304.
10. Carson PL, Moskalik AP, Govil A, et al. The 3D and 2D color flow display of breast masses. *Ultrasound Med Biol* 1997;23(6):837–849.
11. Sheen DM, Collins HD, Gribble RP. Wideband holographic three-dimensional ultrasonic imaging of breast and liver phantoms. *Stud Health Technol Inform* 1996;29:461–470.
12. Davies JD, Chinyama CN, Jones MG, Astley SM, Bates SP, Kulka J. New avenues in 3-D computerized (stereopathological) imaging of breast cancer (review). *Anticancer Res* 1996;16(6C):3971–3981.

13

Ophthalmology

OVERVIEW

This chapter will present an overview of the current clinical utility of three-dimensional ultrasound (3DUS) in the diagnosis and management of patients with orbital diseases. 3DUS has the potential to improve evaluation of orbital diseases because the orbit is small, superficial, easily accessible, can be immobilized during scanning, and the vitreous and aqueous humors have excellent acoustic properties. High-frequency transducers produce excellent images, and transducer coupling is simple. 3DUS can help physicians diagnose some ocular diseases, assess the exact extent and severity of others, plan optimal surgeries, and devise appropriate tumor management plans.

KEY CONCEPTS

- High-quality 3DUS ocular scans are relatively easy to obtain:
  - Anatomic reasons:
    - The orbit is small, superficial, and has excellent acoustic properties; voluntary and involuntary motion rarely degrade the image.
    - Coupling is readily achieved.
    - Scanning access is easy.
  - Technical reasons:
    - High-frequency transducers provide excellent ultrasound data.
- Ocular ultrasound differs from other ultrasound:
  - It is usually performed by ophthalmologists in an office or clinic setting.
  - The machines are smaller, higher frequency, and less expensive.
- Proven clinical benefits:
  - Accurate and repeatable volumetric assessments are clinically necessary and possible.

KEY CONCEPTS, *continued*

- The ability to view different perspectives is helpful during diagnosis and surgical planning.
- Back-scatter collected in 3DUS may provide improved tissue characterization.
- Potential applications in ophthalmology:
  - 3DUS power/color Doppler may be useful to evaluate retinal blood flow.
- In the anterior segment:
  - Evaluation of hypotony, iris tumor, ciliary body disease, hyphema, after surgical screening.
  - Foreign bodies.
- In the posterior segment:
  - Evaluation of retinal detachments, choroidal tumors, and dislocated lens.

Diagnostic ultrasound is the most widely used nonoptical technique for imaging the orbit (1). It plays a vital role in diagnosing ophthalmic diseases (2) and is the imaging modality of choice for intraocular pathology (3). In addition, 2DUS often provides valuable screening information about intraorbital and extraocular pathology quickly and cost-effectively (3).

Amplitude mode (A-mode) ultrasound is used to depict the size and internal characteristics of ocular lesions, while brightness mode (B-mode) ultrasound describes their location and topographic contour (4). Recently, scanning frequencies of up to 100 MHz have been used to produce exquisite 2DUS images, so-called ultrasound biomicroscopy (3). Some ophthalmologists have recently started evaluating orbital blood flow with color Doppler imaging (5,6).

3DUS has tremendous potential to improve the diagnosis and management of patients with orbital diseases (2,4,7,8). It is relatively easy to obtain high-quality 3DUS scans of the eye (4,9). The orbit is superficial, easily accessible, small, and capable of being immobilized during scanning (4). The vitreous and aqueous humor have excellent acoustic properties (1), and high-frequency transducers produce good-quality ultrasound data. Transducer coupling is simple. The transducer may be placed either on the closed eyelid or directly on the globe (4).

3DUS increases the clinical utility of ultrasound by allowing ophthalmologists to view pathologies from different perspectives (Figure 13-1) and to make more accurate and repeated volumetric assessments. 3DUS can help physicians diagnose some ocular diseases (10), assess the exact extent and severity of others (Figure 13-2) (2,4), plan optimal surgeries (10), and devise appropriate tumor management plans (Figure 13-3) (7,11). Despite this potential, application of 3DUS to ocular imaging is maturing slowly, with only a few groups reporting their results to date. If the initial positive results are repeated by others, it is likely to become more popular.

3DUS affords one the luxury of interacting with the virtual orbital data without being present at the examination. Consultations may be done remotely in time (after the patient has left the department) and/or in place (data can be sent over computer networks or phone lines to experts at a remote site) (4).

ADVANTAGES OF 3DUS COMPARED WITH 2DUS IN OPHTHALMOLOGY

Flexible Viewing Perspective

To make the correct diagnosis, ocular pathology sometimes must be viewed from a particular perspective. Hypotony—a disease characterized by decreased intraocular pressure—can be caused by a variety of iridociliary pathologies, and accurate diagnosis is crucial to proper management. Coleman recommends viewing the ciliary body from a mid-vitreous perspective to diagnose correctly the etiology of hypotony (10). This view can be provided only with 3DUS imaging.

The decision to repair retinal detachment surgically depends on a variety of factors,

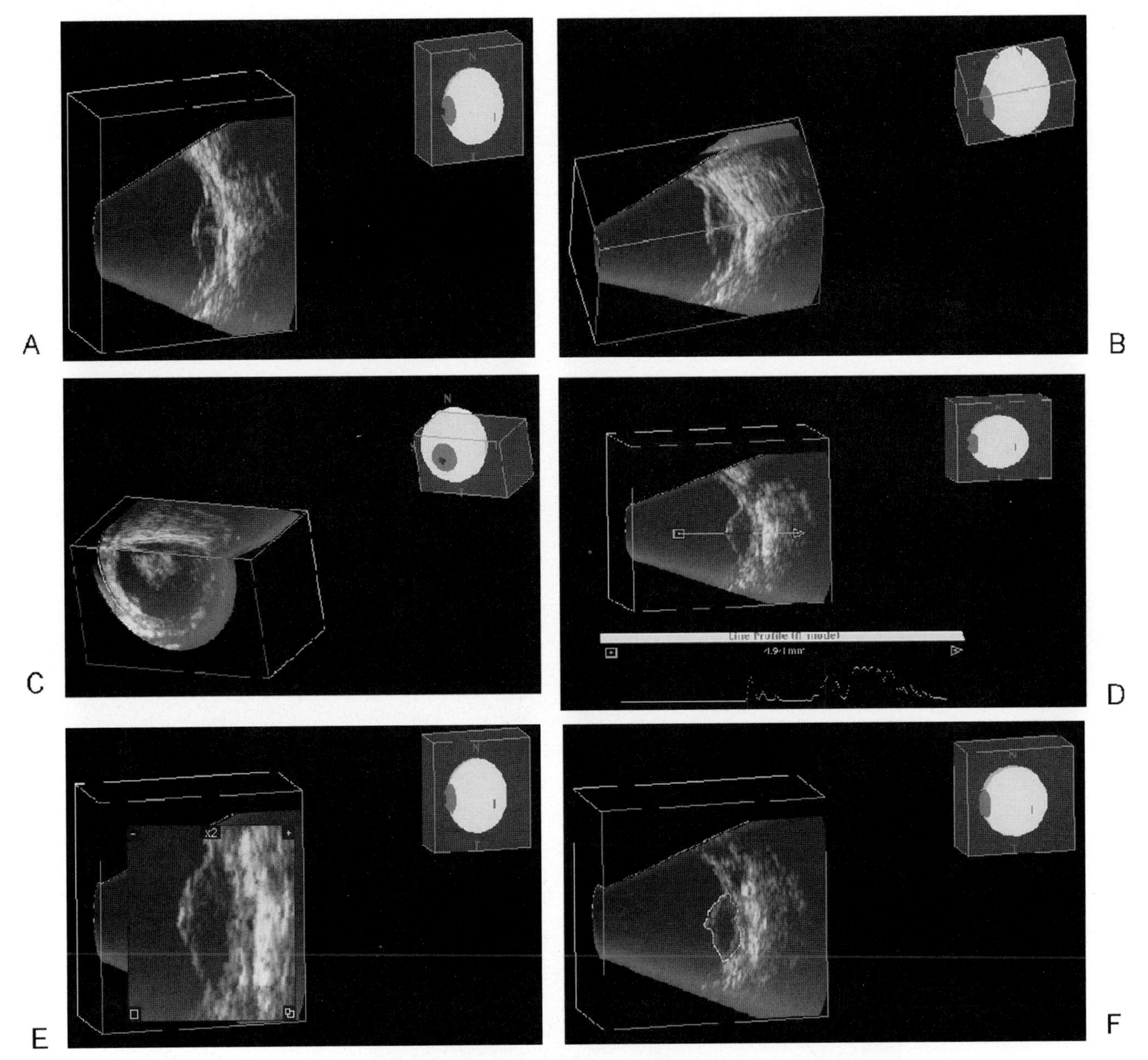

Figure 13-1. **A:** 3-D ultrasound image of choroidal melanoma. Image is sliced perpendicular to the tumor. **B:** Same lesion but now two planes have been sliced into. **C:** Tumor has been cut in two planes, one of which is parallel to the lens. **D:** An A-mode type tracing can be obtained by placing a cursor across the area of interest and an appropriate A-mode type tracing can be obtained (lower part of screen). **E:** Subsegments of the volume can be magnified in appropriate direction. **F:** Tumors can be manually outlined and from these manual outlines more accurate volumetric assessments can be made.

including the size of the detachment and its relationship to other structures, such as the fovea centralis (1). It may be difficult to understand the complex relationship between intraocular structures using only standard 2DUS (9), but by increasing the number of perspectives available 3DUS affords the operator greater opportunity to understand complex disease (7,12).

Improved Volumetric Accuracy and Repeatability

Many *in vivo* and *in vitro* reports have shown that 3DUS measurements are more accurate than 2DUS (13–27). Several studies have shown that 3DUS produces more repeatable data than 2DUS examinations (21,25,26,28,29). 3DUS is likely to improve the management of patients by providing accurate measurements of vitreous hemorrhages and

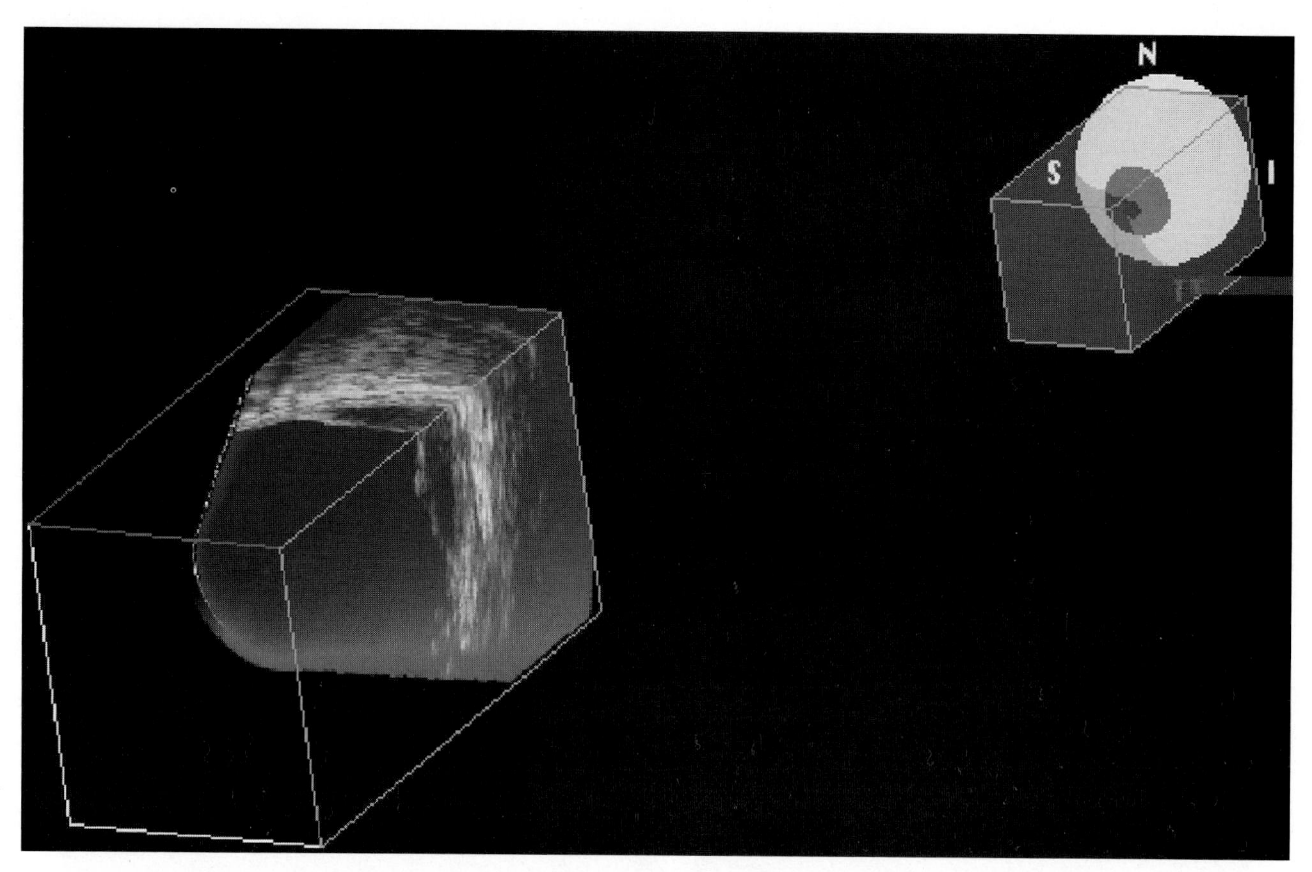

Figure 13-2. 3DUS can be used to accurately assess the extent of a retinal detachment.

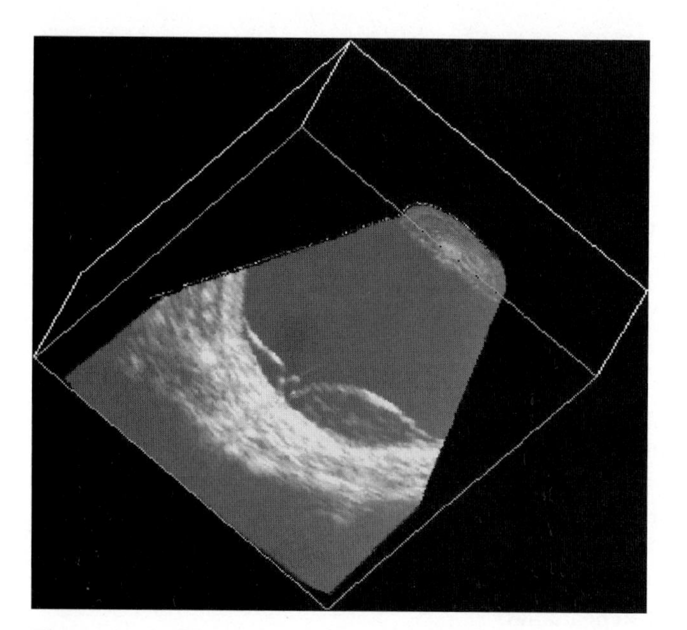

Figure 13-3. Slice from 3DUS study showing both a retinal tumor and detachment in the same eye. Optimizing the viewing plane often will provide a better understanding of the pathology.

ocular tumors, which is extremely important for designing appropriate treatment plans and for assessing patient prognosis (13,30). The decision to resect a choroidal nevus/ melanoma rests heavily on measurement of changes in size and progression from previous assessments.

Improved Scanning Speed

A small number of patients do not tolerate standard 2DUS orbital scanning as well as most people (8). These patients include small children and people with extremely painful globes, as can occur after trauma or postoperatively. If the anatomic area of concern can be scanned into 3DUS volume data set, the patient may be spared discomfort while diagnostic accuracy is preserved.

Improved Characterization of Tissues

Silverman and his colleagues have been developing techniques that combine 3DUS technology with high-frequency transducers and complex spectral analysis algorithms. The back-scatter information is processed to generate tissue characterization information (Figure 13-4) (2, 31). They have reported encouraging results with these techniques, which may help to assess the *in vivo* malignant potential for tumors such as uveal melanomas.

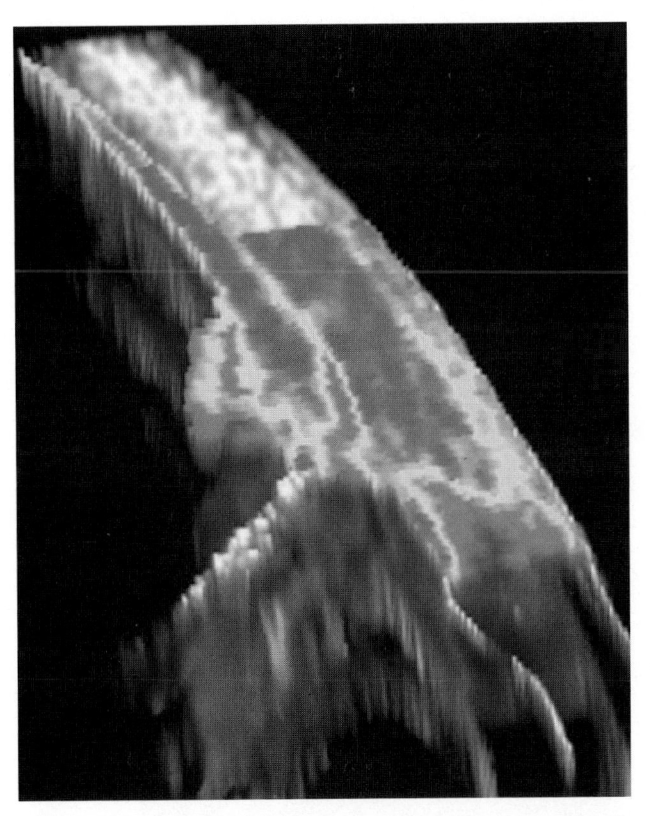

Figure 13-4. 3DUS image of anterior ocular segment obtained from digital 50-MHz ultrasonic data. The image depicts the sclera joining the lower cornea. The thin inner ciliary body and the iris lie to the left. (Image courtesy of Dr. F. Lizzi.) See color version of figure.

Improved Visualization of Vascular Anatomy

3DUS power and color Doppler techniques have the potential to enhance visualization of retinal vascular anatomy (Figure 13-5). Such studies may assist in evaluating retinal diseases and assessing response to therapy. Use of contrast materials may further assist in identifying small vessels in the retina and retro-orbital area.

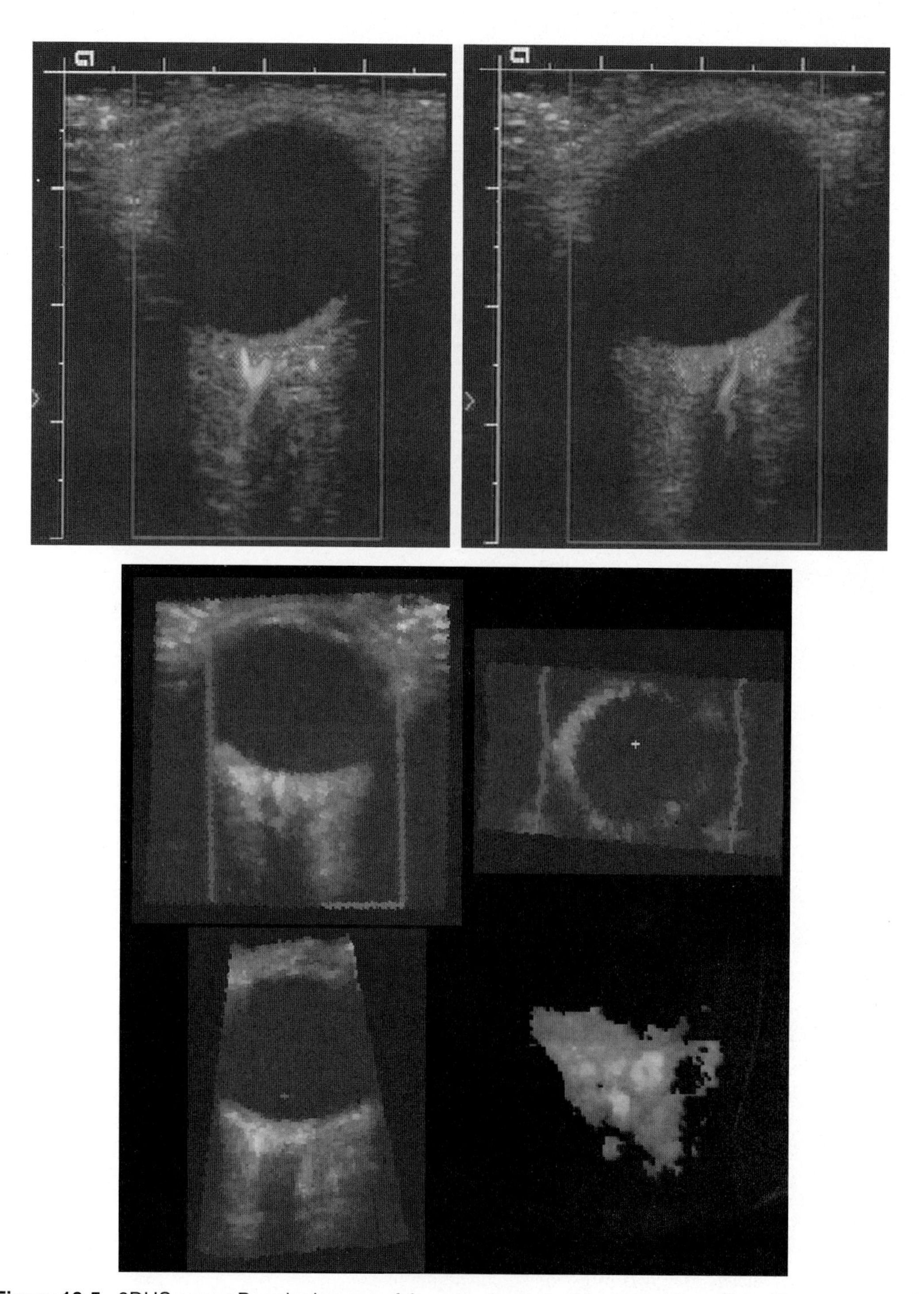

Figure 13-5. 3DUS power Doppler images of the vascular bed of the eye at the retina. The upper panel shows two frames from the 2DUS power Doppler acquisition. The lower panel shows three orthogonal planes through the eye together with a rendered image of the retinal underlying vasculature. The cross marker shows the intersection of the three planes. See color version of figure.

3DUS DISADVANTAGES

Many 3DUS ocular scans have used mechanical positioning systems with either a rotational or a tilting motion. Such devices are heavier and more clumsy than standard transducers, and if not carefully used may increase the pressure on the eye, thereby causing pain. Free-hand techniques with position sensing also can provide satisfactory results and, by following contours of the orbit, may offer a better long-term solution to scanning.

During a standard ocular scan, the patient often is asked to move his or her eye suddenly inward to one side or the other. This simplifies the diagnosis of vitreous floaters and retinal detachments (1). The potential for dynamic assessment of intraocular structures disappears with 3DUS scans, because of the necessity of keeping the eye still (4).

The entire globe may not be included in the 3DUS volume, depending on the scanning geometry used. Although this deficiency can be rectified by doing more than one scan, the entire area of interest may not be covered. As with other types of 3DUS, patient or operator motion can produce artifacts that may mimic disease.

3DUS CLINICAL INDICATIONS

Until now there have been no controlled studies demonstrating that 3DUS is clinically better than 2DUS for any pathology. Positive subjective reports exist on the use of 3DUS in evaluation of anterior and posterior segment tumors, retinal and ciliary body detachments, and a variety of other diseases. These include hyphema, iridocyclic disease, and foreign bodies in the eye. High-frequency 3DUS imaging offers the potential to visualize small structural details of the eye (2). Given these positive initial reports and the theoretical advantage in more accurate and repeatable measurements of tumors, it is reasonable to investigate 3DUS more thoroughly. Provided there are no artifacts in the image, it is also likely to improve the understanding of complex diseases and the planning of some surgeries.

HOW TO DO IT
Practical Clinical Tips for Ophthalmic Imaging

- Gently place the transducer directly on the globe or the closed eyelid.
- Keep the transducer clean and use sterile methyl cellulose or gel.
- Be cognizant of the amount of pressure applied to the globe:
  - Avoid applying excessive pressure to the globe by the transducer and positioning device.
- Scanning is best done supine:
  - Ask patient to focus on a point on the ceiling and not to move the eye during the scan.
- Use the ALARA (As Low As Reasonably Achievable) principle to minimize power deposition in the eye:
  - The orbit is a sound-sensitive part of the body.
- Scan the contralateral eye to clarify unusual pathologic appearances.
- Multiplanar displays are a good way to assess orbital pathology.
- Volume rendering assists in interpreting the organization of vascular structures.

REFERENCES

1. Byrne SF, Green RL. *Ultrasound of the eye and orbit.* St. Louis: Mosby, 1992.
2. Silverman RH, Rondeau MJ, Lizzi FL, Coleman DJ. Three-dimensional high-frequency ultrasonic parameter imaging of anterior segment pathology. *Ophthalmology* 1995;102:837–843.
3. Fledius HC. Ultrasound in ophthalmology. *Ultrasound Med Biol* 1997;23:365–375.
4. Downey DB, Nicolle DA, Fenster A. Three-dimensional orbital ultrasonography. *Can J Ophthalmol* 1995; 30:395–398.

5. Lieb WE, Flaharty PM, Ho A, Sergott RC. Color Doppler imaging of the eye and orbit: a synopsis of a 400 case experience. *Acta Ophthalmol* 1992;[Suppl]:50–54.
6. Munk P, Downey D, Nicolle D, Vellet AD, Rankin R, Lin DT. The role of colour flow Doppler ultrasonography in the investigation of disease in the eye and orbit. *Can J Ophthalmol* 1993;28:171–176.
7. Coleman DJ, Woods S, Rondeau MJ, Silverman RH. Ophthalmic ultrasound. *Radiol Clin North Am* 1992; 30:1105–1114.
8. Downey DB, Nicolle DA, Levin MF, Fenster A. Three-dimensional ultrasound imaging of the eye. *Eye* 1996;10:75–78.
9. Iezzi R, Rosen RB, Tello C, Liebmann J, Walsh JB, Ritch R. Personal computer-based 3-dimensional ultrasound biomicroscopy of the anterior segment. *Arch Ophthalmol* 1996;114:520–524.
10. Coleman DJ. Evaluation of ciliary body detachment in hypotony. *Retina* 1995;15:312–318.
11. Silverman RH, Coleman DJ, Rondeau MJ, Woods SM, Lizzi FL. Measurement of ocular tumor volumes from serial, cross-sectional ultrasound scans. *Retina* 1993;13:69–74.
12. Coleman DJ, Silverman RH, Rondeau MJ, Lizzi FL. New perspectives: 3D volume rendering of ocular tumors. *Acta Ophthalmol* 1992;[Suppl]:22.
13. Baba K, Okai T. Basis and principles of three-dimensional ultrasound. In: Baba K, Jurkovic D, eds. *Three-dimensional ultrasound in obstetrics and gynecology.* New York: Parthenon, 1997:1–20.
14. Buck T, Schon F, Baumgart D, et al. Tomographic left ventricular volume determination in the presence of aneurysm by three-dimensional echocardiographic imaging. I: Asymmetric model hearts. *J Am Soc Echocardiogr* 1996;9:488–500.
15. Elliot TL, Downey DB, Tong S, Mclean CA, Fenster A. Accuracy of prostate volume measurements *in vitro* using three-dimensional ultrasound. *Acad Radiol* 1996;3:401–406.
16. Gilja OH, Smievoll AI, Thune N, et al. *In vivo* comparison of 3D ultrasonography and magnetic resonance imaging in volume estimation of human kidneys. *Ultrasound Med Biol* 1995;21:25–32.
17. Gilja OH, Hausken T, Odegaard S, Berstad A. Three-dimensional ultrasonography of the gastric antrum in patients with functional dyspepsia. *Scand J Gastroenterol* 1996;31:847–855.
18. Gilja OH, Detmer PR, Jong JM, et al. Intragastric distribution and gastric emptying assessed by three-dimensional ultrasonography. *Gastroenterology* 1997;113:38–49.
19. Griewing B, Schminke U, Morgenstern C, Walker ML, Kessler C. Three-dimensional ultrasound angiography (power mode) for the quantification of carotid artery atherosclerosis. *J Neuroimaging* 1997;7:40–45.
20. Kupferwasser I, Mohr-Kahaly S, Menzel T, et al. Quantification of mitral valve stenosis by three-dimensional transesophageal echocardiography. *Int J Card Imaging* 1996;12:241–247.
21. Pauletzki J, Sackmann M, Holl J, Paumgartner G. Evaluation of gallbladder volume and emptying with a novel three-dimensional ultrasound system: comparison with the sum-of-cylinders and the ellipsoid methods. *J Clin Ultrasound* 1996;24:277–285.
22. Pini R, Monnini E, Masotti L, et al. Echocardiographic three-dimensional visualization of the heart. In: Hohne KH, ed. *3D imaging in medicine.* 1990:263–274.
23. Riccabona M, Nelson TR, Pretorius DH, Davidson TE. *In vivo* three-dimensional sonographic measurement of organ volume: validation in the urinary bladder. *J Ultrasound Med* 1996;15:627–632.
24. Riccabona M, Nelson TR, Pretorius DH. Three-dimensional ultrasound: accuracy of distance and volume measurements. *Ultrasound Obstet Gynecol* 1996;7:429–434.
25. von Birgelen C, di Mario C, Li W, et al. Morphometric analysis in three-dimensional intracoronary ultrasound: an *in vitro* and *in vivo* study performed with a novel system for the contour detection of lumen and plaque. *Am Heart J* 1996;132:516–527.
26. von Birgelen C, van der Lugt A, Nicosia A, et al. Computerized assessment of coronary lumen and atherosclerotic plaque dimensions in three-dimensional intravascular ultrasound correlated with histomorphometry. *Am J Cardiol* 1996;78:1202–1209.
27. Wong J, Gerscovich EO, Cronan MS, Seibert JA. Accuracy and precision of *in vitro* volumetric measurements by three-dimensional sonography. *Invest Radiol* 1996;31:26–29.
28. Kyei-Mensah A, Maconochie N, Zaidi J, Pittrof R, Campbell S, Tan SL. Transvaginal three-dimensional ultrasound: reproducibility of ovarian and endometrial volume measurements. *Fertil Steril* 1996;66:718–722.
29. Tong S, Downey DB, McLoughlin RF, et al. Variability and reliability of ultrasound prostate volume measurements. *Can Assoc Radiol J* 1996;47:S8 (abst).
30. Flock M, Gerende JH, Zimmerman LE, The size and shape of malignant melanomas of the choroid and ciliary body in relation to prognosis and histologic characteristics. *Trans Am Acad Ophthalmol Otolaryngol* 1955;59:740.
31. Coleman DJ, Silverman RH, Rondeau MJ, Lizzi FL, McLean IW, Jakobiec FA. Correlations of acoustic tissue typing of malignant melanoma and histopathologic features as a predictor of death. *Am J Ophthalmol* 1990;110:380–388.

14

Interventional Applications

OVERVIEW

This chapter will present an overview of the current clinical utility of three-dimensional ultrasound (3DUS) interventional procedures and discuss possible future roles. Interventional and minimally invasive therapies are becoming increasingly popular with patients and physicians because they tend to reduce the cost of procedures and our ability to use technology wisely is constantly improving. Fast 3DUS implemented in a user-friendly manner will improve the quality of interventional procedures. To date, there have been relatively few reports regarding 3DUS-guided surgery and minimally invasive therapy.

KEY CONCEPTS

- All fields of medicine are experiencing a strong trend toward minimally invasive diagnosis and therapies.
- 3DUS provides useful anatomic information that facilitates the safe performance of these procedures.
- 3DUS compliments fluoroscopic interventional procedures by permitting identification of catheter and needle position and seed placements at or in specific sites.
- 3DUS shortens invasive procedures by providing more accurate information regarding the positional relationships between anatomy and devices (e.g., stent to vessel).

The last decade has seen a strong trend toward less invasive diagnostic and therapeutic procedures. Laparoscopy, endoscopy, arthroscopy, lithotripsy, intraoperative ultrasound, percutaneous biopsies and drainage procedures, and image-guided surgeries have all grown dramatically in popularity. In contrast, the number of laparotomies and radical surgeries has been in relative decline. This trend shows no sign of slowing and, in fact, may be accelerating.

Table 14-1. *Minimally invasive diagnostic and therapeutic procedure requirements*

- The location and volume of tissue to be sampled or treated must be known
- The tissue must be clearly seen, preferably in 3D
- Clear comprehension of the 3D anatomy surrounding the target; including to the skin surface
- Accurate computer mapping of the target 3D anatomy and pathology is highly desirable
- Rapid, flexible, intuitive, and easy-to-learn manner interaction with 3DUS data is essential
- Comprehension of biopsy needle position or the tissue ablative therapy action is required
- Understanding of the different methods by which the needle or therapy can be deployed
- Evaluation of a variety of different treatment scenarios by processing system
- Planned procedure should be delivered accurately and in a timely manner
- Built-in safeguards for the delivery system to ensure patient and operator safety
- Direct monitoring of both the delivery and effects of the procedure by the system

3DUS guidance has considerable clinical utility in the performance of interventional procedures. This chapter will briefly review 3DUS guidance as currently practiced and discuss anticipated advances. The most common requirements for minimally invasive diagnostic and therapeutic procedures are given in Table 14-1.

Most minimally invasive procedures used today are missing some of the items listed in Table 14-1 and most rely heavily on the trained human to fill in the missing steps, which leads to considerable variability in the results obtained, because skills vary between operators. 3DUS imaging promises to help improve overall standards by decreasing the number of steps the operator must perform mentally. To date, some utility has been shown for 3DUS in several minimally invasive and biopsy procedures, and it is likely that the role of 3DUS will increase significantly in the near future.

3DUS provides anatomic information about needle and catheter placement that is not available using 2DUS guidance (Figure 14-1). Before a procedure is performed, the anatomy/pathology can be evaluated with 3DUS to identify the optimal pathway for needle and catheter placement. Visualization of the C-plane (parallel to the skin surface) may assist in understanding the area of interest and the surrounding anatomy. Scanning prior to the invasive procedure permits the actual procedure to be performed more rapidly because the optimal window(s) will have been identified. It is important to determine the optimal window for visualization and to understand the orientation (i.e., anterior, posterior, right, left, cranial, caudal) of each of the orthogonal image planes. Generally, a standard orientation (transverse, sagittal, coronal) is chosen.

During the invasive procedure, the needle/catheter can be displayed in three orthogonal planes and the exact location identified. The needle/catheter can be advanced or repositioned as necessary (see Figure 14-1). The volume data also can be used to guide the direction of the needle pass. 3DUS offers the opportunity to perform invasive procedures more successfully with fewer complications and less procedural time.

Neurosurgery

The neurosurgery/neuroradiology field currently has probably the most advanced 3D-guided interventional program. Three-dimensional (3D) magnetic resonance (MR) and 3D computed tomography (CT) permit accurate planning and performance of a variety of neurosurgical procedures (1). For example, Dormont used 3D MR guidance to plan and implement a process to provide chronic stimulation of the ventralis ventromedial nucleus of the thalamus for the treatment of tremor (2). 3DUS methods provide a 3D representation of the surgical space, which has shown utility in spine surgery (3). Recently, systems have been developed that allow registration of data from different imaging modalities. Specifically, ultrasound and MR have been used in the development of a frameless stereotactic neuronavigation system (4). These systems are becoming more sophisticated, and some have also been used to monitor needle insertions under MR guidance (5).

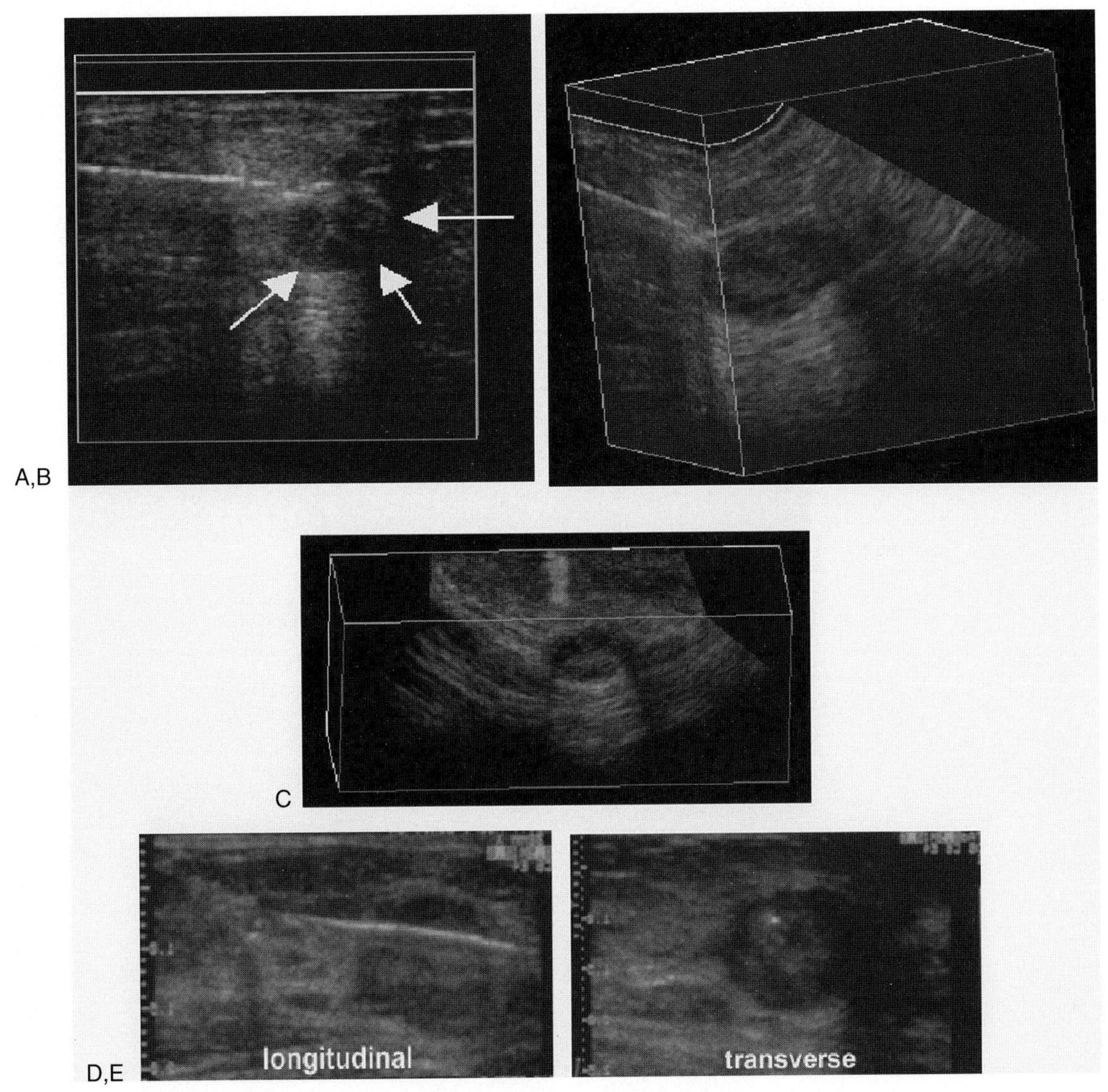

Figure 14-1. 3DUS of a mucinous cancer interventional biopsy. **A:** Longitudinal view of a 14-gauge needle and a mass (*arrows*). Needle tip appears closely approximated with the upper portion of the mass such that after firing it will contain a sample from the mass. **B:** Combined longitudinal and transverse perspective showing that the needle is incorrectly positioned, being too lateral and too superior. **C:** Coronal perspective shows the needle relative to the lesion being too high and too lateral. Note width of echo signal from needle in the coronal plane. **D:** The needle position was repositioned slightly and fired lying in the upper right portion of the mass in both longitudinal and transverse (**E**) perspectives.

Liver

3DUS has been used for planning liver surgery (6). The specific use of the 3DUS in the liver is that the surgery can be planned preoperatively. Van Leeuween stated that the most important use of 3DUS imaging is in clarifying not just which of Couinaud's segments the lesion occupies but also the anatomy of each segment. This is a situation complicated by the high prevalence of anatomic variations, especially in the right hemiliver. Having insight into these variations will help the surgeon accurately plan the procedure preoperatively (6).

3DUS may help assess disease volume more accurately and thereby help the physician use the appropriate therapy dose and best probe position. Kalab and Livraghi comment that alcohol injection appropriately applied to patients with hepatocellular carcinoma is very suitable for palliation (7,8). Kalab comments that the procedure is well tolerated and that the survival of patients, if properly selected, is similar to that with surgical resection

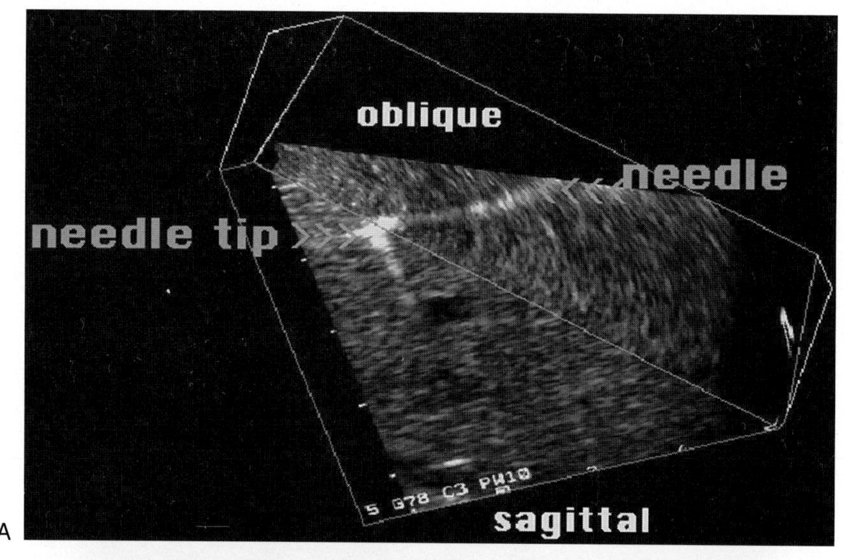

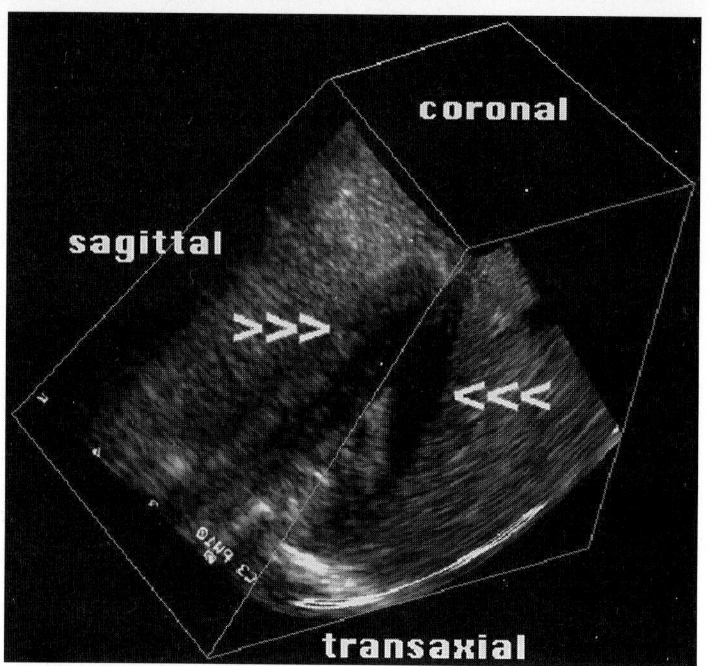

Figure 14-2. 3DUS image of a porcine liver cryoablation procedure. **A:** Oblique slice showing the course of the needle and its tip before the procedure. **B:** Two perspectives after formation of the ice-ball are shown (*arrowheads*).

(7). He emphasizes the importance of titrating the dose to the volume of tumor. Accurate volumetric assessment of tumor is also required to optimally deploy cryosurgery (Figure 14-2) (9) and radiofrequency (10) or microwave (11) ablation techniques.

3DUS also has been used to assist in localizing needles and stents during transjugular intrahepatic portosystemic shunt (TIPS) procedures (Figures 14-3, 14-4) (12). Preliminary results suggest that 3DUS provides numerous positional and directional cues that may improve the efficiency of portal vein access. Information provided by 3DUS included differentiation of liver margin from surrounding ascites, preventing inadvertent perforation of the liver capsule; localization of the angiographic catheter in the middle hepatic vein or inferior right hepatic vein rather than in the anticipated right hepatic vein; selection of an optimized hepatic venous point of needle pass origination; determination of the approximate trajectory of needle passage to puncture a central portal vein; data that allowed the

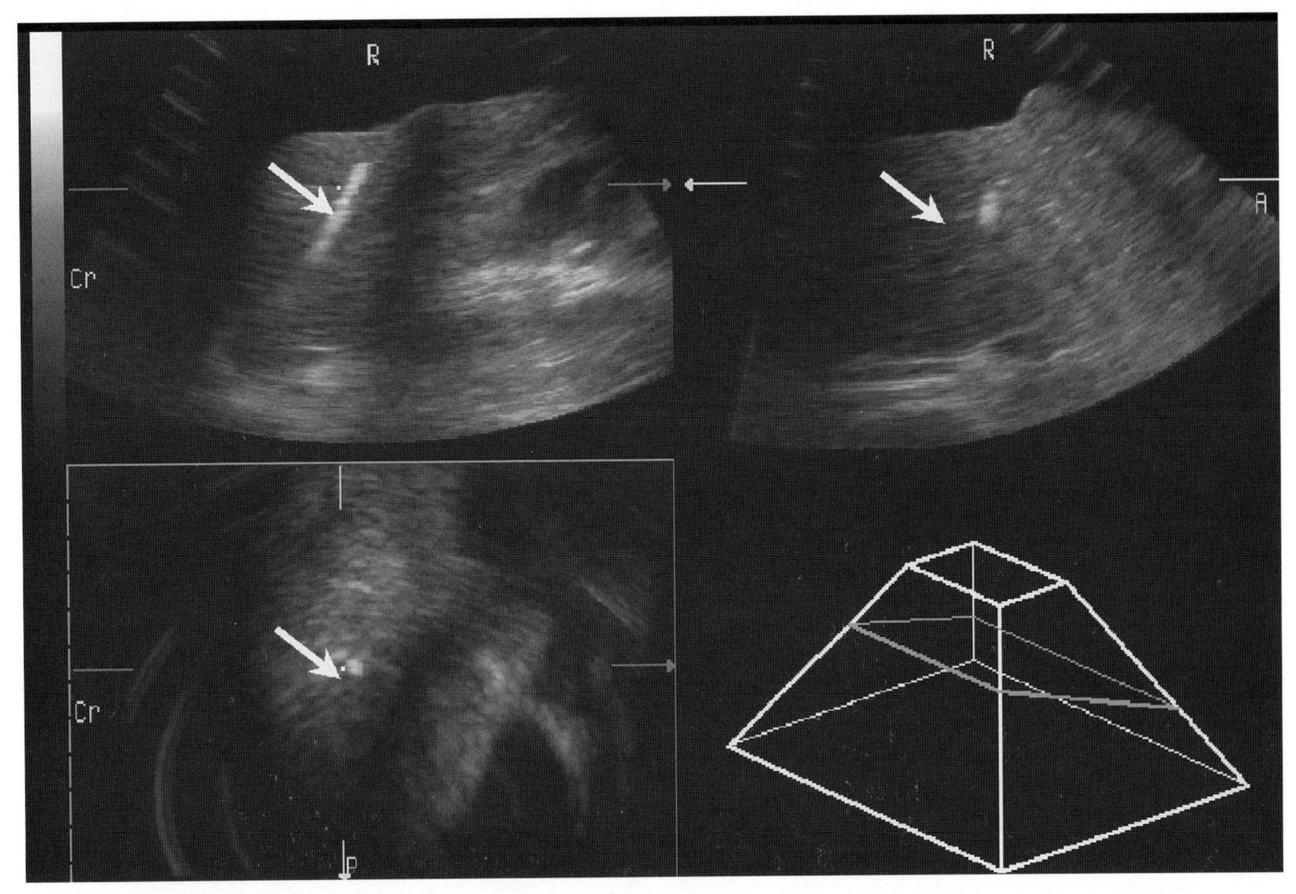

Figure 14-3. 3DUS guidance during TIPS procedure showing location of the needle (*arrow*). The needle is highly echogenic and thus quite visible. Localization is helped by visualization in three orthogonal intersecting planes. Ascites is present. R, right; CR, cranial; A, anterior; P, posterior. (Images courtesy of Dr. S. Rose.)

operators to use a middle hepatic vein origin rather than the right hepatic vein for portal venous access; and identification that the main portal vein puncture was intrahepatic and could safely be used to create a stent channel.

3DUS also may be used to assist in performing transjugular core liver biopsy in patients with focal abnormalities or diffuse disease in which percutaneous biopsy is contraindicated because of hemorrhagic complications (12). 3DUS provides the vascular landmarks in the orthogonal planar images to guide the biopsy needle placement. Historically, these have not been visualized routinely with fluoroscopy in noncatheterized vessels.

Heart and Cardiovascular

Prati et al. have found catheter-based intraluminal technology invaluable for determining key parameters of vascular morphology before and during interventions, and for assessing the accuracy of deployment of device placement (13). They emphasized that the data provided are even more useful when combined with 3DCT reconstruction (13). It is probable that they would be even further assisted if 3DUS was provided immediately, because their system uses off-line reconstruction. Von Birgelen states that 3D intravascular ultrasound (3D-IVUS) permits a more reliable and accurate quantification of interval hyperplasia, as seen in cardiac allograft arteriopathy during interventional procedures (14). This may influence the therapeutic decision because it allows a determination of the longitudinal extent of calcification and frequently discovers atherosclerosis in angiographically normal segments.

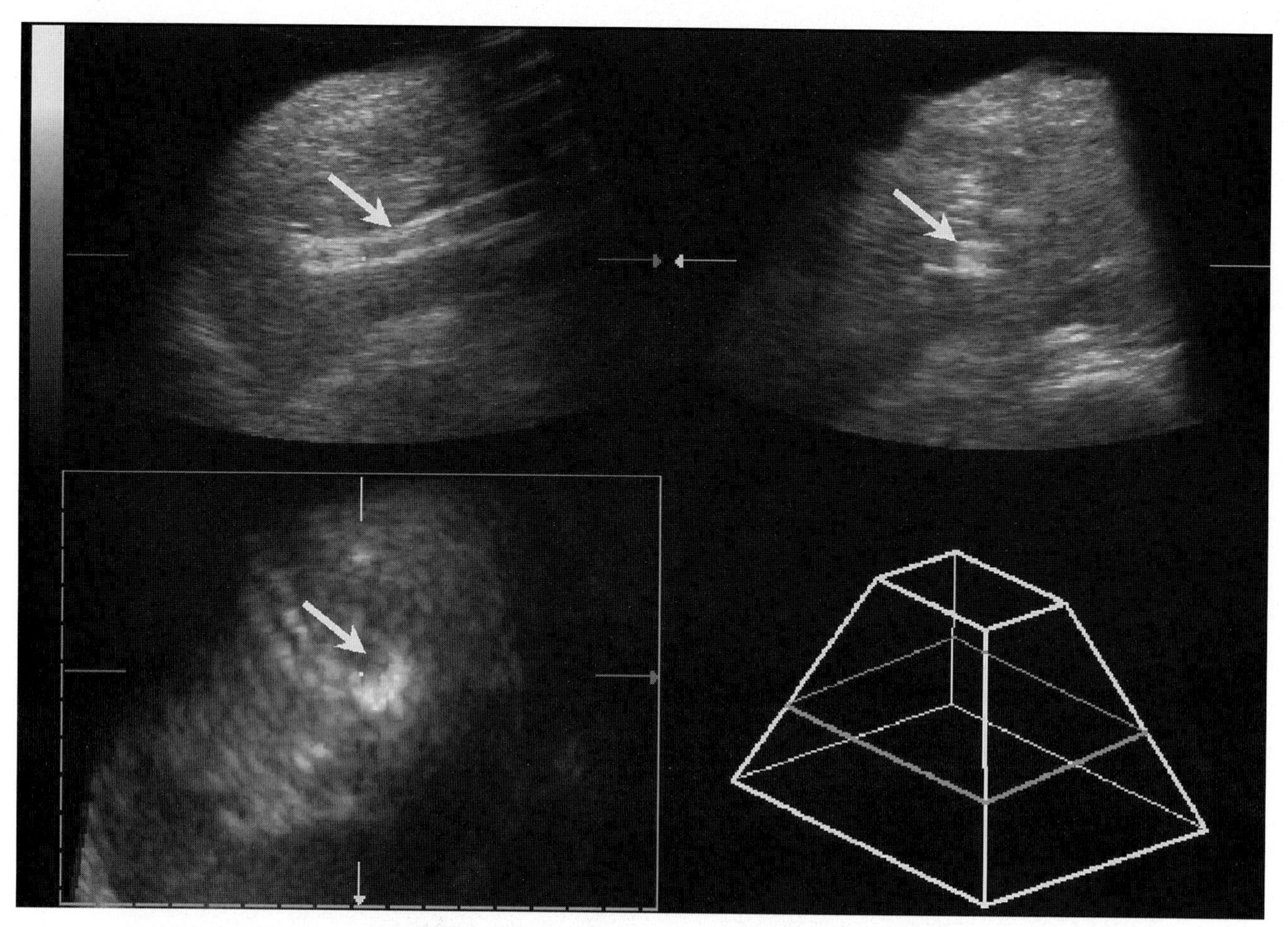

Figure 14-4. 3DUS guidance during the TIPS procedure in Figure 14-8 showing the location of the stent (*arrow*). The stent is highly echogenic and thus quite visible.

Prostate

The prostate is the organ in which 3DUS assistance has been most thoroughly studied, with utility being shown in brachytherapy (Figure 14-5) (15–17), laser therapy (18), stent implantation (19), and cryosurgery (Figures 14-6, 14-7, 14-8) (20,21).

Brachytherapy is a procedure in which radioactive sources are introduced into the body. Ragde et al. devised a procedure that used an externally positioned template to place needles containing radioactive sources in the prostate accurately to deliver a predetermined dose (22). The procedure was reported to be successful, with a 7-year survival of 77% (22). Stock et al. used a similar system, which Stock claims allows him to place seeds three-dimensionally in the prostate in predetermined places (16,17). Martinez et al. use real-time ultrasound guidance and interactive on-line dosimetry to continually update their treatment plan (15). Martinez notes that with this variation patient motion during the procedure becomes insignificant because it can be corrected without increasing the target volume (15).

Chin emphasizes the advantage of seeing the coronal plane from the 3DUS data set to help ensure that needles are appropriately positioned during cryosurgery (20,21). Strasser et al. reported similar opinions when using 3DUS guidance to perform interstitial laser therapy of benign prostatic hyperplasia accurately (18). They commented that it is helpful in placing the fiber exactly in the correct location in the prostate. Parivar et al. demonstrated that 3D MR spectroscopy imaging was superior to standard MR and transrectal ultrasound in detecting persisting or recurrent disease for cryosurgery (23).

In recent years there has been a plethora of new treatments for BPH. These include

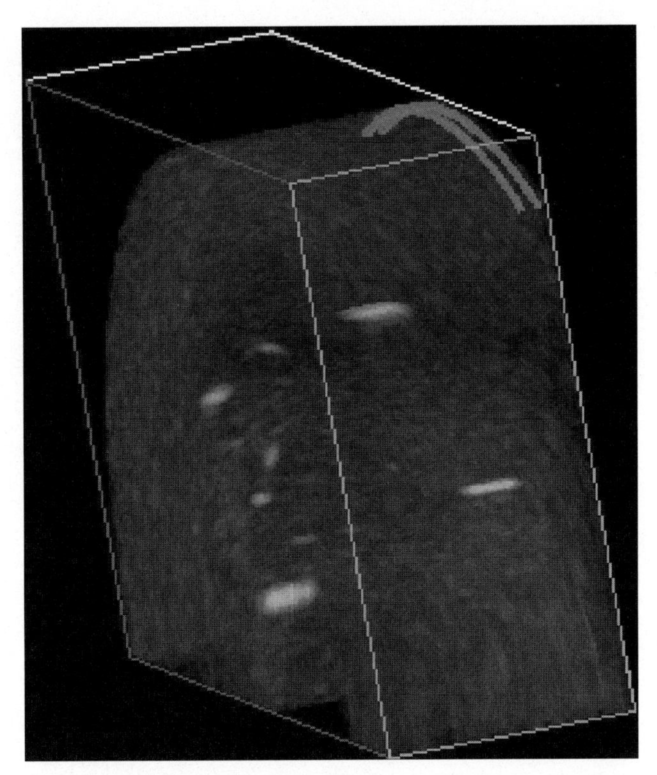

Figure 14-5. 3DUS study evaluating brachytherapy seed placement. Maximum-intensity projection for detecting several brachytherapy seeds within Agar phantom. Maximum-intensity projection shows the seeds clearly, which may significantly aid the detection of both seeds and needles in the future.

A,B

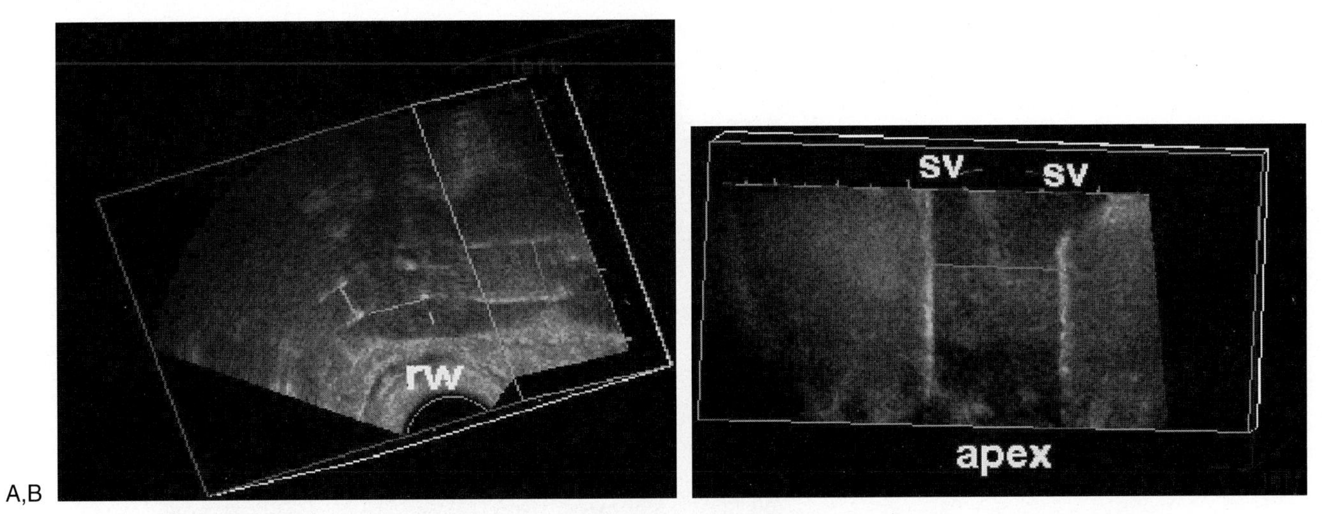

Figure 14-6. 3DUS cryosurgery of the prostate. **A:** Five needles have been inserted in the prostate, showing up as echogenic densities. On the reconstructed images it is possible to measure the exact distance between these needles (*short lines*, distances being measured). **B:** Coronal perspective of the prostate with needles inserted parallel to each other, although the one on the left side needs to be advanced further into the left seminal vesicle (SV) (*short lines,* distances being measured).

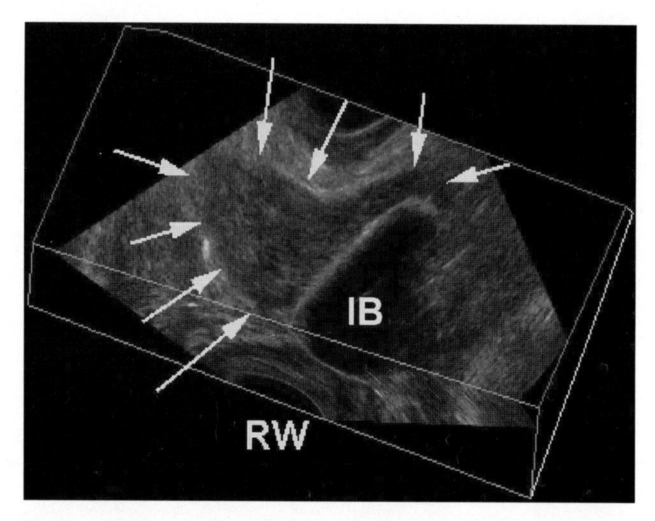

Figure 14-7. 3DUS prostate cryosurgery. Coronal view of the prostate during cryosurgery. The left side of the prostate (prostate outlined by *arrows*) is being frozen with a single cryoprobe. The ice-ball (IB) shows up as a hypoechoic area that is now closely approximated with the rectal wall (RW).

high- and low-energy transurethral microwave thermotherapy, transurethral thermal ablation, high-intensity focused ultrasound, various types of laser ablation, and transurethral electrovaporization (24). There is a great need for objective assessment of their true utility (25). Flannigan strongly recommends we match the dosage of all the newer minimally invasive therapies used for benign prostatic hyperplasia (BPH) with the size of the gland (25). 3DUS is probably the best way of accurately measuring the gland for this purpose.

Bowel

Hunerbein et al. concluded that 3D endosonography with ultrasound-guided biopsy improves diagnosis of extramural recurrence after curative resection (26). 3D image displays allow precise control of the position of the biopsy needle within the target.

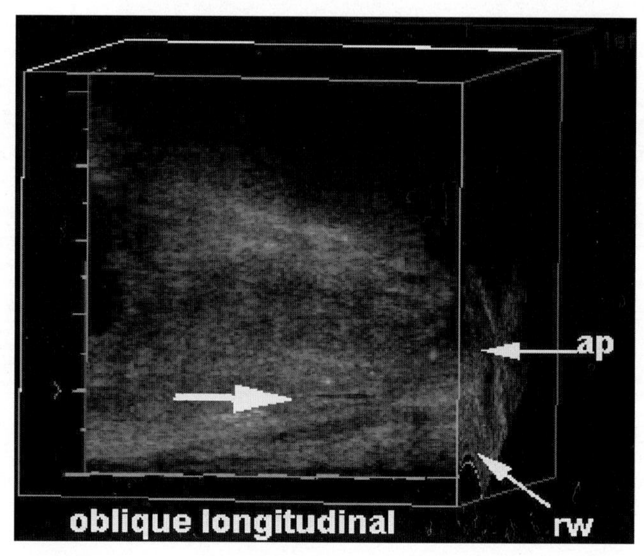

Figure 14-8. 3DUS study following cryosurgery using an oblique coronal plane. The tract of one of the cryoprobes is clearly shown (*arrow*). RW, rectal wall; AP, apex. It is easier to see these tracts by using oblique planes.

Infertility

3DUS also is useful in providing guidance for follicular aspiration and transfer of *in vitro* fertilized embryos. The ability of 3DUS to scroll through a volume in multiple planes simultaneously permits the exact location of the needle or air bubbles to be identified. Feichtinger points out that even the uterine angles and the fallopian tubes can be evaluated to a greater extent than with 2DUS (27). He also suggests that laser surgery under sonographic 3D control may replace several procedures currently performed by hysteroscopic and laparoscopic surgery (28).

Other

3DCT and MR have been used for planning maxillofacial surgery reconstructions (29). Enislidis concludes that computer-assisted 3D guidance systems have the potential to make complex or minimally invasive procedures easier, thereby reducing postoperative morbidity for the patients. Rask and Jensen claim 3DUS is helpful in assessing response to radiation therapy (Figure 14-9) (30).

Intraoperative 3DUS Guidance

In addition to the central nervous system, liver, and prostate, applications of 3DUS to intraoperative procedures are being explored (see Figures 14-1, 14-2, 14-3). Kavic is enthusiastic about the application of 3DUS to the ever-expanding field of endoscopy (31). He believes it may offer a more intuitive and accurate assessment of structures hidden from view during standard laparoscopy. Vogel likes the ability to see cardiac defects prior to opening the chest (32). Fuchs and State are exploring head-mounted display systems that provide an augmented reality system to guide the interventionalist during the procedure without taking eyes off the patient by fusing 3DUS images of the patient interior with patient surface images to assist in guiding needle biopsies (33,34).

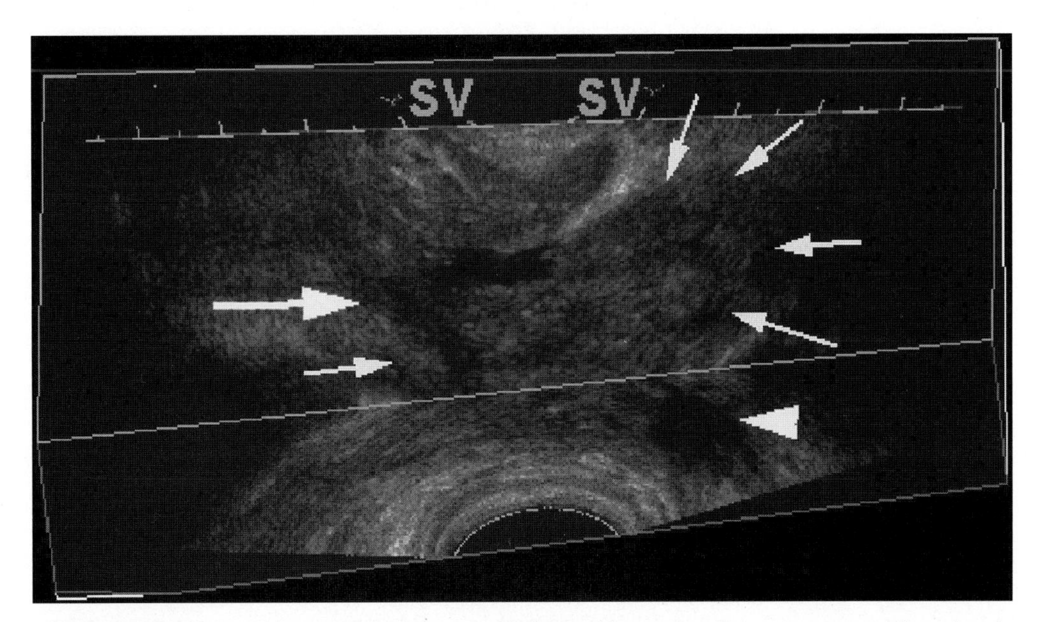

Figure 14-9. 3DUS showing sloughing of prostatic tissue 15 months after cryosurgery. There has been extensive remodeling of the prostate. The coronal plane shows outside of the prostate on the left side and is reasonably well preserved (*white arrows*, edge of prostate). On the right side there is extensive sloughing with fluid in the region of the right prostate bed (*big arrow*). There is also fluid in the region of the left obturator internus muscle (*left arrowhead*).

HOW TO DO IT
Practical Clinical Tips for Interventional Procedures

- Ensure that the 3DUS device is accurately calibrated.
- An interventionalist and imager should be present during interventional procedures:
  - Interventionalist—manipulates catheters.
  - Imager—operates 3DUS system.
- Prescan to find the optimum acoustic window prior to beginning procedure.
- Minimize motion of the patient, examiner, and transducer during acquisition.
- Ensure magnetic positioning systems are not affected by other operating room equipment.
- Locate needle/catheter in three orthogonal views to evaluate position.
- Use orthogonal planes and volume-rendered data to align needle trajectory.

REFERENCES

1. Olivier A, Alonso-Vanegas M, Comeau R, Peters TM. Image-guided surgery of epilepsy. *Neurosurg Clin N Am* 1996;7:229–243.
2. Dormont D, Cornu P, Pidoux B, et al. Chronic thalamic stimulation with three-dimensional MR stereotactic guidance. *Am J Neuroradiol* 1997;18:1093–1107.
3. Foley KT, Smith MM. Image-guided spine surgery. *Neurosurg Clin N Am* 1996;7:171–186.
4. Hata N, Dohi T, Iseki H, Takakura K. Development of a frameless and armless stereotactic neuronavigation. *Neurosurgery* 1997;41:608–613.
5. Leung DA, Debatin JF, Wildermuth S, et al. Real-time biplanar needle tracking for interventional MR imaging procedures. *Radiology* 1995;197:485–488.
6. Van Leeuween MS, Noordzij J, Hennipman A, Fledberg MA. Planning of liver surgery using three dimensional imaging techniques. *Eur J Cancer* 1995;31A:1212–1215.
7. Kalab M. [Treatment of liver tumors with ethanol injection]. *Vnitr Lek* 1997;43:180–183.
8. Livraghi T, Giorgio A, Marin G, et al. Hepatocellular carcinoma and cirrhosis in 746 patients: long-term results of percutaneous ethanol injection. *Radiology* 1995;197:101–108.
9. Lee FT Jr, Mahvi DM, Chosy SG, et al. Hepatic cryosurgery with intraoperative US guidance. *Radiology* 1997;202:624–632.
10. Solbiati L, Goldberg SN, Ierace T, et al. Hepatic metastases: percutaneous radio-frequency ablation with cooled-tip electrodes. *Radiology* 1997;205:367–373.
11. Matsukawa T, Yamashita Y, Arakawa A, et al. Percutaneous microwave coagulation therapy in liver tumors: a 3-year experience. *Acta Radiol* 1997;38:410–415.
12. Rose SC, Pretorius DH, Kinney TB, et al. 3D sonographic guidance of trans venous intrahepatic invasive procedures: feasibility of a new technique. *J Vasc Interv Radiol* 1998 (*in press*).
13. Prati F, di Mario C, Gil R, et al. Usefulness of on-line three-dimensional reconstruction of intracoronary ultrasound for guidance of stent deployment. *Am J Cardiol* 1996;77:455–461.
14. Von Birgelen C, di Mario C, Reimers B, et al. Three-dimensional intracoronary ultrasound imaging: methodology and clinical relevance for the assessment of coronary arteries and bypass grafts. *J Cardiovasc Surg (Torino)* 1996;37:129–139.
15. Martinez A, Gonzalez J, Stromberg J, et al. Conformal prostate brachytherapy: initial experience of a phase I/II dose-escalating trial. *Int J Radiat Oncol Biol Phys* 1995;33:1019–1027.
16. Stock RG, Stone NN, Wesson MF, De Wyngaert JK. A modified technique allowing interactive ultrasound-guided three-dimensional transperineal prostate implantation. *Int J Radiat Oncol Biol Phys* 1995;32:219–225.
17. Stone NN, Stock RG. Brachytherapy for prostate cancer: real-time three-dimensional interactive seed implantation. *Tech Urol* 1995;1:72–80.
18. Strasser H, Janetschek G, Horninger W, Bartsch G. Three-dimensional sonographic guidance for interstitial laser therapy in benign prostatic hyperplasia. *J Endourol* 1995;9:497–501.
19. Koelbl H, Hanzal E. Imaging of the lower urinary tract. *Curr Opin Obstet Gynecol* 1995;7:382–385.
20. Chin JL, Downey DB, Onik G, Fenster A. Three-dimensional prostate ultrasound and its application to cryosurgery. *Tech Urol* 1996;2:187–193.
21. Chin JL, Downey DB, Mulligan M, Fenster A. Three-dimensional transrectal ultrasound guided cryoablation for localized prostate cancer in nonsurgical candidates: a feasibility study and report of early results. *J Urol* 1998;159:910–914.
22. Ragde H, Blasko JC, Grimm PD, et al. Interstitial iodine-125 radiation without adjuvant therapy in the treatment of clinically localized prostate carcinoma. *Cancer* 1997;80:442–453.
23. Parivar F, Hricak H, Shinohara K, et al. Detection of locally recurrent prostate cancer after cryosurgery: evaluation by transrectal ultrasound, magnetic resonance imaging, and three-dimensional proton magnetic resonance spectroscopy. *Urology* 1996;48:594–599.
24. Madersbacher S, Djavan B, Marberger M. Minimally invasive treatment for benign prostatic hyperplasia. *Curr Opin Urol* 1998;8:17–26.
25. Flannigan GM. Editorial: imaging, stone disease. *Curr Opin Urol* 1997;7:U10–U12.

26. Hunerbein M, Dohmoto M, Haensch W, Schlag PM. Evaluation and biopsy of recurrent rectal cancer using three-dimensional endosonography. *Dis Colon Rectum* 1996;39:1373–1378.
27. Feichtinger W. Transvaginal three-dimensional imaging. *Ultrasound Obstet Gynecol* 1993;3:375–378.
28. Feichtinger W, Strohmer H, Feldner-Busztin M. Laser surgery under sonographic control: preliminary experimental investigations. *Ultrasound Obstet Gynecol* 1993;3:264–267.
29. Enislidis G, Wagner A, Ploder O, Ewers R. Computed intraoperative navigation guidance—a preliminary report on a new technique. *Br J Oral Maxillofac Surg* 1997;35:271–274.
30. Jensen PK, Hansen MK. Ultrasonographic, three-dimensional scanning for determination of intraocular tumor volume. *Acta ophthalmologica* 1991;69:178–186.
31. Kavic MS. Three-dimensional ultrasound. *Surg Endosc* 1996;10:74–76.
32. Vogel M, Ho SY, Lincoln C, Yacoub MH, Anderson RH. Three-dimensional echocardiography can simulate intraoperative visualization of congenitally malformed hearts. *Ann Thorac Surg* 1995;60:1282–1288.
33. Fuchs H, State A, Pisano ED, et al. Towards performing ultrasound-guided needle biopsies from within a head-mounted display. Proceedings of visualization in biomedical computing 4th International Conference, VBC '96 1996:591–600.
34. State A, Livingston MA, Garrett WF, et al. Technologies for augmented reality systems: realizing ultrasound-guided needle biopsies. Proceedings of SIGGRAPH '96. *Comput Graph* 1996:439–446.

PART IV

The Future

15

Emerging Clinical Applications

OVERVIEW

This chapter will provide an overview of emerging areas of three-dimensional ultrasound (3DUS) application in clinical medicine. Although the applications of 3DUS are increasing, initial work has been limited. As technology and procedures mature, additional areas will join current clinical techniques.

KEY CONCEPTS

- Clinical applications of 3DUS will continue to expand.
- Clinical acceptance will be determined by equipment availability and ease of use.
- 3DUS will develop in specialized clinical applications, including some studied to date:
  - Vascular—improved by combining 3DUS with contrast agents and harmonic imaging.
  - Musculoskeletal.
  - Neonatal brain—provides orientations not available with two-dimensional ultrasound (2DUS).
  - Motion studies—vascular dynamics, speech pathologies, joint kinematic studies.
  - Dermatology.
  - Thyroid.
  - Psychological impact studies—improved patient understanding and bonding.
- 3DUS will have a positive psychological impact on patient care by permitting patients to more clearly understand their anatomy and pathology.
- 3DUS will open new areas to ultrasound imaging in the future as the technology is distributed through clinical areas and throughout the world.

The clinical applications of 3DUS continue to expand. To a large extent the clinical acceptance and use will be determined by the availability and ease of use of the equipment. Presently, most manufacturers of clinical ultrasound equipment are working on some type of 3DUS product. As these products become more available to clinicians, additional applications will appear as experience grows. The increased development of more sophisticated scanning techniques such as harmonic imaging and contrast materials will logically complement the advantages of 3DUS approaches.

The increasing interest in two-dimensional ultrasound (2DUS) for musculoskeletal imaging will benefit from 3DUS methods, potentially enhancing our understanding of muscle, tendon, and bone relationships. Additionally, 3DUS in combination with joint range-of-motion studies may offer new insight about joint flexibility. Specialized applications of 3DUS in dermatology using higher frequencies may assist in evaluation of skin lesions and tumor volumes. The improved visualization afforded by 3DUS also will contribute to psychological studies such as maternal/fetal bonding and may provide a closer identification between the patient and the underlying processes in his or her body.

3DUS applications also can be expected to expand beyond the traditional areas of 2DUS scanning as specialized devices such as intravascular catheters, higher-frequency probes, and 2D transducer arrays become more common. Such developments already are under development and evaluation in the research setting, as has been discussed in previous chapters. One aspect of 3DUS that has proven true thus far is that the areas of clinical applicability often have exceeded initial projections. As clinical experience grows, that trend is expected to continue. Well-designed clinical trials will be essential to assess the diagnostic and patient management impact of 3DUS to complete utilization and acceptance in medical diagnosis and treatment.

A few areas of clinical exploration that are under way are discussed in the following sections. Although there has been limited 3DUS work in these areas thus far, it is expected that interest will continue to grow.

Vascular Imaging

Imaging of blood flow and vessel anatomy is one of the most important areas of US imaging. 3DUS methods show much promise to provide clear pictures of vascular anatomy and blood flow dynamics, particularly with the advent of contrast materials and harmonic imaging. The clinical principles and applications of contrast material have been studied extensively (1,2) and show increasing promise for a variety of purposes. They significantly enhance blood flow signal and increase visualization of vascular anatomy with power and color Doppler techniques. As a result, 3DUS methods based on power/color Doppler signal enhancements from contrast material offer an important capability to more completely image the 3D vascular anatomy of an organ (Figure 15-1) (3). The increased sensitivity to blood flow afforded with contrast should lead to better quantitative measurements of blood flow and vascular anatomy.

3DUS combined with contrast materials provides an important technique to improve vascular imaging. Current power/color Doppler images suffer from reduced spatial and temporal resolution compared with B-mode images because of the additional processing and signal averaging required to produce high-quality images. As a result, small vessels or rapidly moving structures often can be blurred to appear much larger than actual size. The loss of temporal information makes it difficult to image the heart, for example. Recent developments in harmonic imaging technology offer a solution to these difficulties by producing high-quality images of contrast materials at spatial resolutions and frame rates comparable to B-mode images. Harmonic imaging provides high contrast and spatial resolution that permits recognition of vessels with higher temporal resolution and without blooming artifacts common with color/power Doppler that represents a perfect complement to 3DUS techniques to visualize vascular anatomy (Figure 15-2) (4).

Musculoskeletal Imaging

Musculoskeletal imaging using ultrasound has proven to be a challenging area. The vast differences in bone, cartilage, and soft-tissue acoustic impedances make it difficult

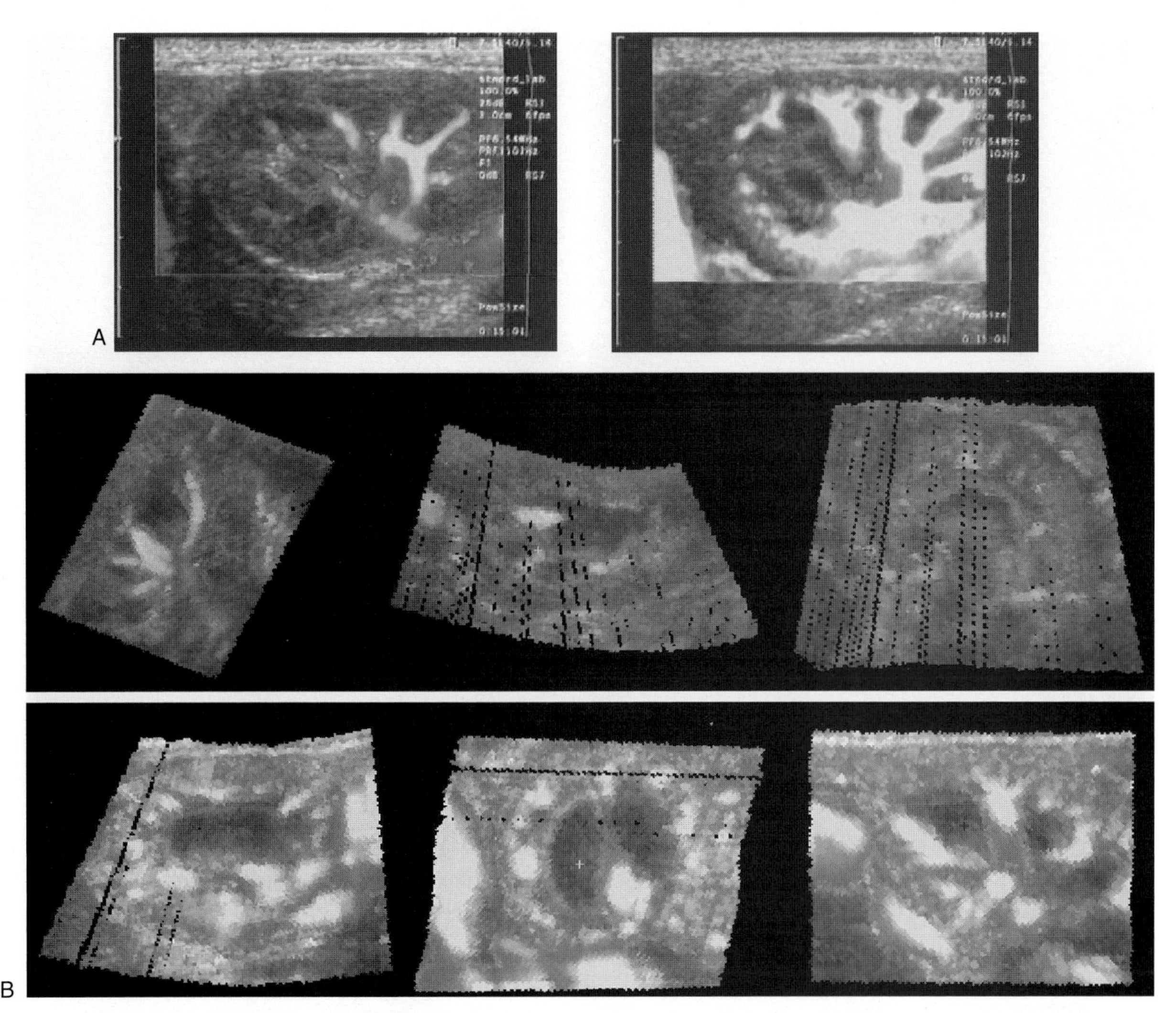

Figure 15-1. 3DUS power Doppler study of a dog kidney imaged using contrast material (Imagent®, Alliance Pharmaceuticals, San Diego, CA). **A:** Two 2DUS images from a free-hand acquisition with position sensing. The left image is without contrast material and the right is with contrast material. **B:** Two panels showing three orthogonal slices through the 3DUS volumes from A. The upper panel is without contrast; the lower is with contrast. 3DUS data provide clear visualization of the entire kidney and an appreciation of the 3D geometry of the organ. Although the kidney is well visualized without contrast material, the smaller vessels and parenchyma are much more clearly seen after contrast. See color version of figure.

to obtain clear views of the spatial relationships between bone, cartilage, tendon, and muscles. Despite this, 3DUS has the potential to improve visualization of joint anatomy by providing an integrated view of the overall spatial relationships (see Figure 15-3).

Early applications of 3DUS to musculoskeletal anatomy have evaluated the pediatric hip. 3DUS provides an improved view of the femoral head and acetabulum for evaluation of hip dysplasia (see Figure 15-4). Gerscovich et al. studied 38 data acquisitions from nine patients with a clinical diagnosis of developmental dysplasia of the hip (5). 3DUS data were acquired predominantly from the coronal plane but reviewed in both sagittal and craniocaudal orientations, which is impossible with 2DUS (6). The capability of viewing improved orientations permitted the relationship of the femoral head to the acetabulum to be more clearly appreciated than with 2DUS. In addition, 3DUS permitted images to be cross-referenced to one another, providing for a more comprehensive understanding of the anatomic relationships. A single 3DUS acquisition can be used to obtain the standard images that are conventionally used to evaluate the pediatric hip. 3DUS studies have shown that a better view of anatomic relationships is possible (5,7,8). 3DUS also has been

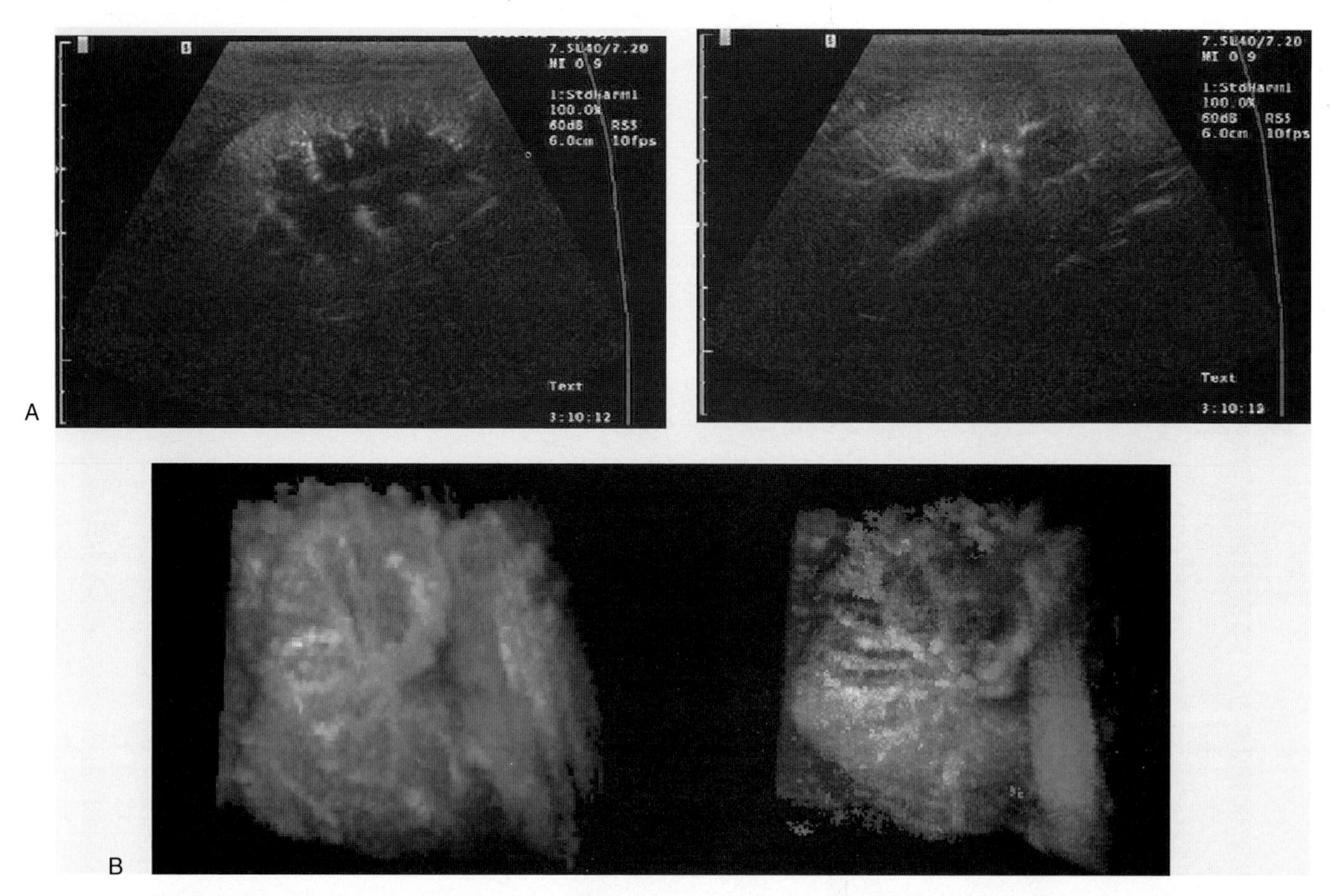

Figure 15-2. 3DUS study of a dog kidney imaged using wide-band harmonic (Elegra, Siemens Ultrasound, Issaquah, WA) imaging and contrast material (Imagent, Alliance Pharmaceuticals, San Diego, CA). **A:** Two 2DUS images from a free-hand acquisition with position sensing. **B:** Two 3DUS volume-rendered images using wide-band harmonic imaging and a contrast agent. The left image was acquired at 5 MHz and the right image was acquired at 7 MHz. Excellent detail regarding vascular anatomy is possible with this approach.

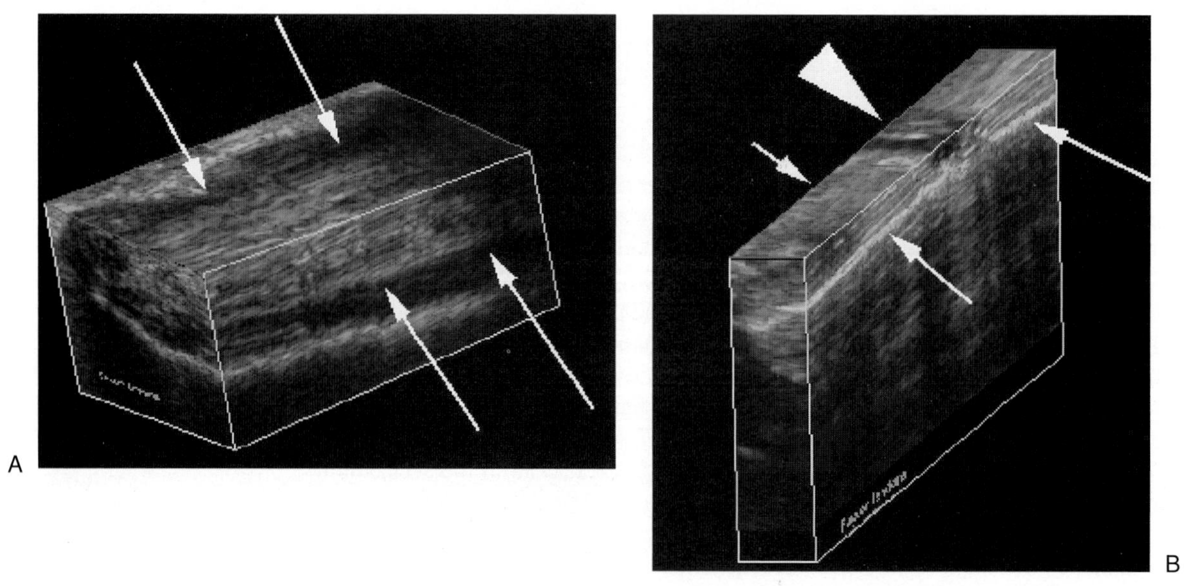

Figure 15-3. **A:** 3DUS multiplanar view of the check tendon of a horse (deceased). (*Arrows*, tendon fibers.) **B:** Multiplanar view of the same tendon after after it has been cut (*arrowhead*).

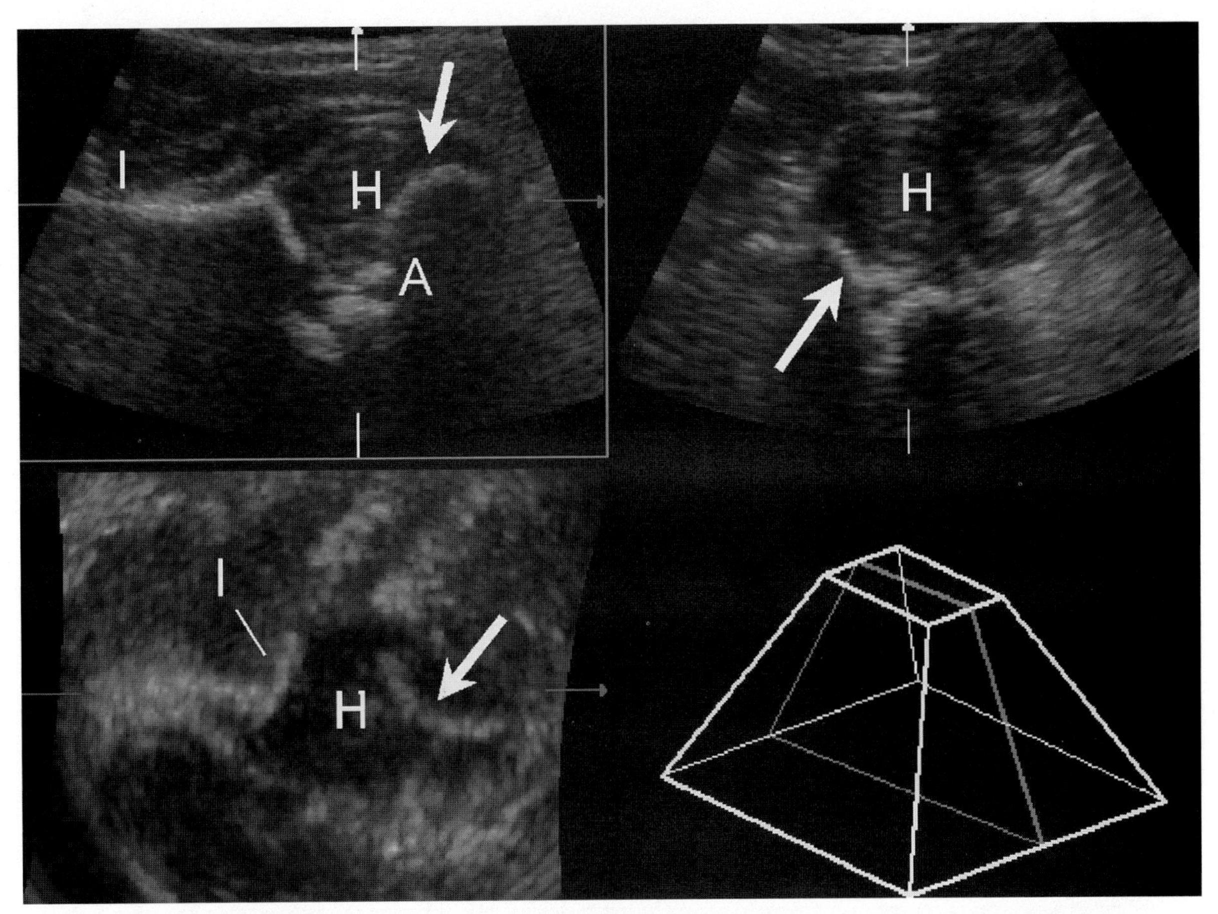

Figure 15-4. 3DUS study of a hip in a 7-week-old infant. Three orthogonal views are displayed of the hip. The upper left is a conventional view; the other two are not normally available. Interactive review of 3DUS data should permit sonographers to better understand neonatal hip anatomy. H, femoral head; I, ilium; A, medial acetabulum; *arrow*, femoral neck.

shown to be useful for calculating the volume of joint effusions and Baker cysts (9). Evaluation of tendons and joint motion also are possible with improving technology.

Neonatal Brain Imaging

3DUS imaging of the neonatal brain offers great promise due to the ability of obtaining axial and oblique images through the brain (Figure 15-5). Conventional 2DUS has been predominantly limited to image planes that are perpendicular to the anterior fontanelle; coronal and sagittal views are the primary planes used to evaluate intraventricular and parenchymal anatomy. In many cases, it is difficult to determine whether echogenicity in the caudothalamic grove is the normal choroid abutting the brain tissue or whether it is hemorrhage. 3DUS should assist in differentiating this situation because planes parallel, and nearly parallel, to the fontanelle can be evaluated, similar to the axial views used in CT and MRI. 3DUS also should be helpful in assessing neonates with ventriculomegaly to clarify the relationship of the dilated ventricles to the remaining parenchyma (Figure 15-6). Evaluation of volume data permits the examiner to move slowly through the ventricular system at optimized orientations to clearly visualize these important relationships. 3DUS also has potential to measure ventricular volumes, which may be important in following patients serially and assessing response to therapy.

Motion Studies

Evaluation of motion as a four-dimensional data set has been studied extensively in cardiology, as discussed in Chapter 11. Other motion applications also show promise.

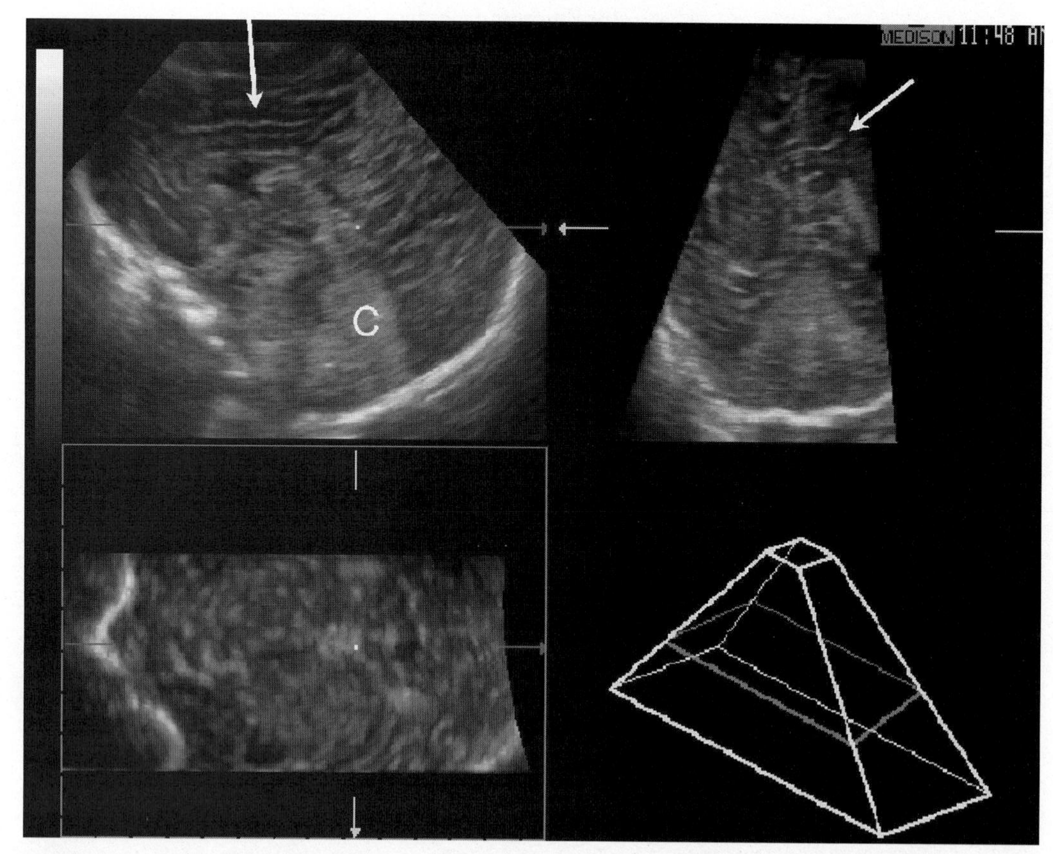

Figure 15-5. Normal neonatal brain in term infant. Multiplanar views show the sagittal view on the upper left, the coronal view on the upper right, and the axial view on the lower left. The axial view can be seen only with 3DUS imaging. Note numerous sulci (*arrows*). (Courtesy of Medison, Korea.)

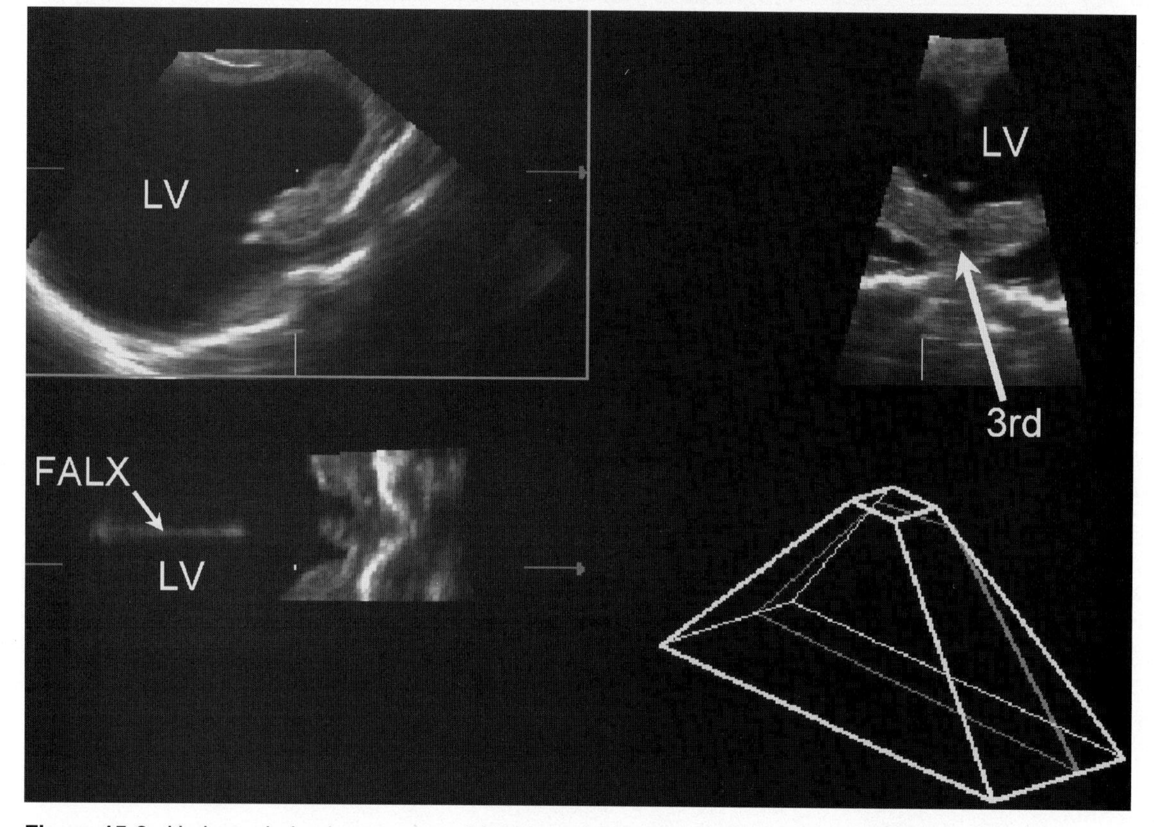

Figure 15-6. Hydrocephalus in a neonate. Multiplanar views show the sagittal view on the upper left, the coronal view on the upper right, and the axial view on the lower left. The axial view can be seen only with 3DUS imaging. The lateral ventricles (LV) and third ventricle are clearly dilated. (Courtesy of Medison, Korea.)

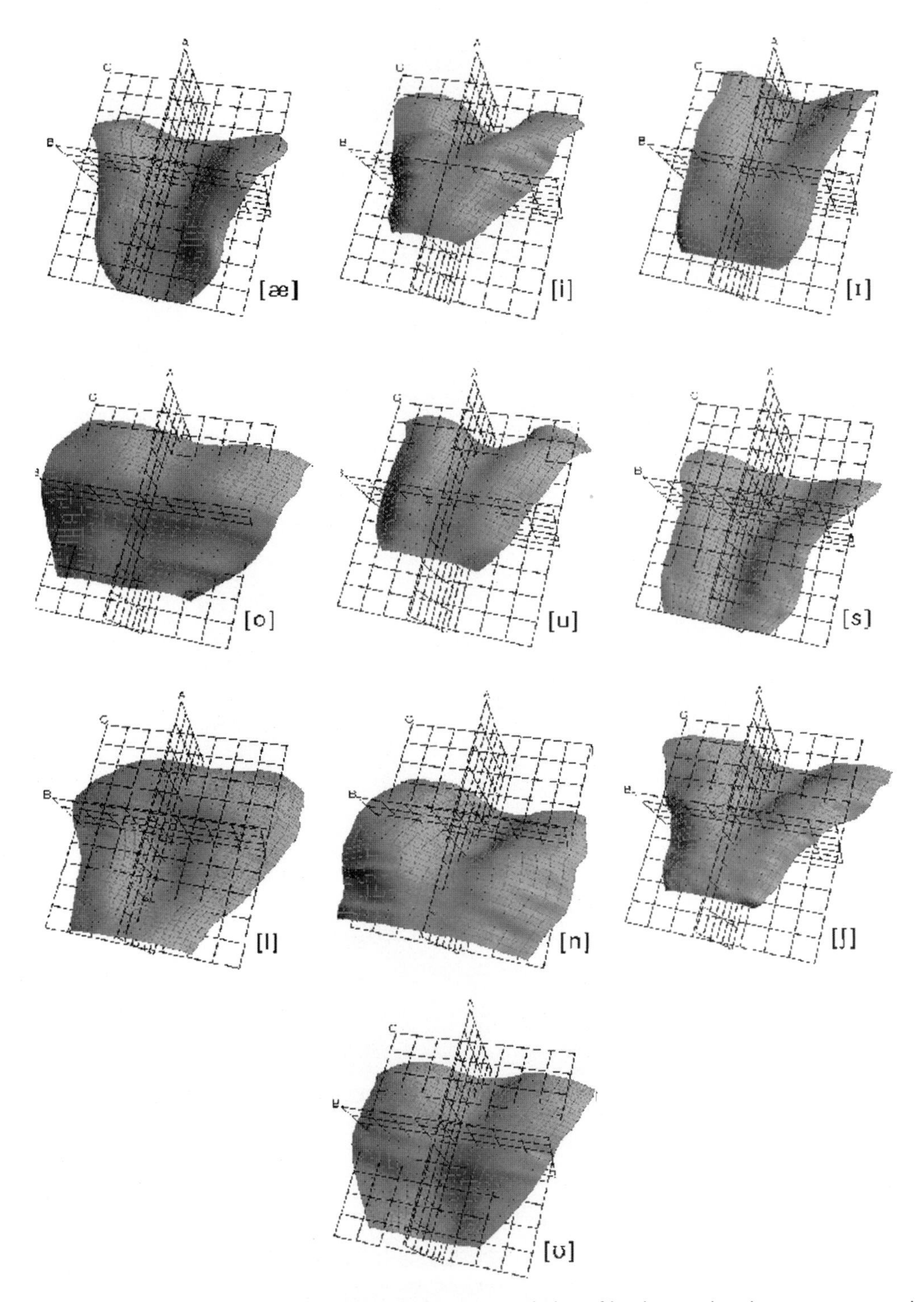

Figure 15-7. 3DUS images of the tongue during pronunciation of basic vowel and consonant sounds. Differences in the surface contour of the tongue with different sounds are clearly seen. (Images courtesy of Dr. M. Stone.)

Monitoring of the shape of the tongue during speech or sustained sound production can be used to gain new insight to speech pathologies (10–13) (Figure 15-7). Musculoskeletal imaging of joint motion is another area of interest that will become more practical when 3DUS imaging is real-time. Also, imaging of blood flow dynamics may ultimately prove to be a valuable means of understanding plaque formation and assessing the impact of flow restrictions.

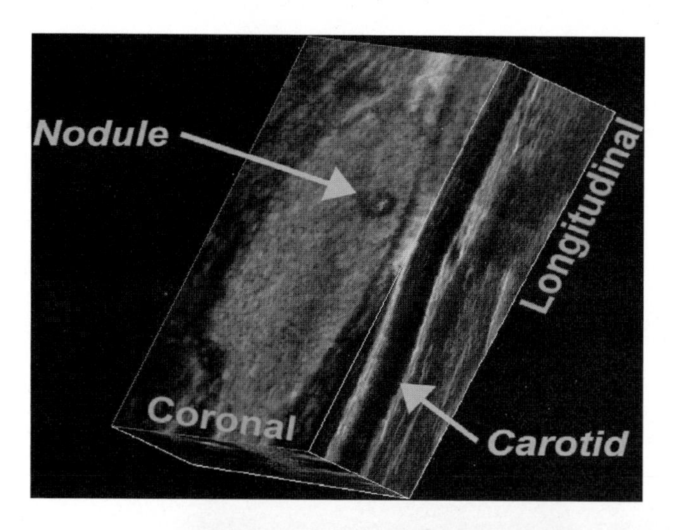

Figure 15-8. 3DUS study of thyroid adenoma. Incidental thyroid adenoma (nonfunctioning) is shown in the right upper lobe. (Courtesy of Dr. D.B. Downey.)

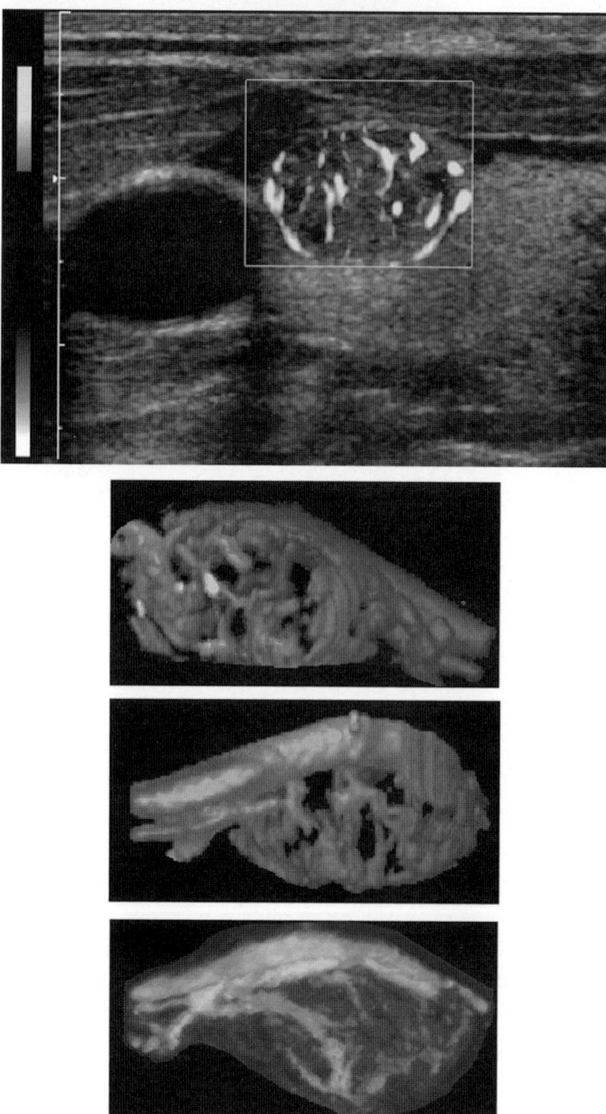

Figure 15-9. 3DUS power Doppler study of thyroid adenoma. The upper image is a 2DUS power Doppler image. The complex vasculature is more clearly appreciated in the 3DUS-rendered images shown below than in the 2DUS power Doppler image. (Image courtesy of Siemens Ultrasound, Issaquah, WA.) See color version of figure.

Dermatologic Imaging

Although 3DUS methods have been predominantly applied using standard clinical imaging equipment, recent work has extended applications to high-frequency (20–100 MHz) imaging in intravascular imaging and dermatology and ophthalmology. Studies in the range of 20 MHz have shown promise for imaging of the superficial layers of the skin for the assessment of skin lesions (14,15). Such methods are valuable for obtaining reproducible measurements of skin tumor volumes that can be useful for assessing responses to therapy (16).

Thyroid Imaging

3DUS evaluation of the thyroid provides a more comprehensive view of the anatomy and pathology of the thyroid (Figure 15-8, Figure 15-9). 3DUS permits the thyroid to be examined in three orthogonal planes simultaneously, which is valuable for assessing the features of thyroid nodules. When there are multiple nodules, each nodule can be identified separately and measured. Currently, 2DUS is used to measure nodules in transverse and longitudinal dimensions. Volume measurements with 3DUS should provide improved measurement accuracy. More accurate volume measurements are important for studies that follow growth. 3DUS will assist in assessing when significant changes have occurred as part of determining and monitoring appropriate therapy (e.g., surgical vs. medical, which often is determined by growth).

Psychological Impact Studies

One of the greatest potentials of 3DUS is to enhance visualization and comprehension of patient anatomy, making it easier to convey information to the patient directly. Such capability permits a clearer impression of spatial relationships that makes sharing information to non-specialist physicians and patients more successful. Certainly, visualization of the fetal face has greatly improved comprehension of developmental anomalies (17,18), helping patients understand management options. The clear visualization of the fetal face provided by 3DUS also can have a very important role in improving maternal–fetal bonding (19) in normal patients, in understanding the severity of congenital anomalies, and in helping parents to understand the management options. 3DUS also may be useful in improving the maternal–fetal bond and thereby lead to a means of reducing abusive behaviors (e.g., smoking) in pregnant mothers.

HOW TO DO IT
Practical Clinical Tips for Emerging Clinical Applications

- Neonatal brain—provides orientations not available with 2DUS. Neonatal brain studies will require transducers with a small footprint and a large sweep angle to display a large portion of the brain, particularly in the plane parallel to the skin. The transducers must be small enough to fit into a neonatal isolette.
- Motion studies—vascular dynamics, joint kinematic studies will require acquisition of multiple volumes, encompassing the extent of motion desired.
- Thyroid imaging of the entire gland requires free-hand techniques, although mechanical positioners or volume transducers may be used to produce volumes from each side.
- Musculoskeletal and joint motion studies are subject to the same type of imaging challenges as traditional 2DUS. The additional challenge is to maintain a stationary position during data acquisition. Joint motion studies require acquisition of multiple volumes in sequence during motion through a proscribed range. Analysis will be similar to that associated with cardiac dynamics.

HOW TO DO It, *continued*

- Dermatology and ophthalmology studies will benefit from high-frequency transducers.
- Psychological studies will be based on evaluating the impact 3DUS imaging has on patients. Study protocols must be developed to examine this impact, particularly in obstetrics.

REFERENCES

1. Burns PN, Powers JE, Simpson DH, et al. Harmonic power mode Doppler using microbubble contrast agents: an improved method for small vessel flow imaging. 1994 IEEE Ultrasonics Symposium Proceedings (Cat. No. 94CH3468–6) 1994;3:1547–1550.
2. Burns PN, Powers JE, Simpson DH, Uhlendorf V, Fritzsch T. Harmonic imaging: principles and preliminary results. *Angiology* 1997;47:S63–S73.
3. Nelson TR, Mattrey RF, Steinbach GC, Lee Y, Lazenby J. 3D harmonic grey-scale ultrasound visualization of vascular anatomy. *JUM* 1998;17(3):55.
4. Nelson TR, Mattrey RF, Steinbach GC, Lee Y, Lazenby J. Gated 3D ultrasound imaging of the coronary arteries using contrast and harmonic imaging. *JUM* 1998;17(3):S56.
5. Gerscovich EO, Greenspan A, Cronan MS, Karol LA, McGahan JP. Three-dimensional sonographic evaluation of developmental dysplasia of the hip: preliminary findings. *Radiology* 1994;190:407–410.
6. Harcke HT, Grissom LE. Performing dynamic sonography of the infant hip. *AJR* 1990;155:837–844.
7. Bohm K, Niethard FU. [Three-dimensional ultrasound image of the infant hip] Dreidimensionale Ultraschalldarstellung der Sauglingshufte. *Bildgebung* 1994;61:126–129.
8. Graf R, Lercher K. Experience with a 3-dimensional sonographic system: infant hip joints, *Ultraschall Med* 1996;17:218–224.
9. Kellner H, Liess H, Zoller WG. [3D-ultrasound of soft tissues and joints] 3D-Sonographie an Weichteilen und Gelenken. *Bildgebung* 1994;61:130–134.
10. Harshman R, Ladefoged P, Goldstein L. Factor analysis of tongue shapes. *J Acoust Soc Am* 1977;62:693–707.
11. Stone M. A three-dimensional model of tongue movement based on ultrasound and x-ray microbeam data. *J Accoust Soc Am* 1990;87(5):2207–2217.
12. Stone M, Lundberg A. Three-dimensional tongue surface shapes of English consonants and vowels. *J Acoust Soc Am* 1996;99:3728–3737.
13. Watkin KL, Rubin JM. Pseudo-three-dimensional reconstruction of ultrasonic images of the tongue. *J Acoust Soc Am* 1989;85:496–499.
14. Fornage BD, McGavran MH, Duvic M, Waldron CA. Imaging of the skin with 20-MHz US. *Radiology* 1993;189:69–76.
15. Stiller MJ, Driller J, Shupack JL, et al. Three-dimensional imaging for diagnostic ultrasound in dermatology. *J Am Acad Dermatol* 1993;29:171–175.
16. Stucker M, Wilmert M, Hoffmann K, et al. [Objectivity, reproducibility and validity of 3D ultrasound in dermatology] Objektivitat, Reproduzierbarkeit und Validitat der 3D-Sonographie in der Dermatologie. *Bildgebung* 1995;62:179–188.
17. Merz E, Weber G, Bahlmann F, et al. Application of transvaginal and abdominal three-dimensional ultrasound for the detection or exclusion of malformations of the fetal face. *Ultrasound Obstet Gynecol* 1997;9(4): 237–243.
18. Pretorius DH, House M, Nelson TR, et al. Evaluation of normal and abnormal lips in fetuses: comparison between three- and two-dimensional sonography. *AJR* 1995;165:1233–1237.
19. Maier B, Steiner H, Wienerroither H, Staudach A. The psychological impact of three-dimensional fetal imaging on the fetomaternal relationship. In: Baba K, Jurkovic D, eds. *Three-dimensional ultrasound in obstetrics and gynecology.* New York: Parthenon, 1997:67–74.

16

Emerging Technologies and Future Developments

OVERVIEW

This chapter will discuss areas of future development in three-dimensional ultrasound (3DUS), including advances in volume data acquisition, analysis, and display. New technology that may have an impact on 3DUS imaging will be reviewed.

KEY CONCEPTS

- New technology will play a vital role in extending the capability of 3DUS.
- Virtual reality methods may enhance 3DUS utility in the clinical setting, particularly for interventional procedures.
- Miniaturization of 3DUS equipment will expand areas of clinical application.

3DUS is an area undergoing rapid development that represents a natural extension of conventional sonography methods that require integration of a series of 2DUS image slices to develop a 3D impression of underlying anatomy or pathology. Interactive 3DUS visualization enhances the diagnostic process by providing better delineation of complex anatomy and pathology. Interactive manipulation of images by rotation and zooming in on localized features or isolating cross-sectional slices greatly assists interpretation by physicians and allows for quantitative measurement of volume or area. However, the interactivity essential to help physicians comprehend patient anatomy and injury and quickly extract vital information currently requires affordable high-performance computer graphics systems, which are now beginning to become available.

Acquisition of 3DUS data affords the possibility of review after the patient has left the medical facility and allows communication of the entire volume via an interactive communications link to a specialist at a tertiary care center. The availability of 3DUS data could reduce the need to refer a patient to a specialized center by permitting the primary physician and the specialist to consult and interactively review the study, thus improving patient care and reducing costs. In addition, network review of volume data could reduce the operator dependence on patient scanning, thereby standardizing examination protocols. Access to 3DUS data at specialization centers affords more sophisticated analysis and review, further augmenting patient diagnosis and treatment. Ultimately, an improved understanding of patient anatomy offered by 3DUS may make it easier for primary care physicians to understand complex patient anatomy.

Real-time 3DUS imaging will be a major advancement for sonography. Transducers will be moved slightly to optimize image quality and to obtain valuable anatomic information as currently done with 2DUS today. Interventional procedures will be performed more rapidly and safely as needles and catheters are advanced under real-time observation. Display techniques will be explored to provide more intuitive understanding of patient anatomy and pathology. Integration of stereo viewing to routine clinical practice may assist in displaying anatomy to physicians and patients more clearly and rapidly, resulting in improved patient care.

FUTURE DEVELOPMENTS

Clinical applications of 3DUS are growing, and further developments will broaden the adoption of 3DUS imaging worldwide. However, researchers must address specialized problems before widespread clinical use can occur as shown in Table 16-1.

3DUS imaging permits obtaining images that are impossible to get with conventional 2DUS methods, because of limitations in patient position or anatomy. Future 3DUS systems will display images of anatomy and organs in an intuitively straightforward manner that will allow physicians to feel as if they are holding a model of the organ in their hands, allowing them to "see" the organ or fetus as it actually is (Figure 16-1) (1–4). Dynamic motion will be reviewed by "slowing down" the images to assess fine movement; the valves or the walls of the heart will be able to be artificially "stopped" to make measurements or to compare specific images, either volume or conventional planar images. Blood flow visualization through the cardiac cycle will be three-dimensional rather than two-dimensional. The 3DUS system of the future will treat sonographic data not as a series of two-dimensional images viewed in either real time or statically but as a volume with which the clinician rapidly interacts as though exploring internal patient anatomy directly. The technology to accomplish this is available today and will benefit from continued performance increases and cost reductions, ultimately providing 3DUS imaging systems as standard equipment at costs comparable to those of conventional systems available today.

Ultimately, 3DUS technology will provide a central integrating focus in ultrasound imaging. 3DUS equipment will provide more conventional 2DUS imaging capability but

Table 16-1. Improvements needed to facilitate primary clinical acceptance of 3DUS

- Produce scanners having faster volume acquisition, analysis, and interactive display
- Improve overall scanning system performance and image quality
  - Including elevational focus, compounding, and shadowing compensation
- Develop smaller 3DUS scanners and smaller and more specialized transducers
- Improve image and position volume acquisition methods
  - Free-hand image-based scanning
  - Intraluminal and endocavitary probe systems
- Improve physiologic gating for cardiac and respiratory motion
  - Eliminate motion artifacts
  - Evaluate dynamic processes
- Improve user interfaces for evaluating volume data and understanding patient anatomy
  - Head-mounted displays and data gloves for interventional procedures
- Improve algorithms for automated data segmentation, classification, and biometry
- Integrate 3DUS imaging, ultrasound contrast agents, and harmonic imaging
  - Enhance visualization of vascular anatomy
- Create real-time volume-imaging systems

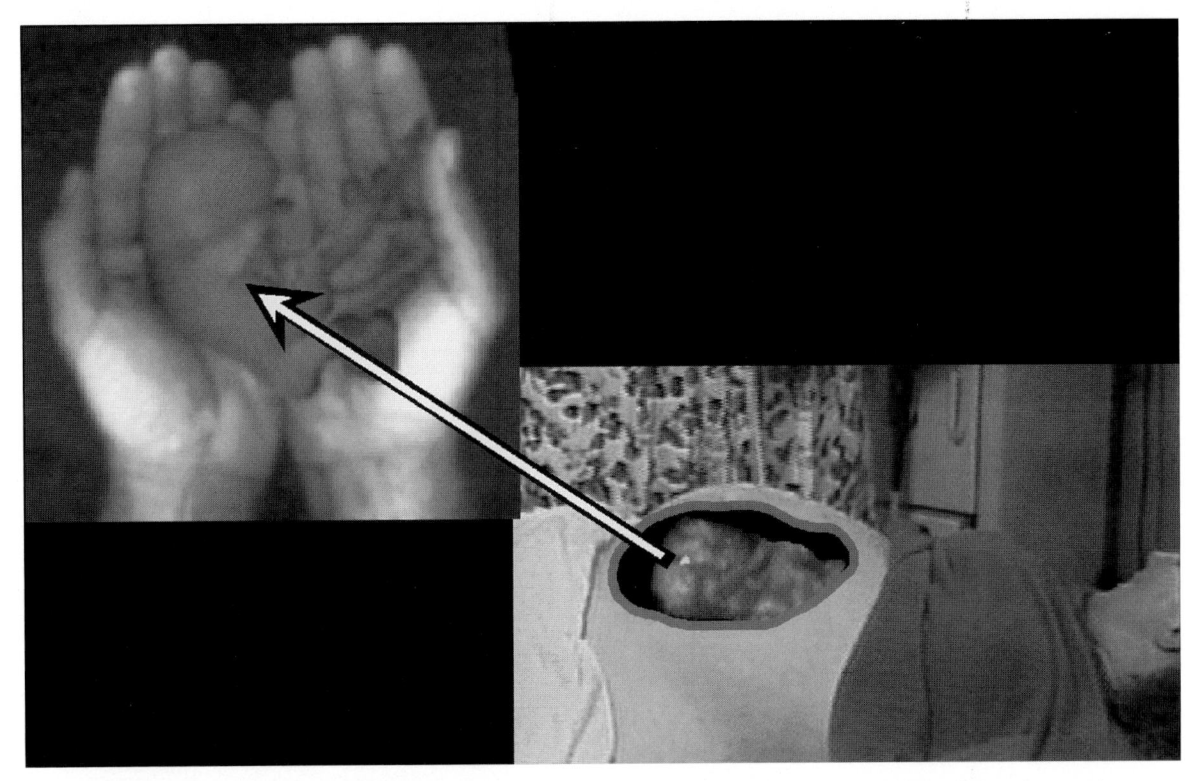

Figure 16-1. Future volume imaging in ultrasound, and other modalities, will offer real-time interactive review and manipulation of patient data. Instead of viewing patient data on a computer or scanner console, the sonographer or physician will directly view the internal patient anatomy and organs in an intuitively straightforward manner that will allow physicians to feel as if they are holding a model of the organ in their hands and allow them to "see" the organ or fetus as it actually is. Direct viewing of internal anatomy would facilitate more rapid comprehension of the patient's condition and provide improved feedback for interventional procedures. See color version of figure.

seamlessly expand imaging to include volume acquisition and display as necessary to obtain the diagnosis. Focused application of 3DUS methods will expand the areas in which ultrasound is the predominant imaging modality beyond those areas in which ultrasound currently is the method of choice. Through rapid transmission of volume data to specialists at distant locations patient care will benefit from improved diagnosis and treatment.

REFERENCES

1. Bajura M, Fuchs H, Ohbuchi R. Merging virtual objects with the real world: seeing ultrasound imagery within the patient. *Comput Graph* 1992;26:203–210.
2. Fuchs H, Levoy M, Pizer SM. Interactive visualization of 3D medical data. *IEEE Comput* 1989;(Aug):46–51.
3. Fuchs H, State A, Pisano ED, et al. Towards performing ultrasound-guided needle biopsies from within a head-mounted display. Proceedings of visualization in biomedical computing 4th International Conference, VBC '96 1996:591–600.
4. State A, Chen DT, Tector C, et al. Observing a volume rendered fetus within a pregnant patient. Proceedings visualization (Cat. No. 94CH35707) 1994;CP41:364–368.

Subject Index

Subject Index

Figure numbers are indicated in italic type.
Table numbers are indicated in bold type followed by a t.

X